Advances in Experimental Medicine and Biology

Volume 1276

Advances in Experimental Medicine and Biology provides a platform for scientific contributions in the main disciplines of the biomedicine and the life sciences. This series publishes thematic volumes on contemporary research in the areas of microbiology, immunology, neurosciences, biochemistry, biomedical engineering, genetics, physiology, and cancer research. Covering emerging topics and techniques in basic and clinical science, it brings together clinicians and researchers from various fields.

Advances in Experimental Medicine and Biology has been publishing exceptional works in the field for over 40 years, and is indexed in SCOPUS, Medline (PubMed), Journal Citation Reports/Science Edition, Science Citation Index Expanded (SciSearch, Web of Science), EMBASE, BIOSIS, Reaxys, EMBiology, the Chemical Abstracts Service (CAS), and Pathway Studio.

2019 Impact Factor: 2.450   5 Year Impact Factor: 2.324

More information about this series at http://www.springer.com/series/5584

Xian-Cheng Jiang
Editor

Lipid Transfer in Lipoprotein Metabolism and Cardiovascular Disease

Springer

Editor
Xian-Cheng Jiang
Department of Cell Biology
SUNY Downstate Health Sciences University
Brooklyn, NY, USA

ISSN 0065-2598 ISSN 2214-8019 (electronic)
Advances in Experimental Medicine and Biology
ISBN 978-981-15-6081-1 ISBN 978-981-15-6082-8 (eBook)
https://doi.org/10.1007/978-981-15-6082-8

This Springer imprint is published by the registered company Springer Nature Singapore Pte Ltd.
The registered company address is: 152 Beach Road, #21-01/04 Gateway East, Singapore 189721, Singapore

Preface

Lipid studies present unique challenges to vascular biology and cardiovascular diseases. Lipid transfer proteins, lipid transporters, and lipid binding proteins have important impact on in lipid metabolism.

This book brings together in a single volume an updated knowledge about lipid transfer proteins, such as phospholipid transfer protein (PLTP), cholesteryl ester transfer protein (CETP) and lipopolysaccharide binding protein (LBP), and microsomal triglyceride transfer protein (MTP); certain important lipid transporters, such as ABC binding cassette (ABC) A1, ABCG1, and ABCG5/ABCG8; certain newly discovered important lipid binding proteins, such as major facilitator superfamily domain containing 2a (Mfsd2a); and apolipoprotein M. We will also discuss recent progresses on modified lipoproteins and their impact on cardiovascular diseases. Moreover, we will summarize the rare diseases related with lipoprotein metabolism and re-evaluate preβ1 high density lipoprotein (HDL) as a pro- or anti-atherosclotic particle.

Our goal in this volume was to compile chapters presenting broad overviews of proteins with lipid interaction, while emphasizing the relationship between lipid or lipoprotein metabolism and cardiovascular diseases.

We hope you enjoy the volume.

Brooklyn, NY, USA Xian-Cheng Jiang

Contents

Impact of Phospholipid Transfer Protein in Lipid Metabolism and Cardiovascular Diseases

1

Xian-Cheng Jiang

Abstract

PLTP plays an important role in lipoprotein metabolism and cardiovascular disease development in humans; however, the mechanisms are still not completely understood. In mouse models, PLTP deficiency reduces cardiovascular disease, while its overexpression induces it. Therefore, we used mouse models to investigate the involved mechanisms. In this chapter, the recent main progresses in the field of PLTP research are summarized, and our focus is on the relationship between PLTP and lipoprotein metabolism, as well as PLTP and cardiovascular diseases.

Keywords

Phospholipid transfer protein · Lipoprotein · VLDL · HDL · Cardiovascular diseases

Abbreviation

Apo	apolipoprotein
BLp	apoB-containing-triglyceride-rich particles
CVD	cardiovascular disease
HDL	high density lipoprotein
KO	gene knockout
PLTP	phospholipid transfer protein
VLDL	very low density lipoprotein

X.-C. Jiang (✉)
Department of Cell Biology, SUNY Downstate Health Sciences University, Brooklyn, NY, USA
e-mail: xian-cheng.jiang@downstate.edu

1.1 Phospholipid Transfer Protein (PLTP)

PLTP is one of the members of lipid transfer protein family, which includes bactericidal/permeability increasing protein (BPI), lipopolysaccharide-binding protein (LBP), and cholesterol ester transfer protein (CETP) [1]. PLTP has two molecular weight, 55 kDa and 81 kDa, which could be due to different glycosylations [2]. PLTP is a nonspecific lipid transfer protein, and it has ability to transfer phospholipids, free cholesterol, α-tocopherol, diacylglycerol, and lipopolysaccharides [3]. Two forms of PLTP protein mass in human serum were discovered. ApoA-I-containing lipoproteins (about 160 kDa in size) is associated with highly active PLTP, while apoE-containing lipoproteins (about 520 kDa in size) is associated with inactivity PLTP [4–6]. So far, the significance for the existence of active and inactive PLTP in the circulation is unknown. It is quite possible that PLTP could have activities independent from its lipid transfer activity. However, no report indicates that there are two forms of PLTP in the blood of mice and rabbits.

Almost all tested tissues express PLTP [2, 7]. Liver, adipose tissues, and macrophages are the important sites for PLTP expression,

X.-C. Jiang (ed.), *Lipid Transfer in Lipoprotein Metabolism and Cardiovascular Disease*, Advances in Experimental Medicine and Biology 1276, https://doi.org/10.1007/978-981-15-6082-8_1

although their levels are not as high as that of the placenta, thymus, ovary, and lung [2, 8–10]. Importantly, PLTP is highly expressed in human atherosclerotic lesions [11, 12].

The liver is one of the major tissues for lipoprotein production and PLTP expression. In order to investigate the effect of liver-derived PLTP on cholesterol and phospholipid metabolism, we prepared liver-specific PLTP expressed mouse model with no PLTP expression in all other tissues. We found the mice have about 25% plasma PLTP activity of that of WT mice [13]. Using Cre-lox system, we prepared liver PLTP knockout (KO) mice, and we found that the mice have significantly lower PLTP activity (about 20%) than that of controls [14]. These results indicated that liver makes about 20% contribution to blood PLTP activity.

Adipose tissues express much higher PLTP mRNA levels than that in the liver [7]. PLTP not only transfers phospholipids but also free cholesterol [15] which is the major cholesterol in the adipose tissues [16]. We established adipose tissue PLTP KO mice which have significantly lower plasma PLTP activity, HDL-phospholipid, HDL-cholesterol, and apoA-I concentrations [17]. In order to investigate the mechanisms, we used adipose tissue explants to measure cholesterol efflux, mediated by apoA-I. We found that exogenous and endogenous PLTP significantly increases cholesterol efflux [17]. Therefore, like liver-derived PLTP [14], adipose tissue-derived PLTP plays an important role in blood PLTP activity and HDL metabolism.

The lung is another important tissue for PLTP expression [18]. In order to explore the impact of lung-derived PLTP in blood PLTP activity and lipoprotein metabolism, we treated PLTP-Flox mice with adenovirus (AdV)-Cre and AdV-GFP (intratracheally) [19]. We found that lung PLTP deficiency resulted in significant reductions of plasma PLTP activity (about 18%), phospholipids (about 20%), and cholesterol (about 23%).

PLTP also produced by the brain; however, the function of PLTP in the brain is still not quite clear [20, 21]. PLTP activity may play an important role in maintaining neuron structural integrity and in conducting signal transduction pathways [22]. PLTP KO mice have significantly lower brain vitamin E concentration, and these mice significantly increase anxiety [23]. Interestingly, Alzheimer's disease patients have significantly higher PLTP levels [20, 21]. PLTP deficiency increased amyloid-β (Aβ)-associated memory defect in mice [24]. PLTP mRNA expression levels were 6.8-fold higher in cerebral vessels [25] than that in the whole brain. PLTP could be important in maintaining blood–brain barrier, and this effect could be mediated by its vitamin E transfer activity and, thus, regulate cerebrovascular oxidative stress [26]. It is possible that PLTP may have an important impact in the brain, physiologically and pathophysiologically.

1.2 PLTP and Cholesteryl Ester Transfer Protein (CETP)

PLTP and CETP have some similarity in structural features [1, 27] and sequence [2], but they have functional overlap. We had prepared CETP transgenic/PLTP KO mice, and we found that the expression of CETP can further reduce HDL cholesterol levels on PLTP deficient background [28]. In fact, PLTP has an interaction with CETP. Although PLTP has no cholesteryl ester transfer activity, purified PLTP promotes HDL cholesteryl ester transfer to VLDL [29]. In addition, PLTP KO/CETP transgenic mice have significantly lower CETP activity than that of CETP transgenic mice [28].

1.3 Regulation of PLTP

Many factors can regulate PLTP activity and mRNA levels. Western-type diets upregulate both PLTP activity and mRNA levels [7]. Lipopolysaccharide treatment can significantly reduce plasma PLTP activity and significantly reduce PLTP mRNA levels in the liver and adipose tissues [7]. Glucose can promote PLTP activity and expression [30], while insulin has opposite effect [31, 32]. PLTP activity can also be regulated by diacylglyceride [33].

Human and mouse PLTP gene promoter regions contain multiple AP2 and Sp1 consensus sequences which are associated with PLTP transcription [34, 35]. Both fenofibrate and chenodeoxycholic acid can upregulate PLTP expression, suggesting that peroxisome proliferator-activated receptor (PPAR) and farnesoid X-activated receptor (FXR) are involved in the regulation [34]. We [8] and other researchers [36] indicated that liver X receptor (LXR) can upregulate PLTP expression. Both human and mouse PLTP promoter contain an LXR response element. LXR/RXR heterodimers can bind on the element, and LXR/RXR transient-transfection can activate the expression [41]. Through activating of SREBP-1,c LXR agonists can activate triglyceride biosynthesis and PLTP transcription [37].

We also found that, in LDL receptor KO mice, profurin (prodomain of furin) overexpression significantly attenuates the development of atherosclerosis and reduces plasma LDL-cholesterol [38]. This effect is related with PLTP degradation in the liver, thus blocking VLDL secretion [39].

1.4 PLTP and HDL

PLTP activity mediates transfer of phospholipids from apoB-containing-triglyceride-rich lipoprotein, such chylomicron and VLDL, into HDL, and also mediates exchange of phospholipids among lipoproteins [40, 41]. Moreover, PLTP can act like a putative fusion factor to influence the size of HDL particles [42]. PLTP-induced phospholipid transfer activity seems to be important in the enlargement of HDL [43], and triglyceride enrichment in the core of HDL might promote HDL fusion [44].

Adenovirus and adenovirus-associated virus (AAV)-mediated overexpression of PLTP in mouse liver caused a dramatic induction of plasma PLTP activity [45, 46]. These mice have a significant reduction in α-HDL but induction in preβ-HDL levels. Adenovirus-associated virus (AAV)-mediated PLTP overexpression in mice resulted in a significant reduction in cholesterol and HDL cholesterol [46]. We also prepared PLTP transgenic mice with low level human PLTP expression and found that the preβ-HDL is significantly increased [47]. High level of PLTP transgenic mice was also generated. These mice showed a significant induction in PLTP activity in the circulation and a reduction in plasma HDL cholesterol levels but an induction in preβ-HDL [48], compared with control mice. Overall, PLTP overexpression causes a significant reduction in plasma HDL but increases preβ-HDL.

Until now, no PLTP gene deficiency or mutation was found in humans. Using PLTP KO mice, we learned a lot about PLTP deficiency. The KO mice show a complete depletion of the transfer activity for following lipids: phosphatidylcholine, sphingomyelin, phosphatidylethanolamine, and phosphatidylinositol. The KO mice also partially lose their activity for free cholesterol transferring [15]. In addition, PLTP KO mice have a defect for [^{3}H] phosphatidylcholine transfer from apoB-containing lipoprotein to HDL in vivo. On normal diet, the KO mice significantly decrease HDL and apoA-I, suggesting that PLTP plays an important role in transferring surface lipid components (phosphatidylcholine, sphingomyelin, and free cholesterol) from triglyceride-rich lipoproteins into HDL, thus maintaining HDL levels in the circulation [15]. Moreover, the HDL from the PLTP KO mice was phosphatidylcholine poor but protein enriched. PLTP deficiency also promotes HDL turnover rate [49, 50]. Overall, PLTP deficiency causes a significant reduction in plasma HDL cholesterol levels. Interestingly, both PLTP overexpression and deficiency result in HDL reduction, and the reason is still unknown.

We compared HDLs, isolated from PLTP transgenic, wild type (WT), and PLTP KO mice. We found that (1) PLTP transgenic mouse has the largest size of HDL, WT mouse has the middle range size of HDL, while the PLTP deficient mouse has the smallest size of HDL [17]; (2) different HDLs have different inflammatory index. HDL from PLTP transgenic mouse has the highest inflammatory index, while HDL from WT mouse is in the middle, and HDL from PLTP KO mouse has the lowest inflammatory

Table 1.1 The influence of PLTP expression on HDL

	PLTP Tg	WT	PLTP KO
HDL size (nm)	9.65 ± 0.15^a	9.25 ± 0.15^b	8.85 ± 0.10^c
HDL inflammatory index	1.22 ± 0.29^a	0.52 ± 0.13^b	0.39 ± 0.19^c
HDL-cholesterol (mg/dl)	57 ± 10^a	92 ± 8^b	35 ± 7^c
HDL-phospholipid (mg/dl)	79 ± 12^a	135 ± 15^b	62 ± 8^a

Value displayed are means ± SD., $n = 5$. Values labeled with different lowercase letters are statistically different ($p < 0.05$). HDL size and HDL inflammatory index were adapted from Jiang H. et al. *Arterioscler. Thromb. Vasc. Biol.* 2015;35: 316–322

index [17]; and (3) the order of HDL cholesterol levels is WT > PLTP transgenic > PLTP KO; the order of HDL total phospholipids is WT > PLTP transgenic = PLTP KO (Table 1.1). Thus, PLTP activity influences HDL particle size, inflammatory index, and cholesterol/phospholipid composition [17]. We also found that hepatocyte PLTP deficiency causes a significant reduction in HDL and apoA-I levels [14].

S1P is a potent lipid mediator composed of one long hydrophobic chain and one phosphoric acid group. S1P exerts potent physiological effects through five S1P receptors (S1PR1–S1PR5) located on cell membranes. S1P is involved in various diseases including atherosclerosis [51] and diabetes [52]. On the one hand, S1P has pro-atherogenic properties. S1P induces inflammation and thrombosis. The S1P gradient facilitates the egress of lymphocytes from lymphoid organs into the circulation and the recruitment of lymphocytes to sites of inflammation [53]. S1P activates NF-κB [54], promotes chemotaxis, and stimulates the production of TNF-α in macrophages and/or monocytes [55]. S1P has been shown to augment the thrombin-induced expression of tissue factor in endothelial cells [56], and S1P has also been proposed to induce the expression of plasminogen activator inhibitor-1 (PAI-1) in adipocytes [57] and hepatocytes [58], suggesting that S1P has a pro-thrombotic property. On the other hand, S1P also has anti-atherogenic properties. S1P promotes the survival and prevents the apoptosis of endothelial cells mainly through S1P1 and S1P3 [59]. Many recent studies have link S1P with HDL, because HDL is a major carrier of S1P in the circulation. In fact, HDL-associated S1P regulates a lot of the physiological and pathological effects in cells and tissues [60–64]. HDL-bound apolipoprotein M (apoM) is a physiologically-relevant S1P chaperone [65]. Despite the potential of the apoM-S1P axis as an endothelium-protective mechanism, the effect of apoM-S1P on atherosclerosis is still controversy [66, 67]. Global apoM deficiency causes only about 25–45% reduction of plasma S1P [65, 67]. There must be some other protein factors that are responsible for assisting the function of S1P transporters or serving as a S1P carrier. PLTP could be one of them. Interesting, we found that PLTP deficiency causes about 60% reduction of plasma S1P which is carried by HDL [68]. Furthermore, PLTP can transfer S1P from red blood cells to HDL, suggesting PLTP is one of determiner for plasma S1P, since red blood cells are the major source for S1P in the circulation. Interesting, PLTP deficiency has no effect on plasma apoM levels [68].

1.5 PLTP and Reverse Cholesterol Transport (RCT)

Macrophage highly expresses PLTP, and, thus, it has been suggested the macrophage PLTP plays an important role potential in cholesterol efflux. However, the role of PLTP in RCT (many studies are mainly based on macrophage cholesterol efflux) is still controversial. There are reports indicating that PLTP has no effect [8] or inhibit [69, 70] or promote [71, 72] cholesterol efflux. The cause of the discrepancy among these studies is still unknown; it could be due to the HDL particles or HDL levels in cell culture medium used in these efflux experiments.

On the one hand, it has been reported that exogenous PLTP accelerates HDL-mediated cholesterol efflux through ABCA1 pathway [71]. We found that recombinant PLTP (50 ng/ml) together with 0.8 nmole/ml HDL promotes HDL-mediated cholesterol efflux (our unpublished result). PLTP interacts and stabilizes ABCA1 which directly mediates lipid efflux [14, 71]. It has been shown that PLTP has an amphipathic helical region of the N-terminal barrel which is critical for ABCA1-mediate cholesterol efflux [72]. Moreover, Lee-Rueckert et al. reported that PLTP KO macrophage has an impairment in cholesterol efflux and that the defect can be corrected by a stimulation of the ABCA1-mediated pathway [10]. These results indicated that PLTP has an ability to help ABCA1 for macrophage cholesterol efflux and such an activity might promote RCT [10].

On the other hand, it has been showed that HDL isolated from PLTP transgenic mice has impaired effect on macrophage cholesterol efflux, compared with control [69]. Furthermore, it has been shown the PLTP might cause the formation a dysfunctional HDL subfraction, which could not be a good cholesterol acceptor [73]. The same group of researchers also found that macrophage cholesterol efflux and reverse cholesterol transport to feces are impaired in PLTP transgenic mice and that higher systemic PLTP activity levels might promote the development of atherosclerosis by reducing the rate of RCT [70]. Based on these observations, PLTP could play an important role in inhibiting macrophage cholesterol efflux or RCT. A recent report indicated that overexpression and depletion of PLTP can reduce HDL mass and cholesterol efflux capacity but has nothing to do with macrophage-mediated RCT [74].

1.6 PLTP and ApoB-Containing-Triglyceride-Rich Lipoprotein (BLp) Secretion

Twenty year ago, we found that PLTP KO mice have a defect in VLDL secretion [75]. Moreover, it has been reported that PLTP overexpression promotes liver VLDL overproduction [76, 77]. Masson et al. [78] found that human PLTP transgenic rabbits have a significant increase in plasma LDL but not of HDL. This observation could be a real situation in humans, because humans and rabbits are LDL mammals. It has been reported that the PLTP activity is positively associated with the triglyceride BLp incorporation rate [37]. Manchekar et al. indicated that PLTP has an important effect on the initiation of BLp assembly in mouse liver [79]. We also found that the major function of liver PLTP is to promote VLDL secretion. Based what we have observed, we proposed a model for PLTP-related BLp lipidation (Fig. 1.1) [13, 14]. More importantly, based on human genome-wide association studies (GWAS), human PLTP levels are positively associated with plasma triglyceride and apoB levels [80, 81].

Blocking VLDL secretion, long recognized as an effective LDL-C lowering strategy, differs from the use of statins which function through a reduction in de novo cholesterol synthesis. However, this approach can have unwanted consequences. VLDL secretion is a hepatic-specific defense against the excessive liver triglyceride (TG) accumulation that occurs in nutritional overload or metabolic syndrome. Blocking VLDL secretion by inhibition of microsomal triglyceride transfer protein (MTP) results in hepatic lipid accumulation and toxicity in mice and humans [82]. A similar response occurs in mice with genetic deletion of methionine adenosyltransferase [83] or superoxide dismutase 1 [84]. Two drugs targeting apoB and MTP have been approved only for treatment of extreme dyslipidemia rather than common lipid disorders due to hepatotoxicity concern. Thus, targeting VLDL secretion without causing hepatic lipid accumulation offers great potential as an alternative treatment method for milder lipid disorders. PLTP deficiency in mice did not cause lipid accumulation in the liver [75]. Thus, potentially, PLTP inhibition in humans could result in reduction of BLp production with no consequence of liver lipid accumulation.

PLTP can transfer vitamin E among the lipoproteins and between lipoprotein and cell

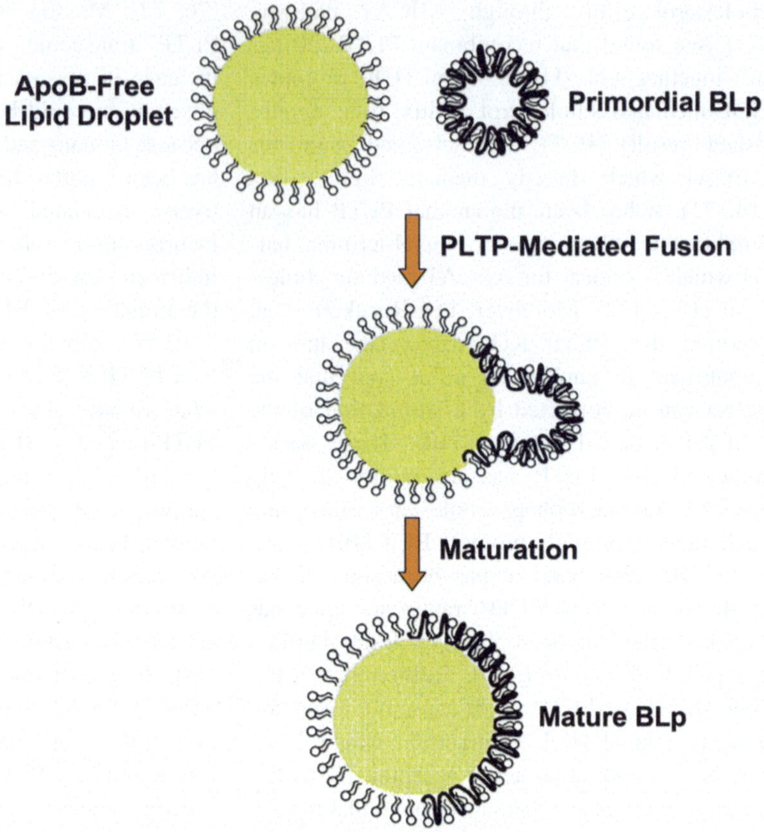

Fig. 1.1 *A model of PLTP-involved BLp lipidation.* ApoB-containing lipoprotein lipidation is involved in the fusion of primordial BLp and apoB-free/TG-rich lipid droplets. PLTP-mediated PL transfer or exchange on both particles' surface would fuse two particles. *BLp* apoB-containing particles. (Adapted from Yazdanyar A and Jiang XC. Hepatology. 2012;56:576–84)

transfer. PLTP activity is important for tissue vitamin E levels. We found that PLTP KO mice have vitamin E-enriched LDL which is resistant to cupper-induced oxidation and have much less activity to induce monocyte chemotactic activity [50, 85], while LDL from PLTP transgenic mice has opposite effects [46]. Many results suggested that the function of PLTP in tissues is different from its role in the circulation. PLTP-deficient macrophages accumulate a lot of cholesterol in the presence of LDL [86]. However, vitamin E supplementation normalizes the cholesterol levels in the macrophages [86]. We found that PLTP-deficient hepatocytes secrete less BLp, and this is related to BLp premature degradation caused by vitamin E deficiency-mediated oxidation stress induction [87]. Thus, a possible effect of PLTP inside cells might be related with changes in cellular vitamin E levels and oxidative stress.

1.7 PLTP and Obesity/Insulin Resistance

PLTP expression is increased in different pathologies associated with increasing risk of cardiovascular diseases, such as obesity [88, 89], insulin resistance [90], and type II diabetes [91]. We evaluate the effect of PLTP deficiency on dietary-induced obesity and insulin resistance. We found that although PLTP KO mice have normal body weight under chow diet, the KO mice are protected from high fat diet-induced obesity and insulin resistance, compared with control mice. To understand the mechanism, we evaluated insulin receptor and Akt activation and found that PLTP deficiency significantly enhanced phosphorylated insulin receptor and Akt levels in mouse livers, adipose tissues, and muscles after insulin stimulation. Moreover, we found that the PLTP deficiency induced significantly more GLUT4 protein in the plasma membranes of adipocytes and muscle cells after insulin stimulation. The conclusion is that PLTP deficiency leads to an improvement in tissue and whole-body insulin sensitivity [92].

1.8 PLTP and Thrombosis

Klein et al. showed that PLTP KO mice have a longer blood clotting time than that of control mice. This phenomenon is related to a reduction of phosphatidylserine externalization through vitamin E reduction in red blood cells [93]. Consistent with these results, same group of the researchers further indicated that PLTP deficiency can reduce thrombotic response to acute intravascular oxidative stress [94]. Thus, PLTP activity seems to be related with hypercoagulation. However, other studies suggested that PLTP has an anticoagulation effect [95, 96]. Therefore, it is still unclear whether PLTP is a factor involved in hypercoagulation or hypocoagulation. Very recently, we found that PLTP promotes phosphatidylserine externalization at the plasma membrane of platelets and accelerates ADP- or collagen-induced platelet aggregation. This effect plays an important role in the initiation of thrombin generation and platelet aggregation under sheer stress conditions. Thus, PLTP is involved in hypercoagulation [97]. Therefore, PLTP inhibition could be a novel approach for countering thrombosis.

1.9 PLTP and Inflammation

It is still controversial whether PLTP is an anti-inflammatory or pro-inflammatory factor. PLTP KO mice reduce plasma interleukin-6 (IL-6) levels [98, 99] and have less expression of IL-6 and infiltrating macrophages in aortic tissue [100], compared with control mice. It has been showed that, in PLTP KO mice, a shift of T helper (Th) lymphocytes toward the anti-inflammatory subset Th2 was observed [101]. On the other hand, other studies (LPS-induced inflammation) indicated that PLTP has an anti-inflammatory property [102–104]. LPS administration causes higher mortality in PLTP deficient mice, compared with WT mice [102]. Decrease in PLTP expression or activity was also associated with enhancing inflammatory responses toward LPS treatment and cigarette smoke exposition [103], since PLTP has binding and neutralizing LPS ability which could explain its anti-inflammatory functions [102, 105]. Moreover, PLTP might also have an anti-inflammatory properties in macrophages through an interaction with the ABCA1 and then JAK2/STAT3 pathway [104].

1.10 PLTP and Cardiovascular Diseases

More than 10 years ago, we indicated that PLTP activity is induced in the patients with cardiovascular diseases (CVD) [106]. We proposed that PLTP could be a target for the treatment of CVD. In the last 10 years, many human studies showed that PLTP in the circulation is positively associated with CVD [81, 107–109]. Using a *PLTP* gene score, constructed by a combination of two *PLTP* tagging single nucleotide polymorphisms (SNPs), Vergeerit et al., by using a *PLTP* gene score (constructed by a combination of two *PLTP* tagging single nucleotide polymorphisms), reported that lower hepatic PLTP transcription and lower plasma PLTP activity result in reduction of CVD among 5 cohorts comprising a total of 4658 cases and 11,459 controls [110]. In a relative recent Framingham Heart Study (comprised a total of 2679 participants with 187 first events being ascertained during 10.4 years of follow-up), Robins et al. showed that higher plasma PLTP activity is positively associated with a first cardiovascular event, defined as fatal or non-fatal coronary heart disease and stroke, among men [111]. Further, PLTP activity is also positively associated with left ventricular systolic dysfunction in human [112, 113]. We examined the long-term prognostic significance of plasma PLTP activity levels in a cohort of 170 high-risk diabetic men with known or suspected CVD who

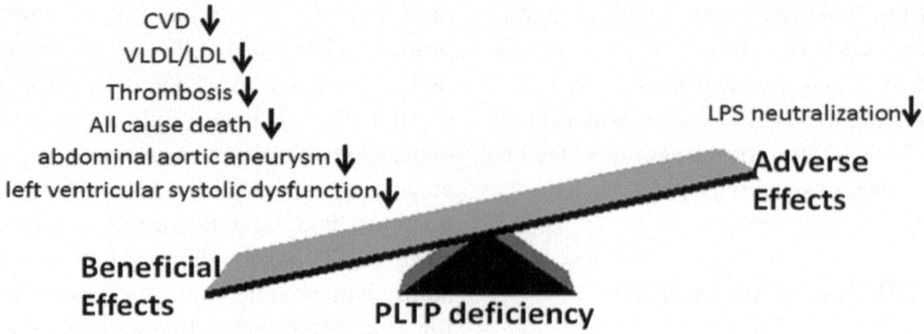

Fig. 1.2 *Beneficial and adverse effects of PLTP inhibition. CVD* cardiovascular disease, *BBB* blood–brain barrier. (Adapted from Jiang XC. Journal Lipid Research. 2018; 59(5):764–771)

were referred for cardiac catheterization. We found that plasma PLTP activity levels were a strong and independent predictor of all-cause mortality in 5 years and higher PLTP activity had higher mortality [114]. One potential mechanism relating PLTP-related CVD is that plasma PLTP activity is positively associated with BLp levels [80, 81]. Contradictorily, PLTP mass was lower in a small group of CVD patients compared to controls [115], although it seems clear that the plasma PLTP protein concentration does not represent the preferred marker of PLTP-associated risk [116, 117]. In addition, reported effects of PLTP on peripheral artery disease are both limited and inconsistent [118, 119].

In mouse models, systemic PLTP deficiency reduces atherosclerosis [75], while its overexpression shows the opposite effect [46, 120, 121]. Systemic PLTP deficiency in mice also is associated with a reduced thrombotic response [94] and a reduced abdominal aortic aneurysm [100]. In rabbits, overexpression of PLTP increases atherosclerotic lesions after a high-fat diet feeding, compared with controls [78]. In general, PLTP is a proven risk factor of atherosclerosis in animal models.

1.11 Conclusion

PLTP plays a role in CVD development, and it might be related with dyslipidemia, inflammation, hypercoagulation, type II diabetes, and metabolic syndrome. However, we still do not have a whole view for PLTP function intracellularly. More epidemiological studies are needed to gain insights into the role of PLTP in human CVD. After about 20 years work, a question is asked: could PLTP inhibition be a treatment for dyslipidemia and CVD? Our answer is "Yes." However, we have to be aware of some adverse effects of such an inhibition, for instance, it could impairment of LPS neutralization. Anyhow, based on our knowledge, so far, PLTP inhibition in human CVD patients should have much more beneficial effects than unwanted effects (Fig. 1.2).

Acknowledgment This chapter was modified from the paper published by Xian-Cheng Jiang in *Journal of Lipid Research* (Jiang XC, 2018, 59(5):764–771). The related contents are re-used with the permission. This work was supported in part by Merit Review Award # I01 RX000900-01) from the United States (US) Department of Veterans Affairs and by National Institutes of Health grant R56HL121409, RO1, HL139582 and RO1, HL149730.

References

1. Bruce C, Beamer LJ, Tall AR (1998) The implications of the structure of the bactericidal/permeability-increasing protein on the lipid-transfer function of the cholesteryl ester transfer protein. Curr Opin Struct Biol 8(4):426–434
2. Day JR, Albers JJ, Lofton-Day CE, Gilbert TL, Ching AF, Grant FJ et al (1994) Complete cDNA encoding human phospholipid transfer protein from human endothelial cells. J Biol Chem 269 (12):9388–9391
3. Massey JB, Hickson D, She HS, Sparrow JT, Via DP, Gotto AM Jr et al (1984) Measurement and prediction of the rates of spontaneous transfer of

phospholipids between plasma lipoproteins. Biochim Biophys Acta 794(2):274–280
4. Oka T, Kujiraoka T, Ito M, Egashira T, Takahashi S, Nanjee MN et al (2000) Distribution of phospholipid transfer protein in human plasma: presence of two forms of phospholipid transfer protein, one catalytically active and the other inactive. J Lipid Res 41 (10):1651–1657
5. Siggins S, Karkkainen M, Tenhunen J, Metso J, Tahvanainen E, Olkkonen VM et al (2004) Quantitation of the active and low-active forms of human plasma phospholipid transfer protein by ELISA. J Lipid Res 45(2):387–395
6. Cheung MC, Albers JJ (2006) Active plasma phospholipid transfer protein is associated with apoA-I- but not apoE-containing lipoproteins. J Lipid Res 47 (6):1315–1321
7. Jiang XC, Bruce C (1995) Regulation of murine plasma phospholipid transfer protein activity and mRNA levels by lipopolysaccharide and high cholesterol diet. J Biol Chem 270(29):17133–17138
8. Cao G, Beyer TP, Yang XP, Schmidt RJ, Zhang Y, Bensch WR et al (2002) Phospholipid transfer protein is regulated by liver X receptors in vivo. J Biol Chem 277(42):39561–39565
9. Valenta DT, Ogier N, Bradshaw G, Black AS, Bonnet DJ, Lagrost L et al (2006) Atheroprotective potential of macrophage-derived phospholipid transfer protein in low-density lipoprotein receptor-deficient mice is overcome by apolipoprotein AI overexpression. Arterioscler Thromb Vasc Biol 26 (7):1572–1578
10. Lee-Rueckert M, Vikstedt R, Metso J, Ehnholm C, Kovanen PT, Jauhiainen M (2006) Absence of endogenous phospholipid transfer protein impairs ABCA1-dependent efflux of cholesterol from macrophage foam cells. J Lipid Res 47(8):1725–1732
11. Desrumaux CM, Mak PA, Boisvert WA, Masson D, Stupack D, Jauhiainen M et al (2003) Phospholipid transfer protein is present in human atherosclerotic lesions and is expressed by macrophages and foam cells. J Lipid Res 44(8):1453–1461
12. O'Brien KD, Vuletic S, McDonald TO, Wolfbauer G, Lewis K, Tu AY et al (2003) Cell-associated and extracellular phospholipid transfer protein in human coronary atherosclerosis. Circulation 108 (3):270–274
13. Yazdanyar A, Jiang XC (2012) Liver phospholipid transfer protein (PLTP) expression with a PLTP-null background promotes very low-density lipoprotein production in mice. Hepatology 56(2):576–584
14. Yazdanyar A, Quan W, Jin W, Jiang XC (2013) Liver-specific phospholipid transfer protein deficiency reduces high-density lipoprotein and non-high-density lipoprotein production in mice. Arterioscler Thromb Vasc Biol 33(9):2058–2064
15. Jiang XC, Bruce C, Mar J, Lin M, Ji Y, Francone OL et al (1999) Targeted mutation of plasma phospholipid transfer protein gene markedly reduces high-density lipoprotein levels. J Clin Invest 103 (6):907–914
16. Krause BR, Hartman AD (1984) Adipose tissue and cholesterol metabolism. J Lipid Res 25(2):97–110
17. Jiang H, Yazdanyar A, Lou B, Chen Y, Zhao X, Li R et al (2015) Adipocyte phospholipid transfer protein and lipoprotein metabolism. Arterioscler Thromb Vasc Biol 35(2):316–322
18. Jiang XC, D'Armiento J, Mallampalli RK, Mar J, Yan SF, Lin M (1998) Expression of plasma phospholipid transfer protein mRNA in normal and emphysematous lungs and regulation by hypoxia. J Biol Chem 273(25):15714–15718
19. DuPage M, Dooley AL, Jacks T (2009) Conditional mouse lung cancer models using adenoviral or lentiviral delivery of Cre recombinase. Nat Protoc 4 (7):1064–1072
20. Vuletic S, Jin LW, Marcovina SM, Peskind ER, Moller T, Albers JJ (2003) Widespread distribution of PLTP in human CNS: evidence for PLTP synthesis by glia and neurons, and increased levels in Alzheimer's disease. J Lipid Res 44(6):1113–1123
21. Vuletic S, Peskind ER, Marcovina SM, Quinn JF, Cheung MC, Kennedy H et al (2005) Reduced CSF PLTP activity in Alzheimer's disease and other neurologic diseases; PLTP induces ApoE secretion in primary human astrocytes in vitro. J Neurosci Res 80(3):406–413
22. Albers JJ, Vuletic S, Cheung MC (2012) Role of plasma phospholipid transfer protein in lipid and lipoprotein metabolism. Biochim Biophys Acta 1821(3):345–357
23. Desrumaux C, Risold PY, Schroeder H, Deckert V, Masson D, Athias A et al (2005) Phospholipid transfer protein (PLTP) deficiency reduces brain vitamin E content and increases anxiety in mice. FASEB J Off Publ Fed Am Soc Exp Biol 19(2):296–297
24. Desrumaux C, Pisoni A, Meunier J, Deckert V, Athias A, Perrier V et al (2013) Increased amyloid-beta peptide-induced memory deficits in phospholipid transfer protein (PLTP) gene knockout mice. Neuropsychopharmacology Off Publ Am Coll Neuropsychopharmacol 38(5):817–825
25. Chirackal Manavalan AP, Kober A, Metso J, Lang I, Becker T, Hasslitzer K et al (2014) Phospholipid transfer protein is expressed in cerebrovascular endothelial cells and involved in high density lipoprotein biogenesis and remodeling at the blood-brain barrier. J Biol Chem 289(8):4683–4698
26. Zhou T, He Q, Tong Y, Zhan R, Xu F, Fan D et al (2014) Phospholipid transfer protein (PLTP) deficiency impaired blood-brain barrier integrity by increasing cerebrovascular oxidative stress. Biochem Biophys Res Commun 445(2):352–356
27. Huuskonen J, Wohlfahrt G, Jauhiainen M, Ehnholm C, Teleman O, Olkkonen VM (1999) Structure and phospholipid transfer activity of human PLTP: analysis by molecular modeling and site-directed mutagenesis. J Lipid Res 40(6):1123–1130

28. Kawano K, Qin SC, Lin M, Tall AR, Jiang XC (2000) Cholesteryl ester transfer protein and phospholipid transfer protein have nonoverlapping functions in vivo. J Biol Chem 275(38):29477–29481
29. Tollefson JH, Ravnik S, Albers JJ (1988) Isolation and characterization of a phospholipid transfer protein (LTP-II) from human plasma. J Lipid Res 29 (12):1593–1602
30. Tu AY, Albers JJ (2001) Glucose regulates the transcription of human genes relevant to HDL metabolism: responsive elements for peroxisome proliferator-activated receptor are involved in the regulation of phospholipid transfer protein. Diabetes 50(8):1851–1856
31. Riemens SC, van Tol A, Sluiter WJ, Dullaart RP (1999) Plasma phospholipid transfer protein activity is lowered by 24-h insulin and acipimox administration: blunted response to insulin in type 2 diabetic patients. Diabetes 48(8):1631–1637
32. Riemens SC, Van Tol A, Stulp BK, Dullaart RP (1999) Influence of insulin sensitivity and the TaqIB cholesteryl ester transfer protein gene polymorphism on plasma lecithin:cholesterol acyltransferase and lipid transfer protein activities and their response to hyperinsulinemia in non-diabetic men. J Lipid Res 40 (8):1467–1474
33. Lalanne F, Motta C, Pafumi Y, Lairon D, Ponsin G (2001) Modulation of the phospholipid transfer protein-mediated transfer of phospholipids by diacylglycerols. J Lipid Res 42(1):142–149
34. Tu AY, Albers JJ (2001) Functional analysis of the transcriptional activity of the mouse phospholipid transfer protein gene. Biochem Biophys Res Commun 287(4):921–926
35. Tu AY, Chen H, Johnson KA, Paigen B, Albers JJ (1997) Characterization of the mouse gene encoding phospholipid transfer protein. Gene 188(1):115–118
36. Laffitte BA, Joseph SB, Chen M, Castrillo A, Repa J, Wilpitz D et al (2003) The phospholipid transfer protein gene is a liver X receptor target expressed by macrophages in atherosclerotic lesions. Mol Cell Biol 23(6):2182–2191
37. Okazaki H, Goldstein JL, Brown MS, Liang G (2010) LXR-SREBP-1c-phospholipid transfer protein axis controls very low density lipoprotein (VLDL) particle size. J Biol Chem 285(9):6801–6810
38. Jin W, Wang X, Millar JS, Quertermous T, Rothblat GH, Glick JM et al (2007) Hepatic proprotein convertases modulate HDL metabolism. Cell Metab 6(2):129–136
39. Yu Y, Lei X, Jiang H, Li Z, Creemers JWM, Zhang M et al (2018) Prodomain of furin promotes phospholipid transfer protein proteasomal degradation in hepatocytes. J Am Heart Assoc 7(9):e008526
40. Tall AR, Hogan V, Askinazi L, Small DM (1978) Interaction of plasma high density lipoproteins with dimyristoyllecithin multilamellar liposomes. Biochemistry 17(2):322–326
41. Tall AR, Krumholz S, Olivecrona T, Deckelbaum RJ (1985) Plasma phospholipid transfer protein enhances transfer and exchange of phospholipids between very low density lipoproteins and high density lipoproteins during lipolysis. J Lipid Res 26 (7):842–851
42. Jauhiainen M, Metso J, Pahlman R, Blomqvist S, van Tol A, Ehnholm C (1993) Human plasma phospholipid transfer protein causes high density lipoprotein conversion. J Biol Chem 268(6):4032–4036
43. Huuskonen J, Olkkonen VM, Ehnholm C, Metso J, Julkunen I, Jauhiainen M (2000) Phospholipid transfer is a prerequisite for PLTP-mediated HDL conversion. Biochemistry 39(51):16092–16098
44. Rye KA, Jauhiainen M, Barter PJ, Ehnholm C (1998) Triglyceride-enrichment of high density lipoproteins enhances their remodelling by phospholipid transfer protein. J Lipid Res 39(3):613–622
45. Foger B, Santamarina-Fojo S, Shamburek RD, Parrot CL, Talley GD, Brewer HB Jr (1997) Plasma phospholipid transfer protein. Adenovirus-mediated overexpression in mice leads to decreased plasma high density lipoprotein (HDL) and enhanced hepatic uptake of phospholipids and cholesteryl esters from HDL. J Biol Chem 272(43):27393–27400
46. Yang XP, Yan D, Qiao C, Liu RJ, Chen JG, Li J et al (2003) Increased atherosclerotic lesions in apoE mice with plasma phospholipid transfer protein overexpression. Arterioscler Thromb Vasc Biol 23 (9):1601–1607
47. Jiang X, Francone OL, Bruce C, Milne R, Mar J, Walsh A et al (1996) Increased prebeta-high density lipoprotein, apolipoprotein AI, and phospholipid in mice expressing the human phospholipid transfer protein and human apolipoprotein AI transgenes. J Clin Invest 98(10):2373–2380
48. van Haperen R, van Tol A, Vermeulen P, Jauhiainen M, van Gent T, van den Berg P et al (2000) Human plasma phospholipid transfer protein increases the antiatherogenic potential of high density lipoproteins in transgenic mice. Arterioscler Thromb Vasc Biol 20(4):1082–1088
49. Qin S, Kawano K, Bruce C, Lin M, Bisgaier C, Tall AR et al (2000) Phospholipid transfer protein gene knock-out mice have low high density lipoprotein levels, due to hypercatabolism, and accumulate apoA-IV-rich lamellar lipoproteins. J Lipid Res 41 (2):269–276
50. Yan D, Navab M, Bruce C, Fogelman AM, Jiang XC (2004) PLTP deficiency improves the anti-inflammatory properties of HDL and reduces the ability of LDL to induce monocyte chemotactic activity. J Lipid Res 45(10):1852–1858
51. Kurano M, Yatomi Y (2018) Sphingosine 1-phosphate and atherosclerosis. J Atheroscler Thromb 25(1):16–26
52. Ng ML, Wadham C, Sukocheva OA (2017) The role of sphingolipid signalling in diabetes associated pathologies (Review). Int J Mol Med 39(2):243–252

53. Swan DJ, Kirby JA, Ali S (2010) Vascular biology: the role of sphingosine 1-phosphate in both the resting state and inflammation. J Cell Mol Med 14 (9):2211–2222
54. Shatrov VA, Lehmann V, Chouaib S (1997) Sphingosine-1-phosphate mobilizes intracellular calcium and activates transcription factor NF-kappa B in U937 cells. Biochem Biophys Res Commun 234 (1):121–124
55. Keul P, Lucke S, von Wnuck LK, Bode C, Graler M, Heusch G et al (2011) Sphingosine-1-phosphate receptor 3 promotes recruitment of monocyte/macrophages in inflammation and atherosclerosis. Circ Res 108(3):314–323
56. Takeya H, Gabazza EC, Aoki S, Ueno H, Suzuki K (2003) Synergistic effect of sphingosine 1-phosphate on thrombin-induced tissue factor expression in endothelial cells. Blood 102(5):1693–1700
57. Lee MH, Hammad SM, Semler AJ, Luttrell LM, Lopes-Virella MF, Klein RL (2010) HDL3, but not HDL2, stimulates plasminogen activator inhibitor-1 release from adipocytes: the role of sphingosine-1-phosphate. J Lipid Res 51(9):2619–2628
58. Iwaki S, Yamamura S, Asai M, Sobel BE, Fujii S (2012) Posttranscriptional regulation of expression of plasminogen activator inhibitor type-1 by sphingosine 1-phosphate in HepG2 liver cells. Biochim Biophys Acta 1819(11–12):1132–1141
59. Kimura T, Sato K, Malchinkhuu E, Tomura H, Tamama K, Kuwabara A et al (2003) High-density lipoprotein stimulates endothelial cell migration and survival through sphingosine 1-phosphate and its receptors. Arterioscler Thromb Vasc Biol 23 (7):1283–1288
60. Sachinidis A, Kettenhofen R, Seewald S, Gouni-Berthold I, Schmitz U, Seul C et al (1999) Evidence that lipoproteins are carriers of bioactive factors. Arterioscler Thromb Vasc Biol 19(10):2412–2421
61. Kimura T, Sato K, Kuwabara A, Tomura H, Ishiwara M, Kobayashi I et al (2001) Sphingosine 1-phosphate may be a major component of plasma lipoproteins responsible for the cytoprotective actions in human umbilical vein endothelial cells. J Biol Chem 276(34):31780–31785
62. Zhang B, Tomura H, Kuwabara A, Kimura T, Miura S, Noda K et al (2005) Correlation of high density lipoprotein (HDL)-associated sphingosine 1-phosphate with serum levels of HDL-cholesterol and apolipoproteins. Atherosclerosis 178(1):199–205
63. Maceyka M, Harikumar KB, Milstien S, Spiegel S (2012) Sphingosine-1-phosphate signaling and its role in disease. Trends Cell Biol 22(1):50–60
64. Pappu R, Schwab SR, Cornelissen I, Pereira JP, Regard JB, Xu Y et al (2007) Promotion of lymphocyte egress into blood and lymph by distinct sources of sphingosine-1-phosphate. Science 316 (5822):295–298
65. Christoffersen C, Obinata H, Kumaraswamy SB, Galvani S, Ahnstrom J, Sevvana M et al (2011) Endothelium-protective sphingosine-1-phosphate provided by HDL-associated apolipoprotein M. Proc Natl Acad Sci U S A 108(23):9613–9618
66. Christoffersen C, Jauhiainen M, Moser M, Porse B, Ehnholm C, Boesl M et al (2008) Effect of apolipoprotein M on high density lipoprotein metabolism and atherosclerosis in low density lipoprotein receptor knock-out mice. J Biol Chem 283(4):1839–1847
67. Bosteen MH, Madsen Svarrer EM, Bisgaard LS, Martinussen T, Madsen M, Nielsen LB et al (2017) Effects of apolipoprotein M in uremic atherosclerosis. Atherosclerosis 265:93–101
68. Yu Y, Guo S, Feng Y, Feng L, Cui Y, Song G et al (2014) Phospholipid transfer protein deficiency decreases the content of S1P in HDL via the loss of its transfer capability. Lipids 49(2):183–190
69. Moerland M, Samyn H, van Gent T, van Haperen R, Dallinga-Thie G, Grosveld F et al (2008) Acute elevation of plasma PLTP activity strongly increases pre-existing atherosclerosis. Arterioscler Thromb Vasc Biol 28(7):1277–1282
70. Samyn H, Moerland M, van Gent T, van Haperen R, Grosveld F, van Tol A et al (2009) Elevation of systemic PLTP, but not macrophage-PLTP, impairs macrophage reverse cholesterol transport in transgenic mice. Atherosclerosis 204(2):429–434
71. Oram JF, Wolfbauer G, Vaughan AM, Tang C, Albers JJ (2003) Phospholipid transfer protein interacts with and stabilizes ATP-binding cassette transporter A1 and enhances cholesterol efflux from cells. J Biol Chem 278(52):52379–52385
72. Oram JF, Wolfbauer G, Tang C, Davidson WS, Albers JJ (2008) An amphipathic helical region of the N-terminal barrel of phospholipid transfer protein is critical for ABCA1-dependent cholesterol efflux. J Biol Chem 283(17):11541–11549
73. Moerland M, Samyn H, van Gent T, Jauhiainen M, Metso J, van Haperen R et al (2007) Atherogenic, enlarged, and dysfunctional HDL in human PLTP/apoA-I double transgenic mice. J Lipid Res 48 (12):2622–2631
74. Kuwano T, Bi X, Cipollari E, Yasuda T, Lagor WR, Szapary HJ et al (2017) Overexpression and deletion of phospholipid transfer protein reduce HDL mass and cholesterol efflux capacity but not macrophage reverse cholesterol transport. J Lipid Res 58 (4):731–741
75. Jiang XC, Qin S, Qiao C, Kawano K, Lin M, Skold A et al (2001) Apolipoprotein B secretion and atherosclerosis are decreased in mice with phospholipid-transfer protein deficiency. Nat Med 7(7):847–852
76. Lie J, de Crom R, van Gent T, van Haperen R, Scheek L, Lankhuizen I et al (2002) Elevation of plasma phospholipid transfer protein in transgenic mice increases VLDL secretion. J Lipid Res 43 (11):1875–1880
77. van Haperen R, Samyn H, van Gent T, Zonneveld AJ, Moerland M, Grosveld F et al (2009) Novel roles of hepatic lipase and phospholipid transfer protein in

VLDL as well as HDL metabolism. Biochim Biophys Acta 1791(10):1031–1036
78. Masson D, Deckert V, Gautier T, Klein A, Desrumaux C, Viglietta C et al (2011) Worsening of diet-induced atherosclerosis in a new model of transgenic rabbit expressing the human plasma phospholipid transfer protein. Arterioscler Thromb Vasc Biol 31(4):766–774
79. Manchekar M, Liu Y, Sun Z, Richardson PE, Dashti N (2015) Phospholipid transfer protein plays a major role in the initiation of apolipoprotein B-containing lipoprotein assembly in mouse primary hepatocytes. J Biol Chem 290(13):8196–8205
80. Teslovich TM, Musunuru K, Smith AV, Edmondson AC, Stylianou IM, Koseki M et al (2010) Biological, clinical and population relevance of 95 loci for blood lipids. Nature 466(7307):707–713
81. Dullaart RP, van Tol A, Dallinga-Thie GM (2013) Phospholipid transfer protein, an emerging cardiometabolic risk marker: is it time to intervene? Atherosclerosis 228(1):38–41
82. Cuchel M, Bloedon LT, Szapary PO, Kolansky DM, Wolfe ML, Sarkis A et al (2007) Inhibition of microsomal triglyceride transfer protein in familial hypercholesterolemia. N Engl J Med 356(2):148–156
83. Cano A, Buque X, Martinez-Una M, Aurrekoetxea I, Menor A, Garcia-Rodriguez JL et al (2011) Methionine adenosyltransferase 1A gene deletion disrupts hepatic very low-density lipoprotein assembly in mice. Hepatology 54(6):1975–1986
84. Uchiyama S, Shimizu T, Shirasawa T (2006) CuZn-SOD deficiency causes ApoB degradation and induces hepatic lipid accumulation by impaired lipoprotein secretion in mice. J Biol Chem 281 (42):31713–31719
85. Jiang XC, Tall AR, Qin S, Lin M, Schneider M, Lalanne F et al (2002) Phospholipid transfer protein deficiency protects circulating lipoproteins from oxidation due to the enhanced accumulation of vitamin E. J Biol Chem 277(35):31850–31856
86. Ogier N, Klein A, Deckert V, Athias A, Bessede G, Le Guern N et al (2007) Cholesterol accumulation is increased in macrophages of phospholipid transfer protein-deficient mice: normalization by dietary alpha-tocopherol supplementation. Arterioscler Thromb Vasc Biol 27(11):2407–2412
87. Jiang XC, Li Z, Liu R, Yang XP, Pan M, Lagrost L et al (2005) Phospholipid transfer protein deficiency impairs apolipoprotein-B secretion from hepatocytes by stimulating a proteolytic pathway through a relative deficiency of vitamin E and an increase in intracellular oxidants. J Biol Chem 280(18):18336–18340
88. Kaser S, Sandhofer A, Foger B, Ebenbichler CF, Igelseder B, Malaimare L et al (2001) Influence of obesity and insulin sensitivity on phospholipid transfer protein activity. Diabetologia 44(9):1111–1117
89. Tzotzas T, Dumont L, Triantos A, Karamouzis M, Constantinidis T, Lagrost L (2006) Early decreases in plasma lipid transfer proteins during weight reduction. Obesity (Silver Spring) 14(6):1038–1045
90. Borggreve SE, De Vries R, Dullaart RP (2003) Alterations in high-density lipoprotein metabolism and reverse cholesterol transport in insulin resistance and type 2 diabetes mellitus: role of lipolytic enzymes, lecithin:cholesterol acyltransferase and lipid transfer proteins. Eur J Clin Investig 33 (12):1051–1069
91. van Tol A (2002) Phospholipid transfer protein. Curr Opin Lipidol 13(2):135–139
92. Song G, Zong C, Shao M, Yu Y, Liu Q, Wang H et al (2019) Phospholipid transfer protein (PLTP) deficiency attenuates high fat diet induced obesity and insulin resistance. Biochim Biophys Acta Mol Cell Biol Lipids 1864(10):1305–1313
93. Klein A, Deckert V, Schneider M, Dutrillaux F, Hammann A, Athias A et al (2006) Alpha-tocopherol modulates phosphatidylserine externalization in erythrocytes: relevance in phospholipid transfer protein-deficient mice. Arterioscler Thromb Vasc Biol 26(9):2160–2167
94. Desrumaux C, Deckert V, Lemaire-Ewing S, Mossiat C, Athias A, Vandroux D et al (2010) Plasma phospholipid transfer protein deficiency in mice is associated with a reduced thrombotic response to acute intravascular oxidative stress. Arterioscler Thromb Vasc Biol 30(12):2452–2457
95. Oslakovic C, Krisinger MJ, Andersson A, Jauhiainen M, Ehnholm C, Dahlback B (2009) Anionic phospholipids lose their procoagulant properties when incorporated into high density lipoproteins. J Biol Chem 284(9):5896–5904
96. Deguchi H, Wolfbauer G, Cheung MC, Banerjee Y, Elias DJ, Fernandez JA et al (2015) Inhibition of thrombin generation in human plasma by phospholipid transfer protein. Thromb J 13:24
97. Zhao XM, Wang Y, Yu Y, Jiang H, Babinska A, Chen XY et al (2018) Plasma phospholipid transfer protein promotes platelet aggregation. Thromb Haemost 118(12):2086–2097
98. Schlitt A, Liu J, Yan D, Mondragon-Escorpizo M, Norin AJ, Jiang XC (2005) Anti-inflammatory effects of phospholipid transfer protein (PLTP) deficiency in mice. Biochim Biophys Acta 1733(2–3):187–191
99. Shelly L, Royer L, Sand T, Jensen H, Luo Y (2008) Phospholipid transfer protein deficiency ameliorates diet-induced hypercholesterolemia and inflammation in mice. J Lipid Res 49(4):773–781
100. Deckert V, Kretz B, Habbout A, Raghay K, Labbe J, Abello N et al (2013) Development of abdominal aortic aneurysm is decreased in mice with plasma phospholipid transfer protein deficiency. Am J Pathol 183(3):975–986
101. Desrumaux C, Lemaire-Ewing S, Ogier N, Yessoufou A, Hammann A, Sequeira-Le Grand A et al (2016) Plasma phospholipid transfer protein (PLTP) modulates adaptive immune functions

through alternation of T helper cell polarization. Cell Mol Immunol 13(6):795–804
102. Gautier T, Klein A, Deckert V, Desrumaux C, Ogier N, Sberna AL et al (2008) Effect of plasma phospholipid transfer protein deficiency on lethal endotoxemia in mice. J Biol Chem 283 (27):18702–18710
103. Brehm A, Geraghty P, Campos M, Garcia-Arcos I, Dabo AJ, Gaffney A et al (2014) Cathepsin G degradation of phospholipid transfer protein (PLTP) augments pulmonary inflammation. FASEB J Off Publ Fed Am Soc Exp Biol 28(5):2318–2331
104. Vuletic S, Dong W, Wolfbauer G, Tang C, Albers JJ (2011) PLTP regulates STAT3 and NFkappaB in differentiated THP1 cells and human monocyte-derived macrophages. Biochim Biophys Acta 1813 (10):1917–1924
105. Yu Y, Cui Y, Zhao Y, Liu S, Song G, Jiao P et al (2016) The binding capability of plasma phospholipid transfer protein, but not HDL pool size, is critical to repress LPS induced inflammation. Sci Rep 6:20845
106. Schlitt A, Bickel C, Thumma P, Blankenberg S, Rupprecht HJ, Meyer J et al (2003) High plasma phospholipid transfer protein levels as a risk factor for coronary artery disease. Arterioscler Thromb Vasc Biol 23(10):1857–1862
107. de Vries R, Dallinga-Thie GM, Smit AJ, Wolffenbuttel BH, van Tol A, Dullaart RP (2006) Elevated plasma phospholipid transfer protein activity is a determinant of carotid intima-media thickness in type 2 diabetes mellitus. Diabetologia 49 (2):398–404
108. Colhoun HM, Scheek LM, Rubens MB, Van Gent T, Underwood SR, Fuller JH et al (2001) Lipid transfer protein activities in type 1 diabetic patients without renal failure and nondiabetic control subjects and their association with coronary artery calcification. Diabetes 50(3):652–659
109. Schlitt A, Blankenberg S, Bickel C, Lackner KJ, Heine GH, Buerke M et al (2009) PLTP activity is a risk factor for subsequent cardiovascular events in CAD patients under statin therapy: the AtheroGene study. J Lipid Res 50(4):723–729
110. Vergeer M, Boekholdt SM, Sandhu MS, Ricketts SL, Wareham NJ, Brown MJ et al (2010) Genetic variation at the phospholipid transfer protein locus affects its activity and high-density lipoprotein size and is a novel marker of cardiovascular disease susceptibility. Circulation 122(5):470–477
111. Robins SJ, Lyass A, Brocia RW, Massaro JM, Vasan RS (2013) Plasma lipid transfer proteins and cardiovascular disease. The Framingham Heart Study. Atherosclerosis 228(1):230–236
112. Cavusoglu E, Marmur JD, Chhabra S, Chopra V, Eng C, Jiang XC (2009) Relation of baseline plasma phospholipid transfer protein (PLTP) activity to left ventricular systolic dysfunction in patients referred for coronary angiography. Atherosclerosis 207 (1):261–265
113. Chen X, Sun A, Zou Y, Ge J, Kamran H, Jiang XC et al (2013) High PLTP activity is associated with depressed left ventricular systolic function. Atherosclerosis 228(2):438–442
114. Cavusoglu E, Marmur JD, Chhabra S, Hojjati MR, Yanamadala S, Chopra V et al (2015) Elevated baseline plasma phospholipid protein (PLTP) levels are an independent predictor of long-term all-cause mortality in patients with diabetes mellitus and known or suspected coronary artery disease. Atherosclerosis 239(2):503–508
115. Yatsuya H, Tamakoshi K, Hattori H, Otsuka R, Wada K, Zhang H et al (2004) Serum phospholipid transfer protein mass as a possible protective factor for coronary heart diseases. Circ J 68(1):11–16
116. Huuskonen J, Ekstrom M, Tahvanainen E, Vainio A, Metso J, Pussinen P et al (2000) Quantification of human plasma phospholipid transfer protein (PLTP): relationship between PLTP mass and phospholipid transfer activity. Atherosclerosis 151(2):451–461
117. Dullaart RP, De Vries R, Scheek L, Borggreve SE, Van Gent T, Dallinga-Thie GM et al (2004) Type 2 diabetes mellitus is associated with differential effects on plasma cholesteryl ester transfer protein and phospholipid transfer protein activities and concentrations. Scand J Clin Lab Invest 64 (3):205–215
118. Ruhling K, Lang A, Richard F, Van Tol A, Eisele B, Herzberg V et al (1999) Net mass transfer of plasma cholesteryl esters and lipid transfer proteins in normolipidemic patients with peripheral vascular disease. Metabolism 48(11):1361–1366
119. Schgoer W, Mueller T, Jauhiainen M, Wehinger A, Gander R, Tancevski I et al (2008) Low phospholipid transfer protein (PLTP) is a risk factor for peripheral atherosclerosis. Atherosclerosis 196(1):219–226
120. van Haperen R, van Tol A, van Gent T, Scheek L, Visser P, van der Kamp A et al (2002) Increased risk of atherosclerosis by elevated plasma levels of phospholipid transfer protein. J Biol Chem 277 (50):48938–48943
121. van Haperen R, van Gent T, van Tol A, de Crom R (2013) Elevated expression of PLTP is atherogenic in apolipoprotein E deficient mice. Atherosclerosis 227 (1):37–42

Cholesteryl Ester Transfer Protein and Lipid Metabolism and Cardiovascular Diseases

2

Helena C. F. Oliveira and Helena F. Raposo

Abstract

In this chapter, we present the major advances in CETP research since the detection, isolation, and characterization of its activity in the plasma of humans and several species. Since CETP is a major modulator of HDL plasma levels, the clinical importance of CETP activity was recognized very early. We describe the participation of CETP in reverse cholesterol transport, conflicting results in animal and human genetic studies, possible new functions of CETP, and the results of the main clinical trials on CETP inhibition. Despite major setbacks in clinical trials, the hypothesis that CETP inhibitors are anti-atherogenic in humans is still being tested.

Keywords

Cholesteryl ester transfer protein (CETP) · Lipoprotein metabolism · Lipid metabolism · CETP inhibitors · Cardiovascular diseases

2.1 Introduction

Cholesteryl ester transfer protein (CETP) was first isolated from plasma and characterized in 1978 [1, 2]. Its activity consists of promoting a net transport of cholesteryl ester (CE) from HDL to VLDL and LDL in exchange for triglycerides (TG). This plasma activity was found in many species, including primates, rabbits, hamsters, reptiles, and fishes, but it is not present in mice, rats, and dogs [3, 4]. The protein purification [5] and cDNA cloning in 1987 [6] and human gene cloning in 1990 [7] fomented investigations on CETP. Since then, a growing interest in CETP research has been observed (Fig. 2.1).

Chemical purification showed that CETP is a plasma hydrophobic glycoprotein consisting of 476 amino acids with an apparent molecular weight of 66–74 kDa [5]. Protein purification and gene cloning allowed for the production of recombinant protein, antibodies, and transgenic mouse models that were very important for further studies on the structure, function, gene expression regulation, human polymorphism screening, and relationship with cardiovascular diseases [8]. However, it took approximately 30 years after the isolation of CETP plasma activity to resolve its molecular structure by Qiu and collaborators in 2007 [9]. This milestone finding allowed for the elucidation of the CETP mechanism of action and a more refined drug design targeting CETP inhibition. In fact, the current view supports the formation of a ternary complex of CETP with donor and acceptor lipoproteins, allowing the transfer of neutral lipids through a hydrophobic tunnel inside its structure [10].

H. C. F. Oliveira (✉) · H. F. Raposo
Department of Structural and Functional Biology, Biology Institute, State University of Campinas, Campinas, SP, Brazil
e-mail: ho98@unicamp.br

X.-C. Jiang (ed.), *Lipid Transfer in Lipoprotein Metabolism and Cardiovascular Disease*, Advances in Experimental Medicine and Biology 1276, https://doi.org/10.1007/978-981-15-6082-8_2

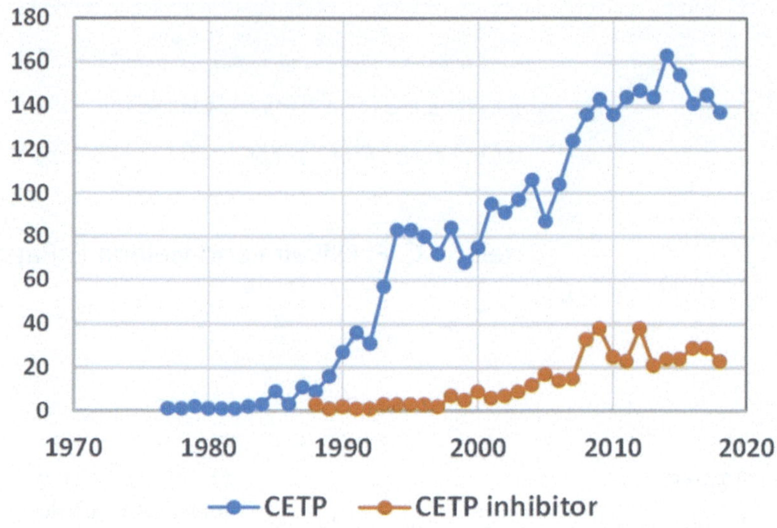

Fig. 2.1 Number of scientific publications through the years with the keywords cholesteryl-ester-transfer-protein or CETP (blue line) and cholesteryl-ester-transfer-protein inhibitor or CETP inhibitor (orange line). PubMed searches were performed on January 13, 2020, limiting the results by the presence of the keywords in the field: Title and Abstract

The human CETP gene is located on chromosome 16 (16q13), spans 25 kb, and contains 17 exons. It is conserved in chimpanzees, rhesus monkeys, chickens, zebrafish, and frogs. Orthologs of the human CETP gene have been described in 217 organisms (Gene ID: 1071, https://www.ncbi.nlm.nih.gov/gene/1071). There is a major CETP gene alternative splicing product that may occur at high frequency [11]. The alternatively spliced product, missing exon 9, is in-frame, and the loss of 60 amino acids in the central region of the protein leads to its inactivation [12]. Cotransfection of wild-type (full length) and exon 9-deleted constructs significantly reduces wild-type secretion and activity [12, 13].

Structurally, CETP belongs to the BPI-like protein family, which includes bactericidal/permeability-increasing protein (BPI), LPS (lipopolysaccharide)-binding protein (LBP), phospholipid transfer protein (PLTP), and CETP. While BPI and LBP are known to be involved in innate immunity against bacteria through their ability to sense lipopolysaccharides, PLTP and CETP are characterized by lipid exchange among plasma lipoproteins [14]. These proteins are also classified within a large family of extracellular and intracellular membrane contact site proteins that contain the tubular lipid binding protein (TULIP) domain [15].

2.2 Reverse Cholesterol Transport

Much of the scientific interest in CETP arises from the fact that it causes a reduction in HDL cholesterol plasma levels. A well-documented negative correlation between plasma concentrations of HDL and cardiovascular risk has been established since the late 1970s [16, 17]. Although the protective mechanisms of HDL were not known at that time, the concept that it was the main player in a process called reverse cholesterol transport (RCT) was emerging. In this process, first proposed by Glomset [18], HDL facilitates the uptake of cholesterol from peripheral tissues and its transport to the liver for catabolism and excretion, thus avoiding accumulation of cholesterol in the plasma compartment and hence in the arteries (Fig. 2.2). In theory, CETP may play a dual role in RCT. It may add an alternative route for delivering cholesterol to the liver via LDL receptor and LDL-receptor related protein (LRP) pathways (often called indirect RCT). However, if these pathways are malfunctioning, CETP activity results in increased LDL cholesterol levels and a risk of atherosclerosis development (Fig. 2.2).

The engineering of CETP transgenic mice in the early 1990s [19–21] was very useful to study its functions, gene expression regulation, and

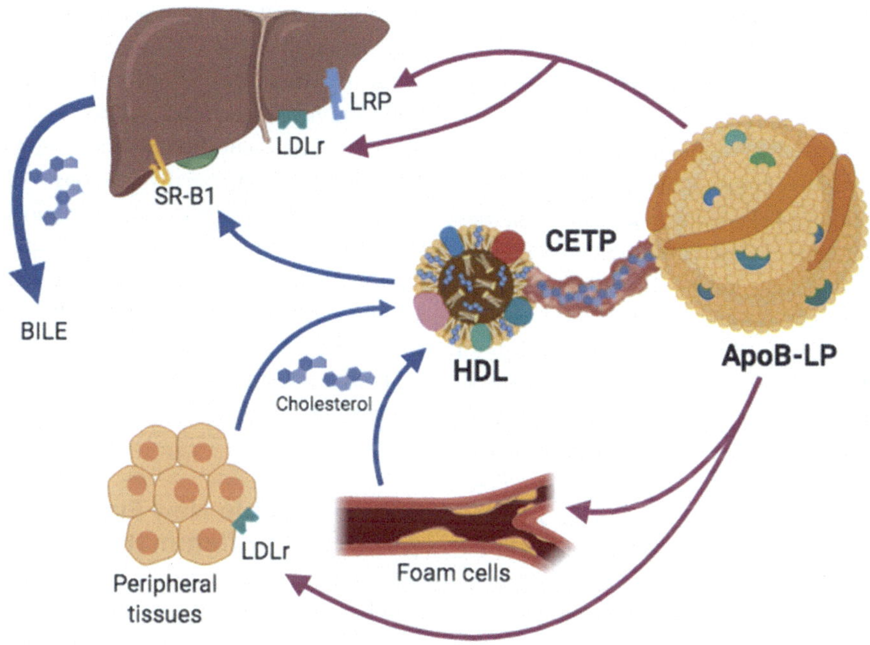

Fig. 2.2 Simplified overview of reverse cholesterol transport (RCT). The first step of RCT is the removal of cholesterol from cell membranes through the interaction of HDL subspecies with ABCA1/G1 membrane transporters. Then, the enzyme lecithin cholesterol acyl transferase (LCAT) esterifies cholesterol on the surface of HDL, which then enters into the hydrophobic core of the HDL particle. HDL-cholesteryl ester has two fates: it is either directly and selectively delivered to steroidogenic tissues (liver, adrenal, and gonadal) via scavenger receptor class B type I (SRBI) or it is transferred to VLDL and LDL (apoB-LP) in exchange for triglycerides (TG) by the action of CETP. The VLDL + LDL enriched in CE are then taken up mainly by the liver through LDL receptors (LDLr) and LDL receptor-related proteins (LRP). The pathway that includes CETP is termed indirect RCT. Once in the liver, cholesterol and cholesterol-derived bile acids are secreted into the bile and excreted from the body via the feces

susceptibility to diseases. In humans, CETP mRNA is mostly expressed in the liver, spleen, and adipose tissue [6]. These data are confirmed in the Human Protein Atlas that analyzed 27 tissue samples from 95 human individuals [22] and included lymph nodes and placenta as high CETP-expressing tissues. Hepatic synthesis is the major source of CETP found in the plasma of primates [23], with Kupffer cells being the predominant source in humans [24]. Human CETP transgenic mice expressing a natural promoter-driven CETP minigene reproduced a similar human pattern of tissue expression [20]. Thus, it was possible to study gene regulation in this model. For instance, CETP gene expression was shown to be upregulated by dietary cholesterol [20, 25, 26], thyroid hormones [27], and fish oil and fibrate treatment [28], while it is downregulated by corticosteroids [29] and hyperinsulinemia [30]. When CETP is expressed in hyperlipidemic atherosclerosis-prone models, such as the LDL receptor knockout mouse or apo E knockout mouse, an acceleration of diet-induced atherosclerosis is observed [31–33]. On the other hand, other experimental mouse model data support the concept that CETP may have a protective role against atherosclerosis in conditions where LDL receptor function is preserved. For instance, CETP expression decreased atherosclerotic lesions in hypertriglyceridemic mice [34], when cholesterol esterification rates were high due to LCAT overexpression [35, 36], in castrated mice [37, 38], in diabetic mice [39, 40], and in SR-BI knockout mice [41, 42].

2.3 Human Genetic Studies

The importance of plasma CETP in lipoprotein metabolism was long ago demonstrated by the discovery of CETP-deficient subjects with marked hyperalphalipoproteinemia in the Japanese population [43]. Among several mutations of the CETP gene, two are common mutations in that population: an intron 14 splicing defect (Int14 + 1 G --> A) and an exon 15 missense mutation (D442G). Although elevated levels of HDL cholesterol are a marker for protection against atherosclerosis, subjects with CETP deficiency show a variety of abnormalities in the composition and function of HDL that impair RCT [44]. Epidemiological studies in Japanese Americans living in Hawaii and Japanese in the Omagari area of Japan, where the intron 14 splicing defect is markedly frequent, have shown a relatively increased incidence of coronary atherosclerosis in CETP deficiency [45, 46]. On the other hand, the TaqIB polymorphism-B2 allele, with low CETP mass and moderate increases in HDL cholesterol, has been associated with a decreased risk for coronary heart disease in many studies, including the Framingham Offspring Study [46, 47]. However, controversy remains, since subsequent reports of the Framingham Heart study and Honolulu Heart Program presented discrepant results [48]. A more recent meta-analysis confirmed an association between the TaqIB polymorphism and the risk of myocardial infarct. This study suggests that the B2B2 genotype of the CETP TaqIB polymorphism is a protective factor against the development of myocardium infarct [49]. However, not all CETP-reducing polymorphisms are protective. A large meta-analysis of selected studies with three common and three uncommon CETP polymorphisms evaluated the associations of CETP genotypes, phenotypes, lipid levels, and coronary risk. The authors concluded that three CETP genotypes that are associated with moderate inhibition of CETP activity (and, therefore, modestly higher HDL-C levels) show weak (but significant) inverse associations with coronary risk, compatible with the expected reductions in risk for equivalent increases in HDL-cholesterol concentration [50]. Another meta-analysis suggests that two (out of seven) CETP polymorphisms (rs708272 [C>T] and rs1800775 [C>A]) may contribute to myocardium infarct susceptibility, especially among Caucasians [51]. Thus, it seems that the associations between CETP levels (due to gene polymorphisms) and cardiovascular diseases may be population-specific, highly dependent on the phenotype (level of CETP activity, LDL and HDL concentration/composition), and highly influenced by environmental factors.

2.4 Possible New Function of CETP: Anti-inflammatory

Similar to its family members, BPI and LBP, CETP may have an anti-inflammatory function that may be relevant not only for atherogenesis but also for immune responses [8]. Experimentally, inhibition of CETP with torcetrapib did not reduce atherosclerosis beyond the statin effect but induced more proinflammatory lesions in hypertriglyceridemic apoE3Leiden/CETP-expressing mice [52]. Normolipidemic human CETP-expressing mice were protected from mortality induced by bacterial lipopolysaccharide acute infusion [53] or polymicrobial sepsis induced by cecal ligation and puncture [54]. Regarding humans, the first large-scale trial with the CETP inhibitor torcetrapib was interrupted because of increased mortality due to several reasons, including infections (22% of noncardiovascular deaths) [55]. In addition, the mortality rate of patients with sepsis correlated with a reduction in plasma CETP concentrations [56]. However, this issue is still enigmatic. A recent report described that a gain-of-function variant (s1800777-A) of CETP promotes a profound reduction in HDL levels and reduced survival in patients with sepsis when compared with noncarriers [57]. Therefore, CETP anti-inflammatory effects may be a direct nonlipid transporting function or indirectly mediated by modulation of HDL size, composition, and concentration [58].

2.5 Does CETP Have an Intracellular Function?

As a lipid-binding and transport protein, hypothetically, CETP may play an intracellular physiological role.

Zhang and collaborators showed that transient transfection of CETP cDNA into COS-7 cells induced higher cholesterol efflux compared with mock-transfected cells, while lipid uptake was not affected. Conversely, the efflux of free cholesterol from macrophages obtained from CETP-deficient subjects was significantly decreased compared with that from normal subjects [59]. These data suggest that local intracellular CETP expression in macrophages plays an anti-atherogenic function, facilitating the removal of cholesterol from the cells.

Because the protein product of the major alternative splice variant of CETP (exon 9 deleted) is retained within the endoplasmic reticulum (ER) [12], Lira and collaborators studied whether the expression of CETP variants could induce ER stress [13]. Transient CETP transfections were performed in a human liposarcoma cell line (SW872) and a human embryonic kidney cell line (HEK293S). Surprisingly, not only the alternative spliced variants of CETP but also the full-length protein expression caused an induction of genes linked to the ER stress response [13]. Thus, although CETP is a secreted protein, intracellular CETP plays a complex role in modulating ER stress or the unfolded protein response.

Morton and collaborators have shown that CETP expression modulates cholesterol and triglyceride homeostasis in the SW872 human liposarcoma cell line. They showed that short-term partial inhibition of CETP in these cells using antisense oligonucleotides induces a phenotype characterized by inefficient mobilization of CE stores leading to CE accumulation [60]. In subsequent studies, CETP was chronically repressed with stably transfected oligonucleotides. CETP-deficient cells had decreased CE and TG biosynthesis and inefficient CETP-mediated translocation of CE and TG from the ER to their site of storage [61, 62]. On the other hand, when they stably overexpressed full-length CETP, SW872 cells accumulated 50% less TG due to a decrease in TG synthesis and a higher TG turnover rate, resulting in the formation of smaller and more metabolically active lipid droplets [63]. In agreement, Zhou and collaborators previously showed that transgenic mice expressing CETP under the control of an adipocyte-specific promoter (aP2) exhibited decreased adipocyte lipid content and size [64]. More recently, Izem et al. showed that exon 9-deleted CETP (the nonsecreted isoform) inhibits full-length CETP synthesis and promotes cellular triglyceride storage [65].

2.6 CETP Inhibition as a Target to Decrease Cardiovascular Diseases

From most human and experimental studies, it is clear that CETP is important clinically and has been a target for drug development. Dozens of review articles on CETP inhibition have been published in the last 5 years, most of them debating the viability of this strategy because of minor or no benefits of the tested molecules in reducing cardiovascular events so far. Despite the setbacks, the hypothesis that CETP inhibitors are anti-atherogenic in humans is still being tested. A summary of the results of the main trials with CETP inhibitors is presented in Box 2.1 and will be discussed below.

The concept that in vivo CETP inhibition could be anti-atherogenic was experimentally demonstrated in rabbits. CETP was inhibited with antisense oligodeoxynucleotides against CETP, reducing CETP mass and increasing HDL cholesterol (HDL-C), resulting in a reduced aortic lesion area [66]. In addition, anti-CETP immunotherapy in rabbits was able to reduce CETP activity, increase HDL, decrease LDL, and diminish atherosclerotic lesions [67].

Box 2.1 Main CETP Inhibitor Trials

Torcetrapib (Pfizer)
Atorvastatin + torcetrapib (vs. atorvastatin only)
Trial **ILLUSTRATE**
24-month follow-up
HDL-C increased by 61%
LDL-C decreased by 20%
Increase in systolic blood pressure of 4.6 mm Hg
From small to no favorable effect on atheroma volume (Nissen et al. 2007) [68]
Trial **ILLUMINATE**
12-month follow-up
HDL-C increased by 72.1%
LDL-C decreased by 24.9%
Increase in systolic blood pressure (5.4 mm Hg)
Increase in apo-CIII and aldosterone levels
Increase in mortality by CVD, cancer, and infections/sepsis (Barter et al. 2007) [55]
Dalcetrapib (Hoffmann-La Roche)
Trial: **dal-OUTCOMES**
Dalcetrapib (vs. placebo)
31-month follow-up
Increased HDL (25%), blood pressure (0.6 mm Hg), and C-reactive protein (hsPCR)
No changes in LDL and aldosterone
No benefits on CVD events/mortality.
Possible benefits in genetically defined population (ADCY9 gene, adenylate cyclase, AA allele) (Schwartz et al. 2012) [69]
Evacetrapib (Eli Lilly)
Trial: **ACCELERATE**
Evacetrapib (vs. placebo)
28-month follow-up
HDL-C increased by 132%
LDL-C decreased by 37%
Increased cholesterol efflux
Increased blood pressure (1.2 mm Hg) and hsPCR
No CVD benefits in 28 months (Lincoff et al. 2017) [70]
Anacetrapib (Merck)
Trial: **REVEAL**
Anacetrapib (vs. placebo) in patients receiving intensive atorvastatin therapy
4.1-year follow-up
HDL-C increased by 104% (midpoint)
LDL-C decreased by 41% (midpoint)
Slightly higher blood pressure (0.7 mm Hg)
No significant differences in the risk of death, cancer, or other serious adverse events
Associated with a lower incidence of new-onset diabetes (11%)
Lower incidence of major coronary events (10.8 vs. 11.8% CVD events) (HPS3/TIMI55–REVEAL Collaborative Group, Bowman L, et al. 2017) [71]
TA-8995 (obicetrapib) – funded by Dezima and undertaken by Xention
Trial: **TULIP**
Nine treatments with TA-8995 alone and combined with statins
Analyses at 12 weeks of treatment
LDL-C reduced by 45.3% (apoB reduced by 33.7%)
HDL-C increased by 179% (apoA-1 increased by 63.4%)
Combined with statins: decrease in LDL-C from 39.8% to 50.2% (Hovingh et al. 2015)
Increased preβ-HDL and cell cholesterol efflux (van Capelleveen et al. 2016) [79]

2.6.1 Torcetrapib

The first pharmacological molecule designed to inhibit CETP was torcetrapib. In two trials, ILLUSTRATE and ILLUMINATE, patients received atorvastatin + torcetrapib or atorvastatin alone and were followed for 12 or 24 months [55, 68]. Compared to atorvastatin monotherapy, torcetrapib + atorvastatin was able to increase HDL-C (61–72%) and decrease LDL-C (20–25%). However, an important side effect was the elevation of systolic blood pressure by approximately 5 mm Hg. There was a small or no favorable effect on atheroma volume [68]. Because torcetrapib-treated patients showed an increased number of cardiovascular events and death from both cardiovascular and noncardiovascular causes [55], the clinical trial was terminated prematurely. The increase in aldosterone levels and blood pressure was considered off-target toxicity rather than a CETP inhibition effect per se, increasing expectations for further generation of CETP inhibitors.

2.6.2 Dalcetrapib

The second molecule to go on to clinical trial was dalcetrapib (dal-OUTCOMES). Although the increase in systolic blood pressure was modest, dalcetrapib was less efficient at increasing HDL-C and reducing LDL-C, so the risk of major cardiovascular outcomes was not significantly altered. Dalcetrapib is considered a relatively weak inhibitor of CETP, meaning that a more potent CETP inhibitor could still be effective regarding clinical benefits in cardiovascular diseases (CVD) [69].

2.6.3 Evacetrapib

A potent CETP inhibitor, evacetrapib, was then evaluated in the ACELLERATE trial [70]. Indeed, evacetrapib caused marked favorable changes in the lipoprotein profile, increasing HDL-C by approximately 130% and reducing LDL-C by 37% compared to placebo. However, there were no significant benefits for CVD risks and events, and the trial was stopped due to futility at 28 months of treatment. Some could raise the possibility that longer treatment could show CVD improvements.

2.6.4 Anacetrapib

The most successful CETP inhibitor to date is anacetrapib. It in fact reduced the incidence of major coronary events [71]. Anacetrapib was added to intensive statin treatment in the REVEAL trial, a much larger and longer trial than the previous ones. Patients were followed up for 4 years, and at midpoint, HDL-C was increased by 104% in the anacetrapib group. The incidence of the primary outcome was reduced in the anacetrapib group (10.8% vs. 11.8% in the placebo group). Although there was no significant difference between groups during the first year of follow-up, the incidence of major coronary events after 1 year was significantly lower in the anacetrapib group (rate ratio, 0.88; 95% CI, 0.81 to 0.95; $P = 0.001$). Anacetrapib is a highly lipophilic drug that accumulates in adipose tissue, explaining its prolonged elimination profile [72]. No serious adverse events were identified, and there was only a slightly higher blood pressure (0.7 mm Hg) in the group of patients receiving anacetrapib. Another unexpected good finding of anacetrapib was its association with a lower incidence of new-onset diabetes cases. A recent meta-analysis of CETP inhibitor trials showed that CETP inhibitor therapy was associated with a significant 12% reduction in the incidence of diabetes and concluded that the improvement in glucose metabolism is at least in part related to the increase in HDL-C concentration [73].

Although CETP inhibitors are expected to increase HDL-C levels, their impact on reducing LDL-C has gained special attention. In the anacetrapib trial, LDL-C levels were reduced by 40%, as indicated by the "direct method" or by 17% when measured by beta-quantification. This discrepancy discloses the importance of understanding what different methods for LDL cholesterol quantification truly quantify. It is important to discover differential inhibitor effects across the whole spectrum of atherogenic apoB-containing lipoproteins [74]. CETP inhibition may also reduce the concentrations of triglyceride-rich remnant lipoproteins rather than affect size-specific LDL particles [75]. Regarding the mechanism of action, anacetrapib reduces LDL-C levels by increasing its catabolism, while the LDL-apoB-100 production rate is unchanged [76].

2.6.5 TA-8995 (Obicetrapib)

Another promising CETP inhibitor compound is TA-8995 (obicetrapib), which is well tolerated and shows beneficial effects on lipid and apolipoprotein profiles in healthy subjects and patients with mild dyslipidemia [77, 78]. TA-8995 was tested in a randomized, double blind, placebo-controlled phase 2 trial (TULIP) [78]. The safety and efficacy of TA-8995 were tested as a monotherapy and combined with statins. This CETP inhibitor reduces LDL cholesterol and apoB levels by 45% and 34%, respectively, conferring an additional decrease in LDL-C in combination with statins. HDL-C and apoA-1 increased up to 179% and 63%, respectively. TA-8995 also reduces lipoprotein(a) ranging from 27% to 37% [78]. TA-8995 is now registered in a new clinical trial (*ClinicalTrials.govidentifier (NCT number): NCT02241772*) to evaluate its effects on subjects with elevated Lp (a). A subsequent study advanced the possible mechanisms of TA-8995. Plasma samples from TULIP trial patients treated with TA-8995 were shown to dose-dependently increase total and ABCA1-specific cholesterol efflux capacity from the J774 macrophage cell line. These findings suggest that TA-8995 not only increases HDL-C and pre-Beta1-HDL particle levels but also promotes the functional properties of these particles. Whether these changes in HDL particle composition and functionality have a beneficial effect on cardiovascular outcome requires formal testing [79]. Thus, a cardiovascular disease outcome trial with TA-8995 is still needed to translate these effects into a reduction in cardiovascular disease events.

2.7 Concluding Remarks

Major advances in understanding CETP biology were obtained after protein purification, gene cloning, and molecular structure resolution. Studies in animal models have elucidated CETP functions, gene expression regulation, and relationships with diseases. Apart from its HDL-reducing systemic action, intriguing reports raise the possibility that CETP may have relevant local and novel nonlipid transfer functions. Animal and human studies suggest that CETP is important clinically and worthy as a target for drug development. The pros and cons of inhibiting CETP were discussed. Despite major setbacks in clinical trials, the hypothesis that CETP inhibitors are anti-atherogenic in humans is still being tested.

Acknowledgments We thank our co-authors on the original studies reviewed here, as well as other groups for their original contributions to the field. HCFO and HFR research was supported by Fundação de Amparo a Pesquisa do Estado de São Paulo (Fapesp grants #2017/17728-8, # 2015/17555-0 and # 2013/07607-8) and Conselho Nacional de Desenvolvimento Científico e Tecnológico (CNPq grants #300937/2018-0)

References

1. Chajek T, Fielding CJ (1978) Isolation and characterization of a human serum cholesteryl ester transfer protein. Proc Natl Acad Sci U S A 75:3445–3449
2. Pattnaik NM, Montes A, Hughes LB, Zilversmit DB (1978) Cholesteryl ester exchange protein in human plasma isolation and characterization. Biochim Biophys Acta 530:428–438
3. Ha YC, Barter PJ (1982) Differences in plasma cholesteryl ester transfer activity in sixteen vertebrate species. Comp Biochem Physiol B 71:265–269
4. Tsutsumi K, Hagi A, Inoue Y (2001) The relationship between plasma high density lipoprotein cholesterol levels and cholesteryl ester transfer protein activity in six species of healthy experimental animals. Biol Pharm Bull 24:579–581
5. Hesler CB, Swenson TL, Tall AR (1987) Purification and characterization of a human plasma cholesteryl ester transfer protein. J Biol Chem 262:2275–2282
6. Drayna D et al (1987) Cloning and sequencing of human cholesteryl ester transfer protein cDNA. Nature 327:632–634
7. Agellon LB et al (1990) Organization of the human cholesteryl ester transfer protein gene. Biochemistry 29:1372–1376
8. Oliveira HCF, de Faria EC (2011) Cholesteryl ester transfer protein: the controversial relation to atherosclerosis and emerging new biological roles. IUBMB Life 63:248–257
9. Qiu X et al (2007) Crystal structure of cholesteryl ester transfer protein reveals a long tunnel and four bound lipid molecules. Nat Struct Mol Biol 14:106–113

10. Charles MA, Kane JP (2012) New molecular insights into CETP structure and function: a review. J Lipid Res 53:1451–1458
11. Inazu A et al (1992) Alternative splicing of the mRNA encoding the human cholesteryl ester transfer protein. Biochemistry 31:2352–2358
12. Quinet E, Yang TP, Marinos C, Tall A (1993) Inhibition of the cellular secretion of cholesteryl ester transfer protein by a variant protein formed by alternative splicing of mRNA. J Biol Chem 268:16891–16894
13. Lira ME, Loomis AK, Paciga SA, Lloyd DB, Thompson JF (2008) Expression of CETP and of splice variants induces the same level of ER stress despite secretion efficiency differences. J Lipid Res 49:1955–1962
14. Alva V, Lupas AN (2016) The TULIP superfamily of eukaryotic lipid-binding proteins as a mediator of lipid sensing and transport. Biochim Biophys Acta 1861:913–923
15. Wong LH, Levine TP (2017) Tubular lipid binding proteins (TULIPs) growing everywhere. Biochim Biophys Acta Mol Cell Res 1864:1439–1449
16. Miller GJ, Miller NE (1975) Plasma-high-density-lipoprotein concentration and development of ischaemic heart-disease. Lancet Lond Engl 1:16–19
17. Gordon T, Castelli WP, Hjortland MC, Kannel WB, Dawber TR (1977) High density lipoprotein as a protective factor against coronary heart disease. The Framingham Study. Am J Med 62:707–714
18. Glomset JA (1968) The plasma lecithins: cholesterol acyltransferase reaction. J Lipid Res 9:155–167
19. Agellon LB et al (1991) Reduced high density lipoprotein cholesterol in human cholesteryl ester transfer protein transgenic mice. J Biol Chem 266:10796–10801
20. Jiang XC, Agellon LB, Walsh A, Breslow JL, Tall A (1992) Dietary cholesterol increases transcription of the human cholesteryl ester transfer protein gene in transgenic mice. Dependence on natural flanking sequences. J Clin Invest 90:1290–1295
21. Marotti KR et al (1993) Severe atherosclerosis in transgenic mice expressing simian cholesteryl ester transfer protein. Nature 364:73–75
22. Fagerberg L et al (2014) Analysis of the human tissue-specific expression by genome-wide integration of transcriptomics and antibody-based proteomics. Mol Cell Proteomics MCP 13:397–406
23. Quinet E, Tall A, Ramakrishnan R, Rudel L (1991) Plasma lipid transfer protein as a determinant of the atherogenicity of monkey plasma lipoproteins. J Clin Invest 87:1559–1566
24. Wang Y et al (2015) Plasma cholesteryl ester transfer protein is predominantly derived from Kupffer cells. Hepatol Baltim Md 62:1710–1722
25. Oliveira HC et al (1996) Human cholesteryl ester transfer protein gene proximal promoter contains dietary cholesterol positive responsive elements and mediates expression in small intestine and periphery while predominant liver and spleen expression is controlled by 5′-distal sequences. Cis-acting sequences mapped in transgenic mice. J Biol Chem 271:31831–31838
26. Luo Y, Tall AR (2000) Sterol upregulation of human CETP expression in vitro and in transgenic mice by an LXR element. J Clin Invest 105:513–520
27. Berti JA et al (2001) Thyroid hormone increases plasma cholesteryl ester transfer protein activity and plasma high-density lipoprotein removal rate in transgenic mice. Metabolism 50:530–536
28. Raposo HF, Patrício PR, Simões MC, Oliveira HCF (2014) Fibrates and fish oil, but not corn oil, up-regulate the expression of the cholesteryl ester transfer protein (CETP) gene. J Nutr Biochem 25:669–674
29. Masucci-Magoulas L et al (1995) Decreased cholesteryl ester transfer protein (CETP) mRNA and protein and increased high density lipoprotein following lipopolysaccharide administration in human CETP transgenic mice. J Clin Invest 95:1587–1594
30. Berti JA et al (2003) Cholesteryl ester transfer protein expression is down-regulated in hyperinsulinemic transgenic mice. J Lipid Res 44:1870–1876
31. Plump AS et al (1999) Increased atherosclerosis in ApoE and LDL receptor gene knock-out mice as a result of human cholesteryl ester transfer protein transgene expression. Arterioscler Thromb Vasc Biol 19:1105–1110
32. Westerterp M et al (2006) Cholesteryl ester transfer protein decreases high-density lipoprotein and severely aggravates atherosclerosis in APOE*3-Leiden mice. Arterioscler Thromb Vasc Biol 26:2552–2559
33. Van Eck M et al (2007) Important role for bone marrow-derived cholesteryl ester transfer protein in lipoprotein cholesterol redistribution and atherosclerotic lesion development in LDL receptor knockout mice. Circ Res 100:678–685
34. Hayek T et al (1995) Decreased early atherosclerotic lesions in hypertriglyceridemic mice expressing cholesteryl ester transfer protein transgene. J Clin Invest 96:2071–2074
35. Föger B et al (1999) Cholesteryl ester transfer protein corrects dysfunctional high density lipoproteins and reduces aortic atherosclerosis in lecithin cholesterol acyltransferase transgenic mice. J Biol Chem 274:36912–36920
36. Berti JA, de Faria EC, Oliveira HCF (2005) Atherosclerosis in aged mice over-expressing the reverse cholesterol transport genes. Braz J Med Biol Res Rev Bras Pesqui Medicas E Biol 38:391–398
37. Cazita PM et al (2003) Cholesteryl ester transfer protein expression attenuates atherosclerosis in ovariectomized mice. J Lipid Res 44:33–40
38. Casquero AC et al (2006) Atherosclerosis is enhanced by testosterone deficiency and attenuated by CETP expression in transgenic mice. J Lipid Res 47:1526–1534

39. Kako Y, Massé M, Huang L-S, Tall AR, Goldberg IJ (2002) Lipoprotein lipase deficiency and CETP in streptozotocin-treated apoB-expressing mice. J Lipid Res 43:872–877
40. MacLean PS et al (2003) Cholesteryl ester transfer protein expression prevents diet-induced atherosclerotic lesions in male db/db mice. Arterioscler Thromb Vasc Biol 23:1412–1415
41. Harder C, Lau P, Meng A, Whitman SC, McPherson R (2007) Cholesteryl ester transfer protein (CETP) expression protects against diet induced atherosclerosis in SR-BI deficient mice. Arterioscler Thromb Vasc Biol 27:858–864
42. El Bouhassani M et al (2011) Cholesteryl ester transfer protein expression partially attenuates the adverse effects of SR-BI receptor deficiency on cholesterol metabolism and atherosclerosis. J Biol Chem 286:17227–17238
43. Koizumi J et al (1985) Deficiency of serum cholesteryl-ester transfer activity in patients with familial hyperalphalipoproteinaemia. Atherosclerosis 58:175–186
44. Yamashita S et al (2000) Molecular mechanisms, lipoprotein abnormalities and atherogenicity of hyperalphalipoproteinemia. Atherosclerosis 152:271–285
45. Zhong S et al (1996) Increased coronary heart disease in Japanese-American men with mutation in the cholesteryl ester transfer protein gene despite increased HDL levels. J Clin Invest 97:2917–2923
46. Nagano M et al (2004) Molecular mechanisms of cholesteryl ester transfer protein deficiency in Japanese. J Atheroscler Thromb 11:110–121
47. Ordovas JM et al (2000) Association of cholesteryl ester transfer protein-TaqIB polymorphism with variations in lipoprotein subclasses and coronary heart disease risk: the Framingham study. Arterioscler Thromb Vasc Biol 20:1323–1329
48. Quintão ECR (2010) Letter by Quintao regarding article, 'Association of circulating cholesteryl ester transfer protein activity with incidence of cardiovascular disease in the community'. Circulation 122:e420; author reply e421
49. Cao M, Zhou Z-W, Fang B-J, Zhao C-G, Zhou D (2014) Meta-analysis of cholesteryl ester transfer protein TaqIB polymorphism and risk of myocardial infarction. Medicine (Baltimore) 93:e160
50. Thompson A et al (2008) Association of cholesteryl ester transfer protein genotypes with CETP mass and activity, lipid levels, and coronary risk. JAMA 299:2777–2788
51. Wang Q et al (2014) Seven functional polymorphisms in the CETP gene and myocardial infarction risk: a meta-analysis and meta-regression. PloS One 9: e88118
52. de Haan W et al (2008) Torcetrapib does not reduce atherosclerosis beyond atorvastatin and induces more proinflammatory lesions than atorvastatin. Circulation 117:2515–2522
53. Cazita PM, Barbeiro DF, Moretti AIS, Quintão ECR, Soriano FG (2008) Human cholesteryl ester transfer protein expression enhances the mouse survival rate in an experimental systemic inflammation model: a novel role for CETP. Shock Augusta Ga 30:590–595
54. Venancio TM et al (2016) CETP lowers TLR4 expression which attenuates the inflammatory response induced by LPS and polymicrobial sepsis. Mediators Inflamm 2016:1784014
55. Barter PJ et al (2007) Effects of torcetrapib in patients at high risk for coronary events. N Engl J Med 357:2109–2122
56. Grion CMC et al (2010) Lipoproteins and CETP levels as risk factors for severe sepsis in hospitalized patients. Eur J Clin Invest 40:330–338
57. Trinder M et al (2019) Cholesteryl ester transfer protein influences high-density lipoprotein levels and survival in sepsis. Am J Respir Crit Care Med 199:854–862
58. Blauw LL, Wang Y, Willems van Dijk K, Rensen PCN (2020) A novel role for CETP as immunological gatekeeper: raising HDL to cure sepsis? Trends Endocrinol Metab. https://doi.org/10.1016/j.tem.2020.01.003
59. Zhang Z et al (2001) Expression of cholesteryl ester transfer protein in human atherosclerotic lesions and its implication in reverse cholesterol transport. Atherosclerosis 159:67–75
60. Izem L, Morton RE (2001) Cholesteryl ester transfer protein biosynthesis and cellular cholesterol homeostasis are tightly interconnected. J Biol Chem 276:26534–26541
61. Izem L, Morton RE (2007) Possible role for intracellular cholesteryl ester transfer protein in adipocyte lipid metabolism and storage. J Biol Chem 282:21856–21865
62. Greene DJ, Izem L, Morton RE (2015) Defective triglyceride biosynthesis in CETP-deficient SW872 cells. J Lipid Res 56:1669–1678
63. Izem L, Greene DJ, Bialkowska K, Morton RE (2015) Overexpression of full-length cholesteryl ester transfer protein in SW872 cells reduces lipid accumulation. J Lipid Res 56:515–525
64. Zhou H et al (2006) Adipose tissue-specific CETP expression in mice: impact on plasma lipoprotein metabolism. J Lipid Res 47:2011–2019
65. Izem L, Liu Y, Morton RE (2020) Exon 9-deleted CETP inhibits full length-CETP synthesis and promotes cellular triglyceride storage. J Lipid Res. https://doi.org/10.1194/jlr.RA120000583
66. Sugano M et al (1998) Effect of antisense oligonucleotides against cholesteryl ester transfer protein on the development of atherosclerosis in cholesterol-fed rabbits. J Biol Chem 273:5033–5036
67. Rittershaus CW et al (2000) Vaccine-induced antibodies inhibit CETP activity in vivo and reduce aortic lesions in a rabbit model of atherosclerosis. Arterioscler Thromb Vasc Biol 20:2106–2112

68. Nissen SE et al (2007) Effect of torcetrapib on the progression of coronary atherosclerosis. N Engl J Med 356:1304–1316
69. Schwartz GG et al (2012) Effects of dalcetrapib in patients with a recent acute coronary syndrome. N Engl J Med 367:2089–2099
70. Lincoff AM et al (2017) Evacetrapib and cardiovascular outcomes in high-risk vascular disease. N Engl J Med 376:1933–1942
71. HPS3/TIMI55–REVEAL Collaborative Group et al (2017) Effects of anacetrapib in patients with atherosclerotic vascular disease. N Engl J Med 377:1217–1227
72. Krishna R et al (2017) Chronic administration of anacetrapib is associated with accumulation in adipose and slow elimination. Clin Pharmacol Ther 102:832–840
73. Masson W et al (2018) Therapy with cholesteryl ester transfer protein (CETP) inhibitors and diabetes risk. Diabetes Metab 44:508–513
74. Holmes MV, Ala-Korpela M (2019) What is 'LDL cholesterol'? Nat Rev Cardiol 16:197–198
75. Kettunen J et al (2019) Lipoprotein signatures of cholesteryl ester transfer protein and HMG-CoA reductase inhibition. PLoS Biol 17:e3000572
76. Millar JS et al (2016) Anacetrapib lowers LDL by increasing ApoB clearance in mildly hypercholesterolemic subjects. J Clin Invest 126:1603–1604
77. Ford J et al (2014) Tolerability, pharmacokinetics and pharmacodynamics of TA-8995, a selective cholesteryl ester transfer protein (CETP) inhibitor, in healthy subjects. Br J Clin Pharmacol 78:498–508
78. Hovingh GK et al (2015) Cholesterol ester transfer protein inhibition by TA-8995 in patients with mild dyslipidaemia (TULIP): a randomised, double-blind, placebo-controlled phase 2 trial. Lancet Lond Engl 386:452–460
79. van Capelleveen JC et al (2016) Effects of the cholesteryl ester transfer protein inhibitor, TA-8995, on cholesterol efflux capacity and high-density lipoprotein particle subclasses. J Clin Lipidol 10:1137.e3–1144.e3

Lipopolysaccharide-Binding Protein and Bactericidal/ Permeability-Increasing Protein in Lipid Metabolism and Cardiovascular Diseases

3

Yang Yu and Guohua Song

Abstract

Lipopolysaccharide-binding protein (LBP) and bactericidal/permeability-increasing protein (BPI) are the main members of BPI-like family based on the similar protein structure and conserved gene homology. Both LBP and BPI participate in lipid metabolism and thereby involve in pathogenesis of certain cardiovascular diseases. This chapter describes four aspects: (1) the loci of BPI and LBP in genome, (2) the characteristics of the cDNAs and expression patterns of LBP and BPI, (3) the structures and functions of LBP and BPI, and (4) the LBP and BPI in lipid metabolism and cardiovascular research.

Keywords

Lipopolysaccharide-binding protein · Bactericidal/permeability-increasing protein · Cholesteryl ester transfer protein · Phospholipid transfer protein · Cardiovascular diseases

Abbreviation

BPI	Bactericidal/permeability-increasing protein
CAD	Coronary artery disease
CETP	Cholesteryl ester transfer protein
LBP	Lipopolysaccharide-binding protein
PLTP	Phospholipid transfer protein

Multicellular organisms are unceasingly challenged by the invasion of microorganisms that flourish in the surroundings. The innate immune system is of dominant importance in keeping such host–microbe homeostasis. Plenty of bio-macromolecules related with the innate immune response exactly interrelate with and responses to the bacterial infection.

Two of the proteins critical to the mediation of signals from the surface of gram-negative bacteria (GNB) are lipopolysaccharide (LPS)-binding protein (LBP) and bactericidal/permeability-increasing protein (BPI) [1]. LBP and BPI both bind the Lipid A component of LPS from the outer envelope of Gram-negative bacteria, although they are normally considered to have opposed functions. LBP brings minute amounts of LPS to trigger the host immune response and can therefore be described as pro-inflammation, whereas BPI not only shows LPS neutralization but also binds ligands from Gram-positive bacteria (GPB) and thereby enhances the pattern recognition molecules in GPB infections [2, 3]. The majority of

Y. Yu (✉) · G. Song
Institute of Atherosclerosis, Shandong First Medical University, Shandong, China

X.-C. Jiang (ed.), *Lipid Transfer in Lipoprotein Metabolism and Cardiovascular Disease*, Advances in Experimental Medicine and Biology 1276, https://doi.org/10.1007/978-981-15-6082-8_3

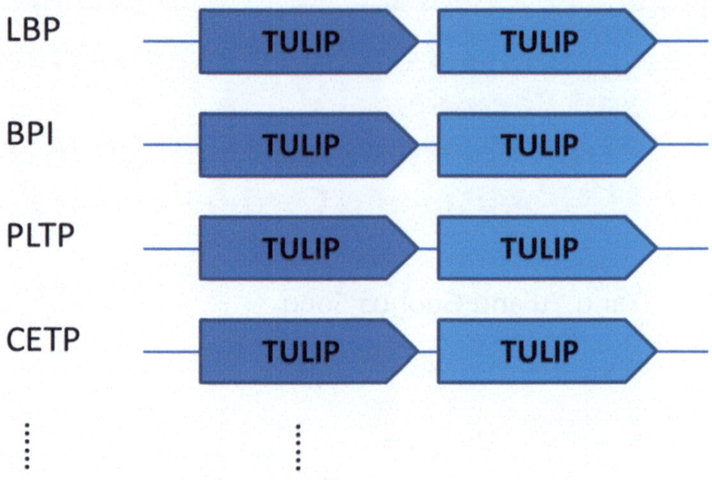

Fig. 3.1 *Domain organization of main members in BPI-like family.* Human BPI proteins contain one or two copies of the TULIP domain and no additional domains. UniProt entry ID for the proteins shown are LBP (P18428), BPI (P17213), PLTP (phospholipid transfer protein) (P55058), and CETP (cholesterol ester transfer protein) (P11597). The information in the picture was from Ref. [4] with brief modification

the involved inflammatory pathways may finally determine the host response to infections.

BPI and LBP are belonging to a family named "tubular lipid-binding protein (TULIP)" based on the similar protein structure and conserved gene homology. There is now evidence that TULIP superfamily includes at least three families: BPI-like, Takeout-like, and SMP (synaptotagmin-like, mitochondrial, and lipid-binding proteins)-like [4]. The genes of LBP, BPI, and phospholipid transfer protein (PLTP) are found on chromosome 20. These three proteins and cholesterol ester transfer protein (CETP) (which is located on chromosome 16) are the main members of BPI-like family. The BPI-like protein family includes LBP, BPI, PLTP, and CETP according to the similar domain organization (Fig. 3.1) and similar functions.

3.1 The Gene Loci and Expression of BPI and LBP

The loci of human LBP and BPI gene cluster were all at chromosome 20 q11.23 (Fig. 3.2). The exon number of LBP is 15. Together with BPI, the LBP expresses in the acute-phase immunologic response to Gram-negative bacterial infections, which might be essential for the quick acute-phase response to LPS. The LBP has restricted high expression in the liver, with very low-level expression in the appendix and endometrium [5]. The BPI gene has 16 exons. The gene of BPI has restricted expression toward the bone marrow and associated with neutrophils and eosinophils [5–7].

3.2 Structures and Functions of LBP and BPI

The crystal structures of LBP and BPI were reported in 2013 and 1997, respectively [8, 9]. Both LBP and BPI have a characteristic, conserved two-domain "boomerang" structure, with an N-terminal domain and a C-terminal domain that share little sequence identity but are very similar in overall architecture (Figs. 3.3 and 3.4) [10].

LBP, a 60 kDa lipid/phospholipid-binding and transfer protein, is synthesized principally by hepatocytes and secreted into the bloodstream [1, 11]. LBP can extract LPS monomers from the out membrane of Gram-negative bacteria (GNB), thereby delivering LPS molecules to

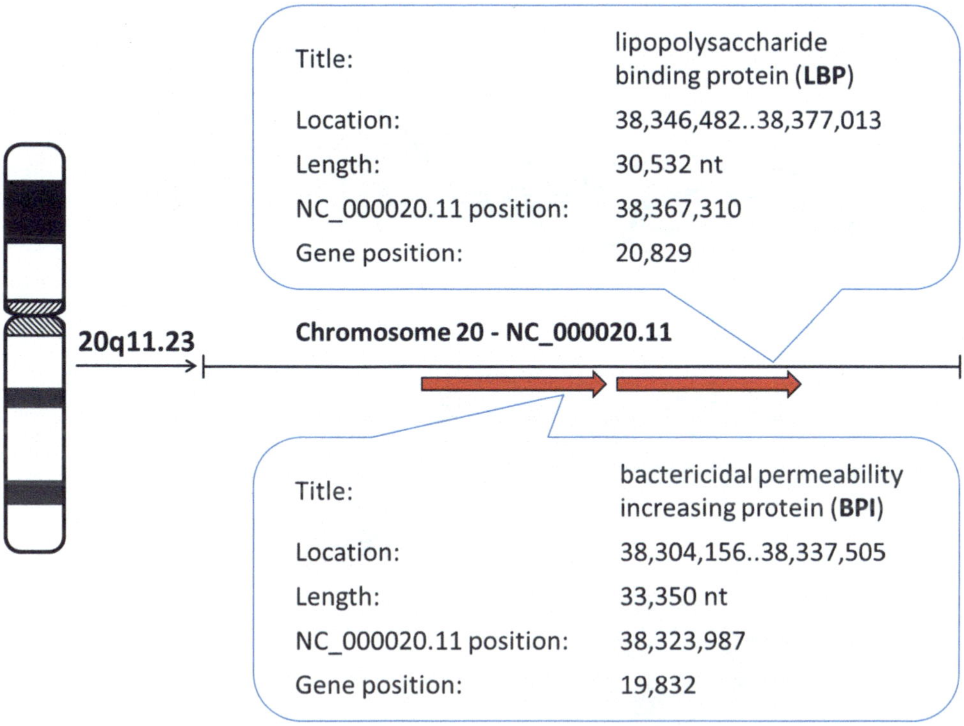

Fig. 3.2 *The loci of the human LBP and BPI.* Human lipopolysaccharide-binding protein (LBP, gene ID: 3929) and bactericidal/permeability-increasing protein (BPI, gene ID: 671) gene cluster on chromosome 20q11.23. The red arrows indicate the direction of transcription. All information of the two proteins was from NCBI Gene

membrane and soluble forms of CD14 receptors and promoting the formation of monomeric LPS-CD14 complex that is a critical intermediate in transport of LPS to MD-2/TLR4 and TLR4-dependent inflammatory cell activation, which markedly increase the host sensitivity to LPS [12]. In endothelial cells which lack of membrane CD14, LBP and soluble CD14 extract LPS from LPS-rich monolayers of GNB to form monomeric endotoxin–sCD14 complex to activate TLR4(+)/MD-2(+) cells [13, 14]. Besides the activation described previously, LBP delivers LPS to lipoproteins leading to hepatic clearance [15, 16]. HDL is the primary mediator to play a major role in the clearance of circulating LPS [15, 17–19].

The LBP-dependent pro-inflammatory effects of LPS are acute mobilization of circulating neutrophils to tissue sites of bacterial infection [20, 21]. The neutrophils and other polymorphonuclear leucocytes play a critical role in the arrest of proliferation and the seclusion of bacteria for disintegrated degradation and clearance. Bactericidal/permeability-increasing protein (BPI), a 55 kD single-chain cationic protein, has higher affinity for LPS and bacteria, is bactericidal, and represses inflammation by preventing LBP from delivering LPS to CD14 [10]. BPI is found mainly in the granules of neutrophils and eosinophils. Additionally, BPI has been detected on the surface of monocytes and colon epithelium, which is possibly due to the secondary secretion from activated neutrophils. Sharing a 45% LBP sequence, the crystal structure of human BPI revealed a boomerang-like shape and two similar domains with a polar pocket on their concave side where phospholipids (or perhaps LPS) can be bound (Fig. 3.4) [9, 22]. The major target cell of BPI–endotoxin aggregates are monocytes, while BPI promotes the CD14-independent delivery of purified LPS aggregates to host cells without apparent cell activation [23–25].

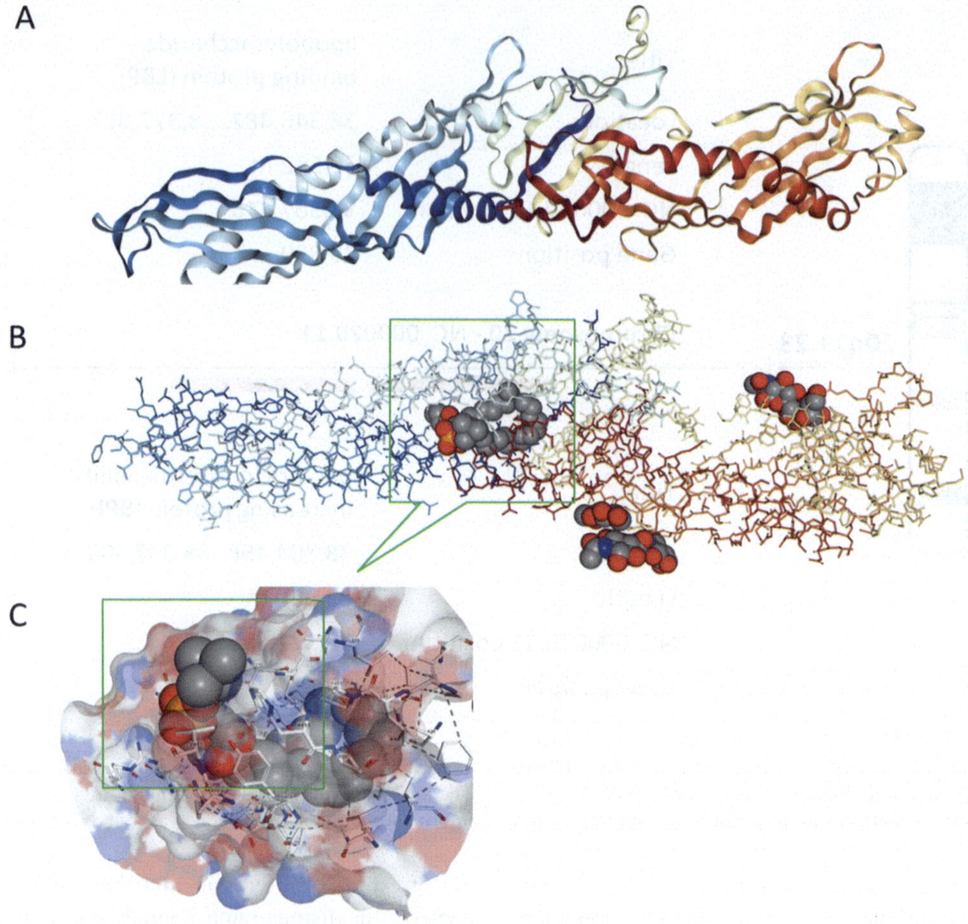

Fig. 3.3 *The structure of lipopolysaccharide-binding protein (LBP) (PDB ID: 4M4D) and its ligand pocket.* The ligand pocket positions of LBP were indicated by green boxes. (**a**), The cartoon style of LBP structure. (**b**), the licorice style of LBP – the ligands were displayed in spacefill mode. (**c**), The view of ligand pocket position of LBP structure. All information was downloaded from Protein Data Bank

The properties of LBP and BPI described above suggest that their coordinated function permits an efficient response to and elimination of invading GNB and a restore to homoeostasis. A small number (approximate 100) of GNB invasion may activate the acute inflammatory response, triggering a quickly mobilization of pro-inflammatory cells (Fig. 3.5). The vast majority of endotoxin delivered to cells via LBP (and CD14) is internalized without cell activation [26]. In fact, if the host cell contains the LPS deacylase acyloxyacyl hydrolase (e.g., macrophages), bulk clearance of LPS is coupled to partial deacylation and detoxification. LBP is also believed to play a role in the handling of LPS, facilitating transport of LPS to lipoproteins, and regulating monocyte/macrophage activation and pro-inflammatory cytokine secretion [27].

3.3 LBP and BPI in Lipid Metabolism and Cardiovascular Research

3.3.1 LBP

LBP is the primary protein to encounter LPS and deliver it to target cells. The serum LBP is a

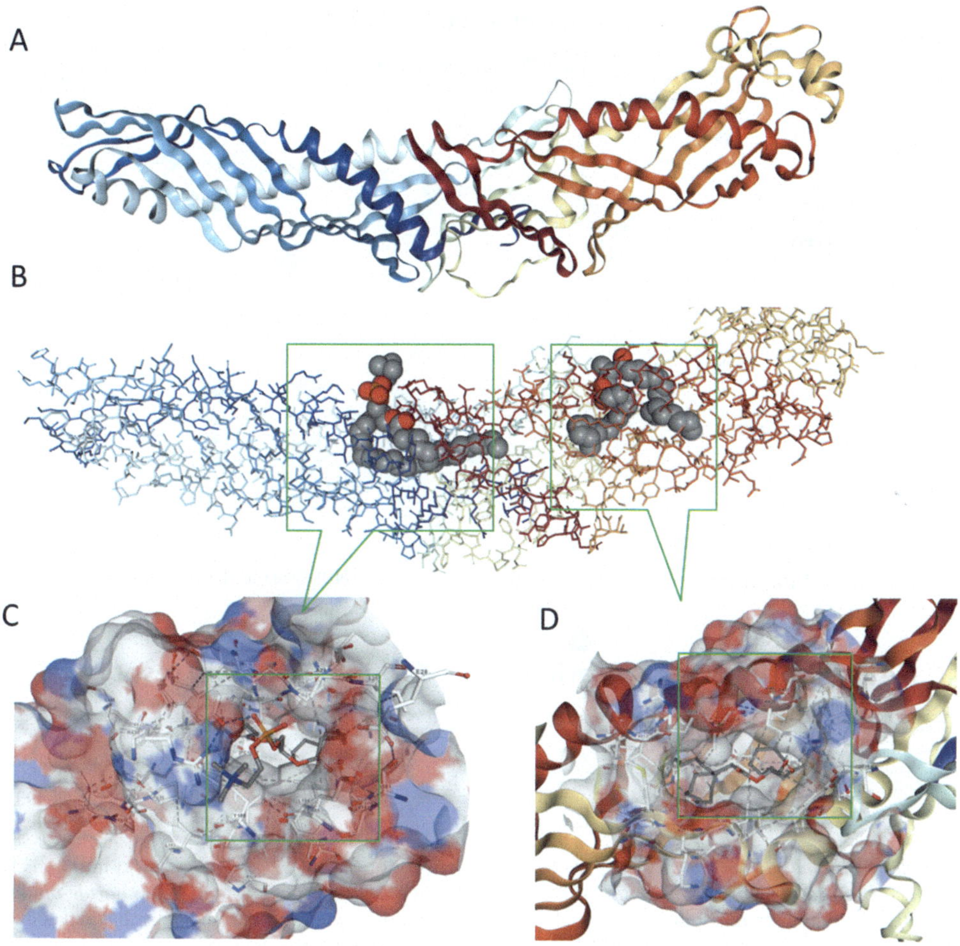

Fig. 3.4 *The structure of bactericidal/permeability-increasing protein (BPI) (PDB ID: 1EWF) and its ligand pockets.* The ligand pocket positions of LBP were indicated by green boxes. (**a**), the cartoon style of BPI structure. (**b**), the licorice style of BPI structure – the ligands were displayed in spacefill style. (**c**), the ligand view of pocket1 position of BPI structure. (**d**), the ligand view of Pocket2 position of BPI structure. All information was downloaded from Protein data bank

useful biomarker that indicates the activation of innate immune responses in the cardiovascular system. Lepper et al. investigated the association of serum LBP level and the risk of coronary artery disease (CAD) and found that LBP is a significant and independent predictor of prevalent CAD and male patients with increased levels of LBP had a fivefold increase in CAD prevalence [28]. LBP was reported as a significant and independent predictor of total and cardiovascular mortality hazard ratio for all-cause mortality [29]. Circulating LBP level is significantly and positively associated with obesity measures, insulin resistance parameters, glycated hemoglobin, fasting glucose, fasting triglycerides, LDL-cholesterol, systolic blood pressure, and inflammatory parameters while negatively associated with high-density lipoprotein-cholesterol. Furthermore, circulating LBP is positively associated with carotid intima media thickness (CIMT) in the internal carotid segments and CIMT in overall carotid segments. The findings reveal that serum LBP is a putative risk factor related to atherosclerosis [30].

A recent 10-year follow-up study revealed that individuals with higher serum LBP levels had a

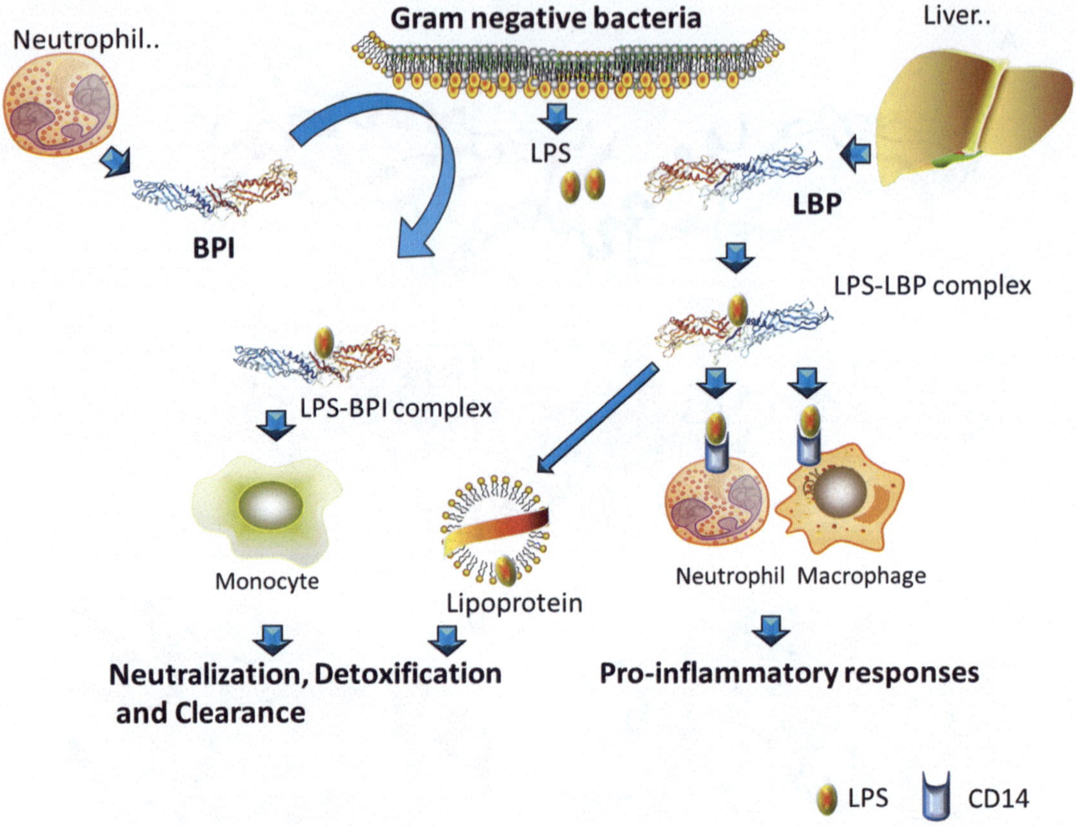

Fig. 3.5 *Role of BPI and LBP in host responses to Gram-negative bacteria. Bactericidal/permeability-increasing protein* (*BPI*) is found mainly in the granules of neutrophils and eosinophils. Additionally, BPI has been detected on the surface of monocytes and colon epithelium, which is possibly due to the secondary secretion from activated neutrophils. BPI has higher affinity for LPS and bacteria, is bactericidal, and represses inflammation by preventing LBP from delivering LPS to CD14. The major target cell of BPI–endotoxin aggregates are monocytes, while BPI promotes the CD14-independent delivery of purified LPS aggregates to host cells without apparent cell activation. *Lipopolysaccharide-binding protein* (*LBP*), synthesized principally by hepatocytes and secreted into the bloodstream, can extract LPS monomers from the out membrane of Gram-negative bacteria (GNB), thereby delivering the LPS molecules to membrane and soluble forms of CD14 receptors, which trigger the pro-inflammatory responses mediated neutrophils, monocytes, and macrophages. Besides the activation described previously, LBP delivers LPS to lipoproteins leading to hepatic clearance

significantly greater risk of the development of cardiovascular disease (CVD) in the general Japanese population CVD after adjusting for conventional cardiovascular risk factors [31]. Furthermore, low-grade endotoxemia might contribute to the pathogenesis of CVD through chronic systemic vascular inflammation. This is the first prospective cohort study to investigate the association between serum LBP levels and the incidence of CVD in a general Japanese population. Further investigations are needed to elucidate the mechanism underlying the association between serum LBP levels and cardiovascular risk. In addition, after assessed by aortic pulse wave velocity (PWV) in patients with type 2 diabetes or obstructive sleep apnea, the serum LBP levels are independently and positively associated with arterial stiffness [32, 33]. Patients with rheumatoid arthritis (RA) have a two- to threefold increased risk of myocardial infarction compared to the general population. Charles-Schoeman C's study found that HDL proteome is abnormal with the increase of HDL-associated LBP in active RA patients.

The treatment of RA may lower the LBP level in HDL fraction, which suggested a decrease in the pro-inflammatory properties of the HDL particle [34]. The study indicated that HDL-associated LBP might be a pro-inflammatory marker of HDL. Moreover, in hemodialysis patients, serum LBP is associated with chronic inflammation and metabolic syndrome [35].

The LPS delivery function of LBP is correlated with lipoprotein metabolism. HDL is the primary acceptor of LPS delivered by LBP. In acute phase of septic shock patients, HDL levels were dramatically decreased with a shift toward large HDL particles, which may reflect a remarkable dysfunction of these lipoproteins. The significantly increased serum LBP level was also observed in this study [36]. Moreover, Wurfel et al. reported that the addition of recombinant LBP enabled prompt binding and neutralization of LPS by recombinant HDL. Thus, LBP appears capable of transferring LPS not only to CD14 but also to lipoprotein particles. In contrast with recombinant HDL, apoA-I-containing lipoproteins isolated from plasma by selected affinity immunosorption on an anti-apoA-I column neutralized LPS without the addition of exogenous LBP. Therefore LBP appears to be physically associated with lipoproteins in plasma; it is positioned to play an important role in the neutralization of LPS [27].

LBP plays an important role in regulating leukocyte responses to LPS via either augmenting these responses at low LBP concentrations or inhibiting them at high concentrations. Richard L. Kitchens and colleagues' investigation of the mechanism of apoA-II activity revealed that LBP promoted the formation of large LPS aggregates with low bioactivity and that apoA-II inhibited the formation of these aggregates without binding and directly inhibiting LPS bioactivity. Their results suggest a novel pro-inflammatory activity of apoA-II that may help maintain sensitive host responses to LPS by suppressing LBP-mediated inhibition [37].

3.3.2 BPI

BPI is the most potent antimicrobial granule protein identified so far and is especially effective against Gram-negative bacteria. In addition to these antimicrobial effects, BPI may play a key role in limiting endotoxin-triggered systemic inflammation by binding with high affinity to the lipid A portion of LPS [25]. Recombinant BPI is a potent inhibitor of LPS-mediated responses in cultured bovine brain microvascular endothelial cells and also inhibited LPS-induced tumor necrosis factor alpha, interleukin-1 beta, and interleukin-6 releases from human whole blood. The findings indicated that BPI treatment is a potent prevention of endotoxemia or endotoxic shock [38]. In rat model with burn and/or sepsis, the recombinant BPI administration could attenuate myocyte cytokine responses to septic challenge and improved contractile function, which suggested that BPI protects myocyte from post-burn infectious inflammation or damage in septic status [39]. A Proteomics study revealed that BPI is decreased dramatically in patients with total coronary atherosclerotic occlusion, suggesting that BPI might be a promising biomarker for severe atherosclerotic coronary stenosis [40].

Moreover, BPI is closely related with diabetes, which is an independent risk factor of atherosclerotic disease. Carme Gubern et al. studied circulating BPI in healthy subjects, in patients with glucose intolerance, and in patients with type 2 diabetes [41]. In subjects with glucose intolerance, the strong associations were observed between plasma BPI and central obesity, glucose metabolism, insulin sensitivity, and components of the metabolic syndrome. Bioactive LPS level was significantly associated with both BPI and LBP. In patients receiving metformin, the improved insulin sensitivity and raised circulating BPI were observed in parallel. The 3-UTR BPI gene polymorphism was associated with both increased BPI and raised insulin sensitivity concomitantly [41]. The decreased circulating BPI concentrations are associated with endothelium-dependent vasodilatation, total LDL, and HDL cholesterol level [42]. In addition, they found that low BPI was associated with increased LPS concentration in healthy volunteers.

Both LBP and BPI participate in lipid metabolism and thereby involve in pathogenesis of certain cardiovascular diseases. Serum LBP level, as a biomarker of pro-inflammatory protein, is highly associated with atherosclerosis, arterial stiffness, and many chronic metabolic dysfunction diseases. BPI, a potential bactericidal protein which fights against infection in vascular system and beyond, is licensed for human use. The association of BPI and CVD needs more evidence. Further mechanistic studies are required to explore the causal relations between LBP/BPI and CVD pathogenesis.

References

1. Weiss J (2003) Bactericidal/permeability-increasing protein (BPI) and lipopolysaccharide-binding protein (LBP): structure, function and regulation in host defence against Gram-negative bacteria. Biochem Soc Trans 31:785–790
2. Schroder NW, Schumann RR (2005) Non-LPS targets and actions of LPS binding protein (LBP). J Endotoxin Res 11:237–242
3. Bulow S, Zeller L, Werner M, Toelge M, Holzinger J, Entzian C, Schubert T, Waldow F, Gisch N, Hammerschmidt S, Gessner A (2018) Bactericidal/permeability-increasing protein is an enhancer of bacterial lipoprotein recognition. Front Immunol 9:2768
4. Alva V, Lupas AN (2016) The TULIP superfamily of eukaryotic lipid-binding proteins as a mediator of lipid sensing and transport. Biochim Biophys Acta 1861:913–923
5. Fagerberg L, Hallstrom BM, Oksvold P, Kampf C, Djureinovic D, Odeberg J, Habuka M, Tahmasebpoor S, Danielsson A, Edlund K, Asplund A, Sjostedt E, Lundberg E, Szigyarto CA, Skogs M, Takanen JO, Berling H, Tegel H, Mulder J, Nilsson P, Schwenk JM, Lindskog C, Danielsson F, Mardinoglu A, Sivertsson A, von Feilitzen K, Forsberg M, Zwahlen M, Olsson I, Navani S, Huss M, Nielsen J, Ponten F, Uhlen M (2014) Analysis of the human tissue-specific expression by genome-wide integration of transcriptomics and antibody-based proteomics. Mol Cell Proteomics 13:397–406
6. Elsbach P, Weiss J, Franson RC, Beckerdite-Quagliata S, Schneider A, Harris L (1979) Separation and purification of a potent bactericidal/permeability-increasing protein and a closely associated phospholipase A2 from rabbit polymorphonuclear leukocytes. Observations on their relationship. J Biol Chem 254:11000–11009
7. Calafat J, Janssen H, Tool A, Dentener MA, Knol EF, Rosenberg HF, Egesten A (1998) The bactericidal/permeability-increasing protein (BPI) is present in specific granules of human eosinophils. Blood 91:4770–4775
8. Eckert JK, Kim YJ, Kim JI, Gurtler K, Oh DY, Sur S, Lundvall L, Hamann L, van der Ploeg A, Pickkers P, Giamarellos-Bourboulis E, Kubarenko AV, Weber AN, Kabesch M, Kumpf O, An HJ, Lee JO, Schumann RR (2013) The crystal structure of lipopolysaccharide binding protein reveals the location of a frequent mutation that impairs innate immunity. Immunity 39:647–660
9. Beamer LJ, Carroll SF, Eisenberg D (1997) Crystal structure of human BPI and two bound phospholipids at 2.4 angstrom resolution. Science 276:1861–1864
10. Krasity BC, Troll JV, Weiss JP, McFall-Ngai MJ (2011) LBP/BPI proteins and their relatives: conservation over evolution and roles in mutualism. Biochem Soc Trans 39:1039–1044
11. Ulevitch RJ, Tobias PS (1999) Recognition of gram-negative bacteria and endotoxin by the innate immune system. Curr Opin Immunol 11:19–22
12. Hailman E, Lichenstein HS, Wurfel MM, Miller DS, Johnson DA, Kelley M, Busse LA, Zukowski MM, Wright SD (1994) Lipopolysaccharide (LPS)-binding protein accelerates the binding of LPS to CD14. J Exp Med 179:269–277
13. Giardina PC, Gioannini T, Buscher BA, Zaleski A, Zheng DS, Stoll L, Teghanemt A, Apicella MA, Weiss J (2001) Construction of acetate auxotrophs of Neisseria meningitidis to study host-meningococcal endotoxin interactions. J Biol Chem 276:5883–5891
14. Gioannini TL, Zhang D, Teghanemt A, Weiss JP (2002) An essential role for albumin in the interaction of endotoxin with lipopolysaccharide-binding protein and sCD14 and resultant cell activation. J Biol Chem 277:47818–47825
15. Read TE, Harris HW, Grunfeld C, Feingold KR, Kane JP, Rapp JH (1993) The protective effect of serum lipoproteins against bacterial lipopolysaccharide. Eur Heart J 14(Suppl K):125–129
16. Van Bossuyt H, De Zanger RB, Wisse E (1988) Cellular and subcellular distribution of injected lipopolysaccharide in rat liver and its inactivation by bile salts. J Hepatol 7:325–337
17. Flegel WA, Wolpl A, Mannel DN, Northoff H (1989) Inhibition of endotoxin-induced activation of human monocytes by human lipoproteins. Infect Immun 57:2237–2245
18. Feingold KR, Funk JL, Moser AH, Shigenaga JK, Rapp JH, Grunfeld C (1995) Role for circulating lipoproteins in protection from endotoxin toxicity. Infect Immun 63:2041–2046
19. Levine DM, Parker TS, Donnelly TM, Walsh A, Rubin AL (1993) In vivo protection against endotoxin by plasma high density lipoprotein. Proc Natl Acad Sci U S A 90:12040–12044
20. Fierer J, Swancutt MA, Heumann D, Golenbock D (2002) The role of lipopolysaccharide binding protein in resistance to Salmonella infections in mice. J Immunol 168:6396–6403
21. Yang KK, Dorner BG, Merkel U, Ryffel B, Schutt C, Golenbock D, Freeman MW, Jack RS (2002)

Neutrophil influx in response to a peritoneal infection with Salmonella is delayed in lipopolysaccharide-binding protein or CD14-deficient mice. J Immunol 169:4475–4480
22. Beamer LJ, Carroll SF, Eisenberg D (1999) The three-dimensional structure of human bactericidal/permeability-increasing protein: implications for understanding protein-lipopolysaccharide interactions. Biochem Pharmacol 57:225–229
23. Iovine N, Eastvold J, Elsbach P, Weiss JP, Gioannini TL (2002) The carboxyl-terminal domain of closely related endotoxin-binding proteins determines the target of protein-lipopolysaccharide complexes. J Biol Chem 277:7970–7978
24. Iovine NM, Elsbach P, Weiss J (1997) An opsonic function of the neutrophil bactericidal/permeability-increasing protein depends on both its N- and C-terminal domains. Proc Natl Acad Sci U S A 94:10973–10978
25. Schultz H, Weiss J, Carroll SF, Gross WL (2001) The endotoxin-binding bactericidal/permeability-increasing protein (BPI): a target antigen of autoantibodies. J Leukoc Biol 69:505–512
26. Gegner JA, Ulevitch RJ, Tobias PS (1995) Lipopolysaccharide (LPS) signal transduction and clearance. Dual roles for LPS binding protein and membrane CD14. J Biol Chem 270:5320–5325
27. Wurfel MM, Kunitake ST, Lichenstein H, Kane JP, Wright SD (1994) Lipopolysaccharide (LPS)-binding protein is carried on lipoproteins and acts as a cofactor in the neutralization of LPS. J Exp Med 180:1025–1035
28. Lepper PM, Schumann C, Triantafilou K, Rasche FM, Schuster T, Frank H, Schneider EM, Triantafilou M, von Eynatten M (2007) Association of lipopolysaccharide-binding protein and coronary artery disease in men. J Am Coll Cardiol 50:25–31
29. Lepper PM, Kleber ME, Grammer TB, Hoffmann K, Dietz S, Winkelmann BR, Boehm BO, Marz W (2011) Lipopolysaccharide-binding protein (LBP) is associated with total and cardiovascular mortality in individuals with or without stable coronary artery disease--results from the Ludwigshafen Risk and Cardiovascular Health Study (LURIC). Atherosclerosis 219:291–297
30. Serrano M, Moreno-Navarrete JM, Puig J, Moreno M, Guerra E, Ortega F, Xifra G, Ricart W, Fernandez-Real JM (2013) Serum lipopolysaccharide-binding protein as a marker of atherosclerosis. Atherosclerosis 230:223–227
31. Asada M, Oishi E, Sakata S, Hata J, Yoshida D, Honda T, Furuta Y, Shibata M, Suzuki K, Watanabe H, Murayama N, Kitazono T, Yamaura K, Ninomiya T (2019) Serum lipopolysaccharide-binding protein levels and the incidence of cardiovascular disease in a general Japanese population: The Hisayama Study. J Am Heart Assoc 8:e013628
32. Sakura T, Morioka T, Shioi A, Kakutani Y, Miki Y, Yamazaki Y, Motoyama K, Mori K, Fukumoto S, Shoji T, Emoto M, Inaba M (2017) Lipopolysaccharide-binding protein is associated with arterial stiffness in patients with type 2 diabetes: a cross-sectional study. Cardiovasc Diabetol 16:62
33. Trojova I, Kozarova M, Petrasova D, Malachovska Z, Paranicova I, Joppa P, Tkacova R (2018) Circulating lipopolysaccharide-binding protein and carotid intima-media thickness in obstructive sleep apnea. Physiol Res 67:69–78
34. Charles-Schoeman C, Gugiu GB, Ge H, Shahbazian A, Lee YY, Wang X, Furst DE, Ranganath VK, Maldonado M, Lee T, Reddy ST (2018) Remodeling of the HDL proteome with treatment response to abatacept or adalimumab in the AMPLE trial of patients with rheumatoid arthritis. Atherosclerosis 275:107–114
35. Lim PS, Chang YK, Wu TK (2019) Serum lipopolysaccharide-binding protein is associated with chronic inflammation and metabolic syndrome in hemodialysis patients. Blood Purif 47:28–36
36. Tanaka S, Diallo D, Delbosc S, Geneve C, Zappella N, Yong-Sang J, Patche J, Harrois A, Hamada S, Denamur E, Montravers P, Duranteau J, Meilhac O (2019) High-density lipoprotein (HDL) particle size and concentration changes in septic shock patients. Ann Intensive Care 9:68
37. Thompson PA, Berbee JF, Rensen PC, Kitchens RL (2008) Apolipoprotein A-II augments monocyte responses to LPS by suppressing the inhibitory activity of LPS-binding protein. Innate Immun 14:365–374
38. Arditi M, Zhou J, Huang SH, Luckett PM, Marra MN, Kim KS (1994) Bactericidal/permeability-increasing protein protects vascular endothelial cells from lipopolysaccharide-induced activation and injury. Infect Immun 62:3930–3936
39. Horton JW, Maass DL, White DJ, Minei JP (1985) Bactericidal/permeability increasing protein attenuates the myocardial inflammation/dysfunction that occurs with burn complicated by subsequent infection. J Appl Physiol 103(2007):948–958
40. Bleijerveld OB, Wijten P, Cappadona S, McClellan EA, Polat AN, Raijmakers R, Sels JW, Colle L, Grasso S, van den Toorn HW, van Breukelen B, Stubbs A, Pasterkamp G, Heck AJ, Hoefer IE, Scholten A (2012) Deep proteome profiling of circulating granulocytes reveals bactericidal/permeability-increasing protein as a biomarker for severe atherosclerotic coronary stenosis. J Proteome Res 11:5235–5244
41. Gubern C, Lopez-Bermejo A, Biarnes J, Vendrell J, Ricart W, Fernandez-Real JM (2006) Natural antibiotics and insulin sensitivity: the role of bactericidal/permeability-increasing protein. Diabetes 55:216–224
42. Esteve E, Castro A, Moreno JM, Vendrell J, Ricart W, Fernandez-Real JM (2010) Circulating bactericidal/permeability-increasing protein (BPI) is associated with serum lipids and endothelial function. Thromb Haemost 103:780–787

Microsomal Triglyceride Transfer Protein: From Lipid Metabolism to Metabolic Diseases

4

Jahangir Iqbal, Zainab Jahangir, and Ali Ahmed Al-Qarni

Abstract

Microsomal triglyceride transfer protein (MTP) was first identified as an endoplasmic reticulum (ER) resident protein that helps in the transfer of neutral lipids to nascent apolipoprotein B (apoB). Its critical role in the assembly and secretion of apoB-containing lipoproteins was identified in abetalipoproteinemia patients who have mutations in MTP and completely lack apoB-containing lipoproteins in the circulation. It has been established now that MTP not only is involved in the transfer of neutral lipids but also plays a role in cholesterol ester and cluster of differentiation 1d (CD1d) biosynthesis. Besides neutral lipids, MTP may also help in the transfer of sphingolipids such as ceramides and sphingomyelin to the apoB-containing lipoproteins. MTP is a multifunctional protein, and its deregulation during pathophysiological conditions gives rise to different metabolic conditions. This book chapter discusses the physiological role and regulation of MTP to maintain the homeostasis of lipids and lipoproteins. It also reviews the regulation of MTP during certain pathophysiological conditions and provides a brief overview of therapeutic interventions that can be possibly used to target its activity or expression to alleviate some of these metabolic diseases.

Keywords

Atherosclerosis · Insulin resistance · Lipid metabolism · Lipoproteins · Metabolic disease · MTP · Obesity · Type 2 diabetes

Abbreviations

ACAT	acyl-CoA:cholesterol acyltransferase
apoB	apolipoprotein B
CCl4	carbon tetrachloride
CD1d	cluster of differentiation 1d
DR1	direct repeat 1
ER	endoplasmic reticulum
fa/fa	Zucker obese
$ERK^{1/2}$	extracellular-signal-regulated kinase 1/2
FFA	free fatty acid
foxA2	forkhead box protein A2
foxO1	forkhead box protein O1
HDL	high-density lipoprotein
HNF-4α	hepatic nuclear factor-4α

J. Iqbal (✉) · A. A. Al-Qarni
King Abdullah International Medical Research Center, King Saud bin Abdulaziz University Health Sciences, King Abdulaziz Medical City, Ministry of National Guard Health Affairs, Al Ahsa, Saudi Arabia
e-mail: iqbalja@ngha.med.sa

Z. Jahangir
John P. Stevens High School, North Edison, NJ, USA

X.-C. Jiang (ed.), *Lipid Transfer in Lipoprotein Metabolism and Cardiovascular Disease*, Advances in Experimental Medicine and Biology 1276, https://doi.org/10.1007/978-981-15-6082-8_4

IRE1β	inositol requiring enzyme 1β
LDL	low-density lipoprotein
miR	microRNA
MTP	microsomal triglyceride transfer protein
NEFA	non-esterified fatty acid
NR2F1	nuclear receptor 2 family 1
OLETF	Otsuka Long-Evans Tokushima Fatty
PI3K	phosphatidylinositol 3-kinase
PTEN	phosphatase and tensin homolog
SHP	small heterodimer partner
SRE/IRE	sterol and insulin response element
SREBP	sterol response element binding protein
VLDL	very low density lipoprotein

4.1 Introduction

Metabolic syndrome is a group of risk factors that includes obesity, insulin resistance, glucose intolerance, endothelial dysfunction, hypertension, and atherogenic dyslipidemia [1, 2]. Access plasma lipids that are carried in apoB-containing lipoproteins are a major risk factor for atherosclerotic cardiovascular disease and other lipid disorders [3, 4]. These apoB-containing lipoproteins require MTP for their assembly and secretion which is present mainly in the liver and intestine. Besides traditional risk factors such as high levels of low-density lipoprotein (LDL) and low levels of high-density lipoprotein (HDL) cholesterol [5, 6], changes in sphingolipids may contribute to the pathogenesis of these disorders [7, 8]. Since MTP plays a central role in the regulation of apoB-containing lipoproteins, it is important to understand how the expression and activity of MTP is controlled during physiological and pathophysiological conditions which can help to develop therapeutic interventions to alleviate some related metabolic diseases.

4.2 Role of Microsomal Triglyceride Transfer Protein in Lipid and Lipoprotein Metabolism

Hydrophobic properties of the lipids make it difficult to transport them in the aqueous environment of the body and tissue fluids. However, nature has evolved different methods to overcome this limitation. To facilitate their transport to enterocyte membranes during dietary absorption, bile acids solubilize lipids in the intestinal lumen along with other hydrophobic molecules [9]. After cellular uptake, several proteins assist in the transport of these lipids between different subcellular compartments for further processing. The absorbed lipids are transported in the extracellular milieu by lipoproteins that possess an amphipathic surface containing a combination of phospholipids, free cholesterol, and amphiphilic proteins surrounded by a hydrophobic core of neutral lipids such as triacylglycerol and cholesteryl esters [10].

The intestine and liver are the predominant tissues that are involved in lipid and lipoprotein metabolism. These tissues were initially thought to be the only organs that express MTP. However, several studies have now revealed that MTP is also expressed in the heart, kidney, brain, eye, macrophages, and adipose tissue [11, 12]. Of the various classes, apoB-containing lipoproteins that transport large amounts of lipids are widely studied due to their association with coronary artery disease and atherosclerosis. Assembly and secretion of these apoB-containing lipoproteins by various tissues are critically dependent on the presence of the lipid transfer activity of MTP [12–18]. MTP exists as a heterodimer consisting of a ubiquitously expressed chaperone protein disulfide isomerase (P-subunit) and a unique M-subunit [19, 20] in the lumen of the ER. Even though the lipid transfer activity resides within the M-subunit of MTP complex, the

P-subunit helps in its structural stabilization and solubilization. Mutations in the M submit of MTP in abetalipoproteinemia subjects were the first indication to suggest that the lipid transfer function of MTP is critical for apoB-lipoprotein assembly and secretion [21–23]. Pharmacological inhibition [24] or missense mutations occurring naturally in the M-subunit [25, 26] alter lipid transfer activity of MTP to decrease the secretion of apoB-containing lipoproteins. Gouda et al. have shown that MTP -493G > T polymorphism may be correlated with the risk of nonalcoholic fatty liver disease and metabolic syndrome [27]. In addition to lipid transfer function, MTP is also known to have apoB-binding and membrane association domains [15, 16]. Several techniques such as yeast two-hybrid, co-immunoprecipitation, and in vitro binding assays have been used to demonstrate interaction of MTP within the N-terminus region of apoB [28, 29]. The importance of these interactions for proper apoB-lipoprotein assembly and secretion has been validated by using small molecule inhibitors against protein-protein interactions in cell culture models [30].

Besides its ability to transfer triglyceride, cholesterol ester, and phospholipids [31, 32], we have shown recently that MTP can also transfer ceramide and sphingomyelin between vesicles [33]. Our data indicated that MTP plays no role in plasma hexosylceramide levels but is critical for determining the plasma ceramide levels and is partially responsible for sphingomyelin levels. We have shown that MTP may regulate plasma ceramide and sphingomyelin levels by transferring these sphingolipids to apoB-containing lipoproteins to facilitate their secretion. The ability of MTP to transfer ceramide and sphingomyelin to nascent apoB-containing lipoproteins or lipid droplets in the lumen of the ER and Golgi to help in their secretion suggests that MTP might be a general lipid transfer protein that can recognize nonspecific hydrophobic motifs [18, 33].

Drosophila MTP ortholog that was shown to transfer only phospholipids provided some insights that triglyceride transfer activity of MTP is not critical for apoB-lipoprotein assembly and secretion [17, 34–36]. Despite being only 50% as efficient as the human MTP, *Drosophila* MTP not only rescued apoB secretion [17] but also responded to the supplementation of oleic acid to increase cellular apoB secretion [35]. Furthermore, overexpression of *Drosophila* MTP in mouse livers has been shown to assemble very low-density lipoproteins (VLDL) that resemble the size of HDL particles [34]. Therefore, the assembly and secretion of primordial lipoproteins only require the phospholipid transfer activity of MTP. However, the presence of triglyceride transfer activity is necessary to increase the association of neutral lipids with apoB-containing lipoproteins.

Although, apoB48 and apoB100 are the major forms used in vivo for lipoprotein assembly, C-terminally truncated apoB has been used in vitro to study the role of MTP in the assembly and lipidation of apoB-containing lipoproteins [37, 38]. Furthermore, it has been suggested that the interaction of lipids with the N-terminal fragments of apoB on the inner leaflet of the ER membrane may initiate the formation of nucleation sites independent of MTP [13, 39–41]. MTP renders naturally occurring hydrophobic peptides such as apoB48 and apoB100 secretion competent by bringing lipids to these peptides. The type and amount of endogenous lipids and the length of the apoB peptide determine the degree of lipidation of apoB-lipoproteins, and the absence of MTP renders these larger peptides to proteasomal degradation [42, 43]. Furthermore, in the absence of the synthesis of larger apoB, cells abort lipoprotein assembly and secrete smaller peptides associated with fewer lipids. Studies have shown that compared to apoB48, secretion of apoB100 is more sensitive to MTP inhibitors [44]. Similarly, a study reported that hepatic ablation of MTP decreased plasma apoB48 levels to a lesser extent [45]. However, intestine-specific deletion of MTP has been shown to reduce apoB48 secretion dramatically by around 80% by enterocytes [46]. However, abetalipoproteinemia patients depicted absence of both apoB100 and apoB48 suggesting that secretion of both peptides requires MTP.

A physiological ratio of cholesterol and phospholipids is necessary to maintain membrane fluidity in the cellular membrane. However, excess free cholesterol is toxic to cells, and several pathways are involved in regulating its levels [47]. One of the pathways adapted by cells to avoid such toxicity is by storing excess free cholesterol in its ester form which is accomplished by acyl-CoA:cholesterol acyltransferase (ACAT) enzymes ACAT1 and ACAT2 [48–52]. ACAT1, expressed ubiquitously [53–56], and ACAT2, expressed mainly in the intestine and liver [53, 56–58], are integral ER membrane proteins with several transmembrane domains [59–61]. Newly synthesized cholesterol esters are either transported for storage or transferred by MTP to apoB-containing lipoproteins for secretion in the intestine and liver. Since MTP is involved in the transfer of cholesterol esters to apoB-lipoproteins, it is expected that inhibition/deficiency of MTP should increase cellular cholesterol esters. However, we have reported that inhibition or genetic ablation of MTP activity increases cellular free cholesterol levels by decreasing microsomal cholesterol ester biosynthesis without effecting ACAT1 and ACAT2 mRNA and protein levels [62]. These observations suggest that MTP regulates cholesterol ester biosynthesis by mechanisms independent of transcriptional or translational control of ACAT1 and ACAT2. It is plausible to surmise that MTP circumvents product inhibition by removing cholesterol esters from the site of synthesis on ACATs and depositing them into apoB-lipoproteins. Not only MTP but also apoB48 co-expression in the cells were shown to increase cholesterol ester biosynthesis and secretion suggesting that biogenesis of lipoproteins by MTP and synthesis of apoB acts in concert to increase the biosynthesis of cholesterol ester.

4.3 Regulation of Microsomal Triglyceride Transfer Protein

Transcription factors play an important role in regulating lipid homeostasis. To maintain this homeostasis, the activity of MTP and the stability of apoB are important factors [12, 63]. MTP gene expression is controlled by a variety of cellular regulators that include several transcription factors, and these factors are sometimes tissue specific. Different types of diet also play an important role in modulating the expression and activity of MTP. Long-term feeding of high-fat diet in hamsters increases MTP mRNA in the liver and intestine [64]. Similarly, a high-fructose diet increases MTP expression in both the liver and intestine [65]. In contrast, a sucrose-enriched diet increases MTP mRNA expression only in the liver [64], whereas diets rich in saturated fat [64, 66] and cholesterol [67] increase the expression of MTP only in the intestine. Sterol response element binding protein (SREBP)-2 has been shown to interact directly with the sterol and insulin response element (SRE/IRE) in the MTP promoter to decrease MTP expression by sterol depletion and pravastatin in HepG2 cells [68]. However, upregulation of MTP by oleic acid in HepG2 cells does not involve SRE/IRE [65]. The potential binding to the SRE/IRE element in the promoter of MTP by SREBPs to change its expression by saturated fat and cholesterol has not been demonstrated in vivo. We have shown that inositol-requiring enzyme 1β (IRE1β) may play a role in the regulation of intestinal MTP during the feeding of high-cholesterol and Western diets [67]. Deletion of IRE1β in mice results in increased intestinal expression of MTP upon Western diet feeding and makes these mice more prone to develop hyperlipidemia than wild-type mice. The increased level of intestinal lipoproteins due to IRE1β deficiency leads to the development of increased atherosclerotic plaques in apoE knockout background mice [69]. Even though macronutrients have been shown to affect MTP expression, further studies are required to investigate how they regulate MTP in vitro and in vivo.

Studies have shown that insulin affects lipoprotein formation either by regulating the amount of fatty acids in the circulation [70] or by direct suppression of the VLDL production [71]. Insulin reduces VLDL secretion most probably through the reduction in MTP activity. Expression of MTP in HepG2 cells is regulated in a dose- and time-dependent manner by insulin [72, 73]. Several studies have shown that insulin regulates MTP expression through mitogen-activated protein kinase and extracellular signal-regulated kinase ($MAPK^{erk}$) cascade and not through phosphatidylinositol 3-kinase (PI3K) signaling pathway that involves phosphorylation of AKT [74, 75]. Increased insulin-mediated MTP suppression by $MAPK^{p38}$ inhibition involves the $MAPK^{erk}$ cascade by phosphorylating and translocating extracellular signal-regulated kinase 1/2 ($ERK^{1/2}$) to the nucleus. Upon translocation, these proteins activate several transcription factors that bind to the *cis*-elements in the promoter regions of various genes [76, 77]. However, the transcription factors activated by $ERK^{1/2}$ that interact with the MTP promoter need to be identified.

Under a non-phosphorylated state, forkhead box protein A2 (FoxA2) and FoxO1 transcription factors are known to bind MTP promoter to increase its expression [78, 79], and this increased expression of MTP is prevented by insulin. Insulin signaling activates phosphorylation of FoxA2 and FoxO1 [78, 79] and prevents their translocation to the nucleus. Furthermore, silencing of FoxO1 in normal and *db/db* mice using RNAi decreased the expression of liver MTP which in turn reduced VLDL production [79]. These combined studies demonstrate the involvement of insulin in lipoprotein metabolism through MTP regulation. Insulin prevents binding of Fox transcription factors to MTP promoter and requires the SRE/IRE element to mediate suppression of MTP in cultured hepatoma cell lines. However, acute administration of insulin into fasted *Apobec1*$^{-/-}$ mice that synthesize apoB100 only did not affect MTP expression but reduced plasma glucose and hepatic FoxO1 levels after 2 h [80].

Even though MTP expression is negatively regulated by insulin, the expression is not reduced in hyperinsulinemic animals. MTP expression is increased in high-sucrose [64] and fructose-fed hamsters [81, 82] that are normalized after the treatment with rosiglitazone by improving insulin sensitivity [83]. Kuriyama et al. have shown that Young Otsuka Long-Evans Tokushima Fatty (OLETF) rats have more hepatic MTP in the absence of hyperinsulinemia during young stage that persists in the adult stage after the development of hyperinsulinemia [84]. Similarly, *ob/ob* mice [85] and Zucker obese (*fa/fa*) rats [86] with hyperinsulinemia have higher MTP expression levels in the intestine and liver. On the other hand, intestinal MTP of alloxan-treated rabbits [87] or streptozotocin-treated rats [88] and mice [85] is increased with no change in liver MTP expression. These studies suggest that hyperinsulinemia is generally associated with increased MTP expression as a result of insulin resistance rather than insulin sensitivity. Regulation of hepatic MTP expression by insulin in vivo remains to be determined, and further studies are needed to understand the role insulin plays in the regulation of MTP during insulin resistance.

In one of the studies, we have shown that intestinal cells express leptin receptors that respond to leptin signaling to regulate MTP levels [89]. We used various mouse models to demonstrate that global deficiency of leptin receptors decreased intestinal MTP expression but not hepatic expression. This regulation of intestinal MTP expression by leptin is independent of central leptin signaling in the hypothalamus. The mechanisms that differentially regulate MTP expression in intestinal and hepatic cells by leptin are unknown and need further investigation.

In early development in mice, MTP expression is initially detected at day 7.5 after gestation mainly in the liver [90]. However, as the embryo matures, the intestine expression of MTP increases and reaches levels higher than the liver [91]. MTP plays a pivotal role in embryonic development as the mice deficient globally in MTP are not viable [92]. Similarly, intestine-specific MTP knockouts do not result in viable progeny after crossing MTP-floxed mice with *Villin-Cre* transgenic mice [46]. Contrary to global or intestine-specific MTP knockouts, liver-specific deletion of MTP using *Alb-Cre* is viable [34] which may be due to delayed expression of albumin during development. As intestine is important in the transport of dietary lipids only after birth, it is unclear why intestine-specific deletion of MTP is critical for embryonic development in mice. Similar to global MTP knockout, apoB knockout mice are also not viable indicating that these lipoproteins may be critical during embryonic development and may be used to nurture the embryo [93]. It is more likely that apoB-containing lipoproteins are involved in the development of the embryo since the inner layer of visceral endodermal cells that line the yolk sac is derived from the embryo [94]. Survival of abetalipoproteinemia subjects during embryonic stage of life suggests that the requirement of MTP during embryo development differs in vertebrates since it is required for yolk lipid utilization in zebrafish larvae [95, 96].

One of the important adaptations that are necessary to uptake dietary fats by the villus cells is the differentiation of stem cells into villus cells to acquire the absorptive phenotype [97]. Human colon adenocarcinoma Caco-2 cells, which can transform into enterocyte-like cells [98], have been extensively used to study differentiation and various intestinal functions [99, 100]. These cells synthesize and secrete apoB-containing lipoproteins only when they are differentiated [99, 101–103] which is dependent on the expression of MTP, and not apoB synthesis [98]. Dai et al. showed that during the undifferentiated stage, cells are inactive due to the binding of nuclear receptor 2 family 1 (NR2F1) repressor to the direct repeat 1 (DR1) element of the MTP promoter, and as the differentiation progresses, NR2F1 expression declines, leading to increased expression of MTP [98]. Besides transcriptional suppression by NR2F1, lower expression of MTP in undifferentiated intestinal cells is also under a posttranscriptional suppression involving IRE1β. Studies have shown that IRE1β cleaves *Mttp* mRNA posttranscriptionally to initiate its degradation [67]. IRE1β, a homolog of ubiquitously expressed IRE1α, is primarily expressed in the intestine and plays a critical role in unfolded protein response [104]. Similar to NR2F1 expression, the level of IRE1β is high in undifferentiated cells which decline during differentiation. Furthermore in mouse intestine, there is a reciprocal expression pattern of MTP with IRE1β and NR2F1 along the jejunum to colon axis and villus to crypt axis in the jejunum [98]. High-cholesterol and Western diets have been shown to enhance the expression of MTP by reducing the expression of IRE1β in the jejunum [67]. These studies suggest that induction of MTP expression involves transcriptional mechanisms that may be dependent on various factors involved in the differentiation of enterocytes. Some studies have shown that MTP may be regulated by posttranscriptional and posttranslational mechanisms also. Activity and protein levels of MTP were lower in mice with liver-specific deletion of phosphatase and tensin homolog (PTEN) with only modest reductions in MTP mRNA [105]. Furthermore, lower MTP activity and protein levels were observed in HepG2 cells overexpressed with a dominant negative form of PTEN [105]. Pan et al. have shown that carbon tetrachloride (CCl4) induces posttranslational ER-associated proteasomal degradation of MTP after covalently modifying it through ubiquitination [106]. These studies indicate that MTP might also be regulated by posttranscriptional and posttranslational mechanisms.

Circadian rhythmicity in humans and rodents maintains a narrow range of plasma lipids by balancing lipoprotein production and catabolism [107–111]. These daily variations in plasma lipids are attributed to synchronized circadian fluctuations in MTP expression in rats and mice [112]. These circadian fluctuations in MTP and plasma lipids were abrogated in *Clock* mutant mice. The clock regulates diurnal expression of MTP by changing the expression of small heterodimer partner (SHP), a repressor of MTP [113]. Expression of MTP is also negatively regulated by bile acids. Chenodeoxycholate has been shown to increase SHP expression that results in suppression of HNF-4α activity leading to a decrease in MTP expression in HepG2 cells [114]. These studies demonstrate that circadian regulation and bile acids may affect MTP expression by modulating the expression of its repressor.

Degradation of mRNAs by microRNAs (miRs) has emerged as a novel mode of posttranscriptional mechanism to control expression of genes in the cells. Large arrays of miRs play a critical role in regulating lipid and lipoprotein metabolism by targeting proteins and enzymes that are involved in these pathways. A recent study by Soh et al. has shown that miR-30c targets hepatic MTP mRNA and modulates lipid substrate availability for VLDL biogenesis [115]. There is ample evidence to suggest modulation of lipid and lipoproteins by the transcriptional and posttranscriptional regulation of MTP, with a very little knowledge about its translational control that needs further investigation.

4.4 Lipids and Lipoproteins in Metabolic Syndrome

Lipids contribute to the risks associated with the complex pathogenesis of metabolic syndrome. Alterations in both atherogenic (VLDL and LDL) and anti-atherogenic (HDL) lipoproteins cause lipid abnormalities that lead to dyslipidemia [116–119] which sometimes may also be caused by a decrease in the clearance of triglyceride-rich lipoproteins [120]. Furthermore, increased prevalence of sedentary life style and obesity has given rise to increased incidence of insulin resistance accompanied with dyslipidemia [121, 122]. Insulin resistance is a critical characteristic of the metabolic syndrome that leads to the development of type 2 diabetes [123, 124] and abnormalities in lipoprotein metabolism contributing to increased cardiovascular risk [123]. Patients with insulin resistance [123] show lipid abnormalities that mainly originate from the overproduction of hepatic VLDL [124, 125] or increased production of chylomicrons [126]. Increased hepatic VLDL production is of key importance in the formation of small dense LDL and is a central feature of dyslipidemia associated with insulin resistance and type 2 diabetes [124, 127]. The presence of small dense LDL particles has been shown to be associated with increased cardiovascular risk that precedes diagnosis of type 2 diabetes [128, 129].

The liver plays a major role in the lipid and lipoprotein metabolism. Under normal physiological conditions, insulin suppresses the gene expression of enzymes involved in triglyceride biosynthesis and reduces the synthesis and secretion of VLDL in the liver [130]. Fatty acids for hepatic lipid and VLDL triglycerides are either synthesized directly by liver or come from dietary fatty acids transported via chylomicrons or plasma non-esterified fatty acid (NEFA) pool originating from adipose tissue [131]. The uptake of fatty acids by the liver is not regulated and is directly related to the concentration of plasma NEFA [132]. However, in insulin-resistant states, inefficiency of insulin signaling results in enhanced lipolysis and flux of fatty acids from adipocytes for increased triglyceride synthesis causing excess triglyceride to be secreted as VLDL [133]. This increased production of VLDL has been implicated to be the major metabolic defect in atherogenic dyslipidemia [134].

During postprandial state, chylomicrons contribute to the large triglyceride-rich lipoproteins, which are critically dependent on apoB48 and MTP [9, 63]. Chylomicrons can accumulate in the circulation and influence overall lipid and lipoprotein turnover during metabolic syndrome [135, 136]. Patients with metabolic syndrome [135, 136] have a significant delay in the clearance of chylomicron remnants due to abnormal postprandial lipoprotein metabolism, thereby leading to impaired glucose and lipid metabolism [137]. In insulin-resistant patients, free fatty acids (FFAs) and inflammatory cytokines are mainly responsible to drive the overproduction of triglyceride-rich lipoproteins [138]. Furthermore, increased activity of lipoprotein lipase that catalyzes the release of fatty acids from the chylomicrons leads to the increase in the plasma fatty acid pool [131]. The dietary fat content determines the overall influence of dietary fatty acids entering the circulation through chylomicrons to hepatic triglycerides. The higher-saturated fatty acid composition in the diet may lead to obesity and insulin resistance [139].

Individuals with insulin resistance have elevated levels of several lipid species, but the causative association between insulin resistance and accumulation of specific lipid metabolites is controversial [140, 141]. Some studies have suggested that hepatic fatty acid biosynthesis pathways are sensitive to the high levels of portal insulin flux, and any imbalance in insulin signaling may lead to the development of a fatty liver [142, 143]. Other studies have shown that hepatic insulin resistance may be due to inflammation and suggested that hepatic steatosis and insulin resistance may be separate manifestations of metabolic disorder [144]. During inflammation various cytokines play an important role in the development of insulin resistance [145]. Obesity is characterized as a state of chronic low-grade inflammation that results in decreased insulin sensitivity [146] and causes lipid accumulation in adipocytes which activates various molecular pathways responsible for increased production of pro-inflammatory cytokines [147].

Insulin resistance in adipose tissue may be an essential aspect for the pathophysiology of the metabolic syndrome. The rising incidence of insulin resistance in the past few decades may be mainly due to increased prevalence of obesity [148]. Hyperinsulinemia during metabolic syndrome correlates with the abdominal adiposity [149] and is a cause of increased hepatic VLDL production that leads to elevated triglycerides in the circulation [150]. The inability of insulin to suppress lipolysis and increase mobilization of FFAs from adipose tissue leads to the increase of FFA flux [148]. Besides increased supply of FFAs to the liver, increased MTP expression and reduced apoB degradation link insulin resistance to increased VLDL secretion [134]. Majority of studies have shown that insulin resistance leads to increased MTP expression [for review, [151]]. Furthermore, insulin-resistant subjects with hyperinsulinemia show significantly higher apoB48 levels [152] contrary to decreased circulating apoB48-containing lipoproteins after insulin administration [153].

4.5 Therapeutic Intervention of MTP to Alleviate Dyslipidemia in Metabolic Syndrome

Besides obesity and insulin resistance, hyperlipidemia is one of the most critical risk factors contributing to metabolic diseases such as atherosclerosis and type 2 diabetes. MTP has been a favorite target to lower production of apoB-containing lipoproteins and treat lipid disorders such as hypercholesterolemia and hypertriglyceridemia and thereby decrease atherosclerosis [154–159]. Several MTP antagonists that decrease lipoprotein production and plasma lipids have been identified [24, 158, 160] and tested in humans after promising preclinical studies provided proof of concept that inhibition of MTP may be an effective therapeutic target to alleviate hyperlipidemia [158–162]. However, some of the subjects in these studies experienced elevated hepatic lipids and increased plasma transaminases [161, 162]. Due to these reasons, the use of MTP inhibitors has been restricted to conditions where alternative therapeutic options to decrease plasma lipids are either not present or are associated with high risks [163, 164].

Novel strategies need to be developed to decrease MTP activity without affecting transaminases in the plasma or lipids in the liver. A recent study by Wang et al. showed that metformin improved lipid metabolism in OLETF rats by reducing the expression of MTP [165]. It has been suggested that inhibition of MTP specifically in the intestine might be useful to avoid hepatic toxicity [157, 166]. However, before promoting this approach, long-term consequences need careful evaluation since intestinal MTP deficiency may be associated with gastrointestinal disturbances [160]. Several natural compounds such as flavanoids [167], isoflavones [168], and garlic extracts [169] have been shown to inhibit MTP activity and should be explored as an alternate therapeutic treatment to lower lipids in the plasma. Another option is to specifically inhibit triglyceride transfer activity of MTP to lower plasma lipids since phospholipid transfer activity is sufficient for lipoprotein assembly and secretion [34, 35]. A combined treatment involving inhibitors that reduce hepatic lipid accumulation along with MTP antagonists may also be beneficial. The discovery of miRs and their critical roles in controlling cellular lipid and lipoprotein metabolism has opened new possibilities to use miRs or their inhibitors as potential therapeutic agents to reduce hyperlipidemia, obesity, diabetes, and atherosclerosis [117, 170–172]. Researchers are now targeting the expression of miRs that regulates lipid metabolism genes as a therapeutic intervention to treat metabolic diseases [115, 173, 174]. Due to their multi-targeting essence, miRs may become a very powerful tool that may influence the pathophysiological process of metabolic diseases.

4.6 Conclusions

Over the years, many researchers have made evident the important role of MTP in the regulation of lipid and lipoprotein metabolism. In this book chapter, we have summarized some of the key functions of MTP and discussed how it is regulated by various factors involving transcriptional, posttranscriptional, and posttranslational mechanisms. We also discussed regulation of MTP and lipoprotein metabolism during pathophysiological conditions that highlights its role in the development of certain metabolic diseases. Pharmaceutical interventions to inhibit MTP to treat these metabolic conditions have so far been unsuccessful due to the adverse effects, such as hepatic steatosis, that are associated with these treatments. However, efforts are ongoing to find the alternative approaches that could be successfully used in the future to treat lipid and lipoprotein metabolism-related disorders. In this respect, miRNA-based therapies have been envisioned to serve as a possible novel approach for the treatment of metabolic diseases although further investigation and refining is necessary.

References

1. Meshkani R, Adeli K (2009) Hepatic insulin resistance, metabolic syndrome and cardiovascular disease. Clin Biochem 42:1331–1346
2. Alberti KG, Eckel RH, Grundy SM, Zimmet PZ, Cleeman JI, Donato KA, Fruchart JC, James WP, Loria CM, Smith SC Jr (2009) Harmonizing the metabolic syndrome: a joint interim statement of the International Diabetes Federation Task Force on Epidemiology and Prevention; National Heart, Lung, and Blood Institute; American Heart Association; World Heart Federation; International Atherosclerosis Society; and International Association for the Study of Obesity. Circulation 120:1640–1645
3. Shapiro MD, Fazio S (2017) Apolipoprotein B-containing lipoproteins and atherosclerotic cardiovascular disease. F1000 Res 6:134
4. Brown WV (1994) Lipoprotein disorders in diabetes mellitus. Med Clin North Am 78:143–161
5. Glass CK, Witztum JL (2001) Atherosclerosis. The road ahead. Cell 104:503–516
6. Lusis AJ (2000) Atherosclerosis. Nature 407:233–241
7. Borodzicz S, Czarzasta K, Kuch M, Cudnoch-Jedrzejewska A (2015) Sphingolipids in cardiovascular diseases and metabolic disorders. Lipids Health Dis 14:55
8. Iqbal J, Walsh MT, Hammad SM, Hussain MM (2017) Sphingolipids and lipoproteins in health and metabolic disorders. Trends Endocrinol Metab 28:506–518
9. Iqbal J, Hussain MM (2009) Intestinal lipid absorption. Am J Physiol Endocrinol Metab 296:E1183–E1194
10. Segrest JP, Jones MK, De LH, Dashti N (2001) Structure of apolipoprotein B-100 in low density lipoproteins. J Lipid Res 42:1346–1367
11. Walsh MT, Hussain MM (2016) Targeting microsomal triglyceride transfer protein and lipoprotein assembly to treat homozygous familial hypercholesterolemia. Crit Rev Clin Lab Sci 54:1–23
12. Hussain MM, Rava P, Walsh M, Rana M, Iqbal J (2012) Multiple functions of microsomal triglyceride transfer protein. Nutr Metab (Lond) 9:14
13. Hussain MM, Bakillah A, Jamil H (1997) Apolipoprotein B binding to microsomal triglyceride transfer protein decreases with increases in length and lipidation: implications in lipoprotein biosynthesis. Biochemistry 36:13060–13067
14. Hussain MM (2000) A proposed model for the assembly of chylomicrons. Atherosclerosis 148:1–15
15. Hussain MM, Iqbal J, Anwar K, Rava P, Dai K (2003) Microsomal triglyceride transfer protein: a multifunctional protein. Front Biosci 8:s500–s506
16. Hussain MM, Shi J, Dreizen P (2003) Microsomal triglyceride transfer protein and its role in apoB-lipoprotein assembly. J Lipid Res 44:22–32
17. Sellers JA, Hou L, Athar H, Hussain MM, Shelness GS (2003) A Drosophila microsomal triglyceride transfer protein homolog promotes the assembly and secretion of human apolipoprotein B. Implications for human and insect transport and metabolism. J Biol Chem 278:20367–20373
18. Sirwi A, Hussain MM (2018) Lipid transfer proteins in the assembly of apoB-containing lipoproteins. J Lipid Res 59:1094–1102
19. Wetterau JR, Zilversmit DB (1985) Purification and characterization of microsomal triglyceride and cholesteryl ester transfer protein from bovine liver microsomes. Chem Phys Lipids 38:205–222

20. Wetterau JR, Combs KA, Spinner SN, Joiner BJ (1990) Protein disulfide isomerase is a component of the microsomal triglyceride transfer protein complex. J Biol Chem 265:9800–9807
21. Wetterau JR, Aggerbeck LP, Bouma ME, Eisenberg C, Munck A, Hermier M, Schmitz J, Gay G, Rader DJ, Gregg RE (1992) Absence of microsomal triglyceride transfer protein in individuals with abetalipoproteinemia. Science 258:999–1001
22. Shoulders CC, Brett DJ, Bayliss JD, Narcisi TM, Jarmuz A, Grantham TT, Leoni PR, Bhattacharya S, Pease RJ, Cullen PM et al (1993) Abetalipoproteinemia is caused by defects of the gene encoding the 97 kDa subunit of a microsomal triglyceride transfer protein. Hum Mol Genet 2:2109–2116
23. Sharp D, Blinderman L, Combs KA, Kienzle B, Ricci B, Wager-Smith K, Gil CM, Turck CW, Bouma ME, Rader DJ et al (1993) Cloning and gene defects in microsomal triglyceride transfer protein associated with abetalipoproteinaemia. Nature 365:65–69
24. Wetterau JR, Gregg RE, Harrity TW, Arbeeny C, Cap M, Connolly F, Chu CH, George RJ, Gordon DA, Jamil H et al (1998) An MTP inhibitor that normalizes atherogenic lipoprotein levels in WHHL rabbits. Science 282:751–754
25. Rehberg EF, Samson-Bouma ME, Kienzle B, Blinderman L, Jamil H, Wetterau JR, Aggerbeck LP, Gordon DA (1996) A novel abetalipoproteinemia genotype. Identification of a missense mutation in the 97-kDa subunit of the microsomal triglyceride transfer protein that prevents complex formation with protein disulfide isomerase. J Biol Chem 271:29945–29952
26. Ohashi K, Ishibashi S, Osuga J, Tozawa R, Harada K, Yahagi N, Shionoiri F, Iizuka Y, Tamura Y, Nagai R et al (2000) Novel mutations in the microsomal triglyceride transfer protein gene causing abetalipoproteinemia. J Lipid Res 41:1199–1204
27. Gouda W, Ashour E, Shaker Y, Ezzat W (2017) MTP genetic variants associated with non-alcoholic fatty liver in metabolic syndrome patients. Genes Dis 4:222–228
28. Mann CJ, Anderson TA, Read J, Chester SA, Harrison GB, Kochl S, Ritchie PJ, Bradbury P, Hussain FS, Amey J et al (1999) The structure of vitellogenin provides a molecular model for the assembly and secretion of atherogenic lipoproteins. J Mol Biol 285:391–408
29. Hussain MM, Bakillah A, Nayak N, Shelness GS (1998) Amino acids 430-570 in apolipoprotein B are critical for its binding to microsomal triglyceride transfer protein. J Biol Chem 273:25612–25615
30. Bakillah A, Nayak N, Saxena U, Medford RM, Hussain MM (2000) Decreased secretion of ApoB follows inhibition of ApoB-MTP binding by a novel antagonist. Biochemistry 39:4892–4899
31. Athar H, Iqbal J, Jiang XC, Hussain MM (2004) A simple, rapid, and sensitive fluorescence assay for microsomal triglyceride transfer protein. J Lipid Res 45:764–772
32. Rava P, Athar H, Johnson C, Hussain MM (2005) Transfer of cholesteryl esters and phospholipids as well as net deposition by microsomal triglyceride transfer protein. J Lipid Res 46:1779–1785
33. Iqbal J, Walsh MT, Hammad SM, Cuchel M, Tarugi P, Hegele RA, Davidson NO, Rader DJ, Klein RL, Hussain MM (2015) Microsomal triglyceride transfer protein transfers and determines plasma concentrations of ceramide and sphingomyelin but not glycosylceramide. J Biol Chem 290:25863–25875
34. Khatun I, Zeissig S, Iqbal J, Wang M, Curiel D, Shelness GS, Blumberg RS, Hussain MM (2012) Phospholipid transfer activity of microsomal triglyceride transfer protein produces apolipoprotein B and reduces hepatosteatosis while maintaining low plasma lipids in mice. Hepatology 55:1356–1368
35. Rava P, Ojakian GK, Shelness GS, Hussain MM (2006) Phospholipid transfer activity of microsomal triacylglycerol transfer protein is sufficient for the assembly and secretion of apolipoprotein B lipoproteins. J Biol Chem 281:11019–11027
36. Rava P, Hussain MM (2007) Acquisition of triacylglycerol transfer activity by microsomal triglyceride transfer protein during evolution. Biochemistry 46:12263–12274
37. Gretch DG, Sturley SL, Wang L, Lipton BA, Dunning A, Grunwald KA, Wetterau JR, Yao Z, Talmud P, Attie AD (1996) The amino terminus of apolipoprotein B is necessary but not sufficient for microsomal triglyceride transfer protein responsiveness. J Biol Chem 271:8682–8691
38. Liang J, Ginsberg HN (2001) Microsomal triglyceride transfer protein binding and lipid transfer activities are independent of each other, but both are required for secretion of apolipoprotein B lipoproteins from liver cells. J Biol Chem 276:28606–28612
39. Shelness GS, Thornburg JT (1996) Role of intramolecular disulfide bond formation in the assembly and secretion of apolipoprotein B-100-containing lipoproteins. J Lipid Res 37:408–419
40. Burch WL, Herscovitz H (2000) Disulfide bonds are required for folding and secretion of apolipoprotein B regardless of its lipidation state. J Biol Chem 275:16267–16274
41. Bakillah A, Jamil H, Hussain MM (1998) Lysine and arginine residues in the N-terminal 18% of apolipoprotein B are critical for its binding to microsomal triglyceride transfer protein. Biochemistry 37:3727–3734
42. Gordon DA, Jamil H, Sharp D, Mullaney D, Yao Z, Gregg RE, Wetterau J (1994) Secretion of apolipoprotein B-containing lipoproteins from HeLa cells is dependent on expression of the microsomal

triglyceride transfer protein and is regulated by lipid availability. Proc Natl Acad Sci U S A 91:7628–7632
43. Nicodeme E, Benoist F, McLeod R, Yao Z, Scott J, Shoulders CC, Grand-Perret T (1999) Identification of domains in apolipoprotein B100 that confer a high requirement for the microsomal triglyceride transfer protein. J Biol Chem 274:1986–1993
44. Kulinski A, Rustaeus S, Vance JE (2002) Microsomal triacylglycerol transfer protein is required for lumenal accretion of triacylglycerol not associated with ApoB, as well as for ApoB lipidation. J Biol Chem 277:31516–31525
45. Raabe M, Veniant MM, Sullivan MA, Zlot CH, Bjorkegren J, Nielsen LB, Wong JS, Hamilton RL, Young SG (1999) Analysis of the role of microsomal triglyceride transfer protein in the liver of tissue-specific knockout mice. J Clin Invest 103:1287–1298
46. Xie Y, Newberry EP, Young SG, Robine S, Hamilton RL, Wong JS, Luo J, Kennedy S, Davidson NO (2006) Compensatory increase in hepatic lipogenesis in mice with conditional intestine-specific Mttp deficiency. J Biol Chem 281:4075–4086
47. Tabas I (2002) Cholesterol in health and disease. J Clin Invest 110:583–590
48. Buhman KF, Accad M, Farese RV (2000) Mammalian acyl-CoA:cholesterol acyltransferases. Biochim Biophys Acta 1529:142–154
49. Chang TY, Chang CC, Cheng D (1997) Acyl-coenzyme A:cholesterol acyltransferase. Annu Rev Biochem 66:613–638
50. Chang TY, Chang CC, Lin S, Yu C, Li BL, Miyazaki A (2001) Roles of acyl-coenzyme A:cholesterol acyltransferase-1 and -2. Curr Opin Lipidol 12:289–296
51. Farese RV Jr (1998) Acyl CoA:cholesterol acyltransferase genes and knockout mice. Curr Opin Lipidol 9:119–123
52. Rudel LL, Lee RG, Cockman TL (2001) Acyl coenzyme A: cholesterol acyltransferase types 1 and 2: structure and function in atherosclerosis. Curr Opin Lipidol 12:121–127
53. Lee RG, Willingham MC, Davis MA, Skinner KA, Rudel LL (2000) Differential expression of ACAT1 and ACAT2 among cells within liver, intestine, kidney, and adrenal of nonhuman primates. J Lipid Res 41:1991–2001
54. Smith JL, Rangaraj K, Simpson R, Maclean DJ, Nathanson LK, Stuart KA, Scott SP, Ramm GA, de JJ (2004) Quantitative analysis of the expression of ACAT genes in human tissues by real-time PCR. J Lipid Res 45:686–696
55. Miyazaki A, Sakashita N, Lee O, Takahashi K, Horiuchi S, Hakamata H, Morganelli PM, Chang CC, Chang TY (1998) Expression of ACAT-1 protein in human atherosclerotic lesions and cultured human monocytes-macrophages. Arterioscler Thromb Vasc Biol 18:1568–1574
56. Chang CC, Sakashita N, Ornvold K, Lee O, Chang ET, Dong R, Lin S, Lee CY, Strom SC, Kashyap R et al (2000) Immunological quantitation and localization of ACAT-1 and ACAT-2 in human liver and small intestine. J Biol Chem 275:28083–28092
57. Sakashita N, Miyazaki A, Chang CC, Chang TY, Kiyota E, Satoh M, Komohara Y, Morganelli PM, Horiuchi S, Takeya M (2003) Acyl-coenzyme A: cholesterol acyltransferase 2 (ACAT2) is induced in monocyte-derived macrophages: in vivo and in vitro studies. Lab Investig 83:1569–1581
58. Parini P, Davis M, Lada AT, Erickson SK, Wright TL, Gustafsson U, Sahlin S, Einarsson C, Eriksson M, Angelin B et al (2004) ACAT2 is localized to hepatocytes and is the major cholesterol-esterifying enzyme in human liver. Circulation 110:2017–2023
59. Lin S, Cheng D, Liu MS, Chen J, Chang TY (1999) Human acyl-CoA:cholesterol acyltransferase-1 in the endoplasmic reticulum contains seven transmembrane domains. J Biol Chem 274:23276–23285
60. Lin S, Lu X, Chang CC, Chang TY (2003) Human acyl-coenzyme A:cholesterol acyltransferase expressed in chinese hamster ovary cells: membrane topology and active site location. Mol Biol Cell 14:2447–2460
61. Joyce CW, Shelness GS, Davis MA, Lee RG, Skinner K, Anderson RA, Rudel LL (2000) ACAT1 and ACAT2 membrane topology segregates a serine residue essential for activity to opposite sides of the endoplasmic reticulum membrane. Mol Biol Cell 11:3675–3687
62. Iqbal J, Rudel LL, Hussain MM (2008) Microsomal triglyceride transfer protein enhances cellular cholesteryl esterification by relieving product inhibition. J Biol Chem 283:19967–19980
63. Hussain MM, Fatma S, Pan X, Iqbal J (2005) Intestinal lipoprotein assembly. Curr Opin Lipidol 16:281–285
64. Lin MC, Arbeeny C, Bergquist K, Kienzle B, Gordon DA, Wetterau JR (1994) Cloning and regulation of hamster microsomal triglyceride transfer protein. The regulation is independent from that of other hepatic and intestinal proteins which participate in the transport of fatty acids and triglycerides. J Biol Chem 269:29138–29145
65. Qiu W, Taghibiglou C, Avramoglu RK, Van Iderstine SC, Naples M, Ashrafpour H, Mhapsekar S, Sato R, Adeli K (2005) Oleate-mediated stimulation of microsomal triglyceride transfer protein (MTP) gene promoter: implications for hepatic MTP overexpression in insulin resistance. Biochemistry 44:3041–3049
66. Hernandez Vallejo SJ, Alqub M, Luquet S, Cruciani-Guglielmacci C, Delerive P, Lobaccaro JM, Kalopissis AD, Chambaz J, Rousset M, Lacorte JM (2009) Short term adaptation of postprandial lipoprotein secretion and intestinal gene expression to a high-fat diet. Am J Physiol Gastrointest Liver Physiol 296:G782–G792

67. Iqbal J, Dai K, Seimon T, Jungreis R, Oyadomari M, Kuriakose G, Ron D, Tabas I, Hussain MM (2008) IRE1beta inhibits chylomicron production by selectively degrading MTP mRNA. Cell Metab 7:445–455
68. Sato R, Miyamoto W, Inoue J, Terada T, Imanaka T, Maeda M (1999) Sterol regulatory element-binding protein negatively regulates microsomal triglyceride transfer protein gene transcription. J Biol Chem 274:24714–24720
69. Iqbal J, Queiroz J, Li Y, Jiang XC, Ron D, Hussain MM (2012) Increased intestinal lipid absorption caused by Ire1beta deficiency contributes to hyperlipidemia and atherosclerosis in apolipoprotein E-deficient mice. Circ Res 110:1575–1584
70. Lewis GF, Uffelman KD, Szeto LW, Steiner G (1993) Effects of acute hyperinsulinemia on VLDL triglyceride and VLDL apoB production in normal weight and obese individuals. Diabetes 42:833–842
71. Malmstrom R, Packard CJ, Caslake M, Bedford D, Stewart P, Yki-Jarvinen H, Shepherd J, Taskinen MR (1998) Effects of insulin and acipimox on VLDL1 and VLDL2 apolipoprotein B production in normal subjects. Diabetes 47:779–787
72. Hagan DL, Kienzle B, Jamil H, Hariharan N (1994) Transcriptional regulation of human and hamster microsomal triglyceride transfer protein genes. Cell type-specific expression and response to metabolic regulators. J Biol Chem 269:28737–28744
73. Lin MC, Gordon D, Wetterau JR (1995) Microsomal triglyceride transfer protein (MTP) regulation in HepG2 cells: insulin negatively regulates MTP gene expression. J Lipid Res 36:1073–1081
74. Au WS, Kung HF, Lin MC (2003) Regulation of microsomal triglyceride transfer protein gene by insulin in HepG2 cells: roles of MAPKerk and MAPKp38. Diabetes 52:1073–1080
75. Allister EM, Borradaile NM, Edwards JY, Huff MW (2005) Inhibition of microsomal triglyceride transfer protein expression and apolipoprotein B100 secretion by the citrus flavonoid naringenin and by insulin involves activation of the mitogen-activated protein kinase pathway in hepatocytes. Diabetes 54:1676–1683
76. Rose BA, Force T, Wang Y (2010) Mitogen-activated protein kinase signaling in the heart: angels versus demons in a heart-breaking tale. Physiol Rev 90:1507–1546
77. Mebratu Y, Tesfaigzi Y (2009) How ERK1/2 activation controls cell proliferation and cell death: is subcellular localization the answer? Cell Cycle 8:1168–1175
78. Wolfrum C, Stoffel M (2006) Coactivation of Foxa2 through Pgc-1beta promotes liver fatty acid oxidation and triglyceride/VLDL secretion. Cell Metab 3:99–110
79. Kamagate A, Qu S, Perdomo G, Su D, Kim DH, Slusher S, Meseck M, Dong HH (2008) FoxO1 mediates insulin-dependent regulation of hepatic VLDL production in mice. J Clin Invest 118:2347–2364
80. Sparks JD, Chamberlain JM, O'Dell C, Khatun I, Hussain MM, Sparks CE (2011) Acute suppression of apo B secretion by insulin occurs independently of MTP. Biochem Biophys Res Commun 406:252–256
81. Taghibiglou C, Carpentier A, Van Iderstine SC, Chen B, Rudy D, Aiton A, Lewis GF, Adeli K (2000) Mechanisms of hepatic very low density lipoprotein overproduction in insulin resistance. Evidence for enhanced lipoprotein assembly, reduced intracellular ApoB degradation, and increased microsomal triglyceride transfer protein in a fructose-fed hamster model. J Biol Chem 275:8416–8425
82. Haidari M, Leung N, Mahbub F, Uffelman KD, Kohen-Avramoglu R, Lewis GF, Adeli K (2002) Fasting and postprandial overproduction of intestinally derived lipoproteins in an animal model of insulin resistance. Evidence that chronic fructose feeding in the hamster is accompanied by enhanced intestinal de novo lipogenesis and ApoB48-containing lipoprotein overproduction. J Biol Chem 277:31646–31655
83. Carpentier A, Taghibiglou C, Leung N, Szeto L, Van Iderstine SC, Uffelman KD, Buckingham R, Adeli K, Lewis GF (2002) Ameliorated hepatic insulin resistance is associated with normalization of microsomal triglyceride transfer protein expression and reduction in very low density lipoprotein assembly and secretion in the fructose-fed hamster. J Biol Chem 277:28795–28802
84. Kuriyama H, Yamashita S, Shimomura I, Funahashi T, Ishigami M, Aragane K, Miyaoka K, Nakamura T, Takemura K, Man Z et al (1998) Enhanced expression of hepatic acyl-coenzyme A synthetase and microsomal triglyceride transfer protein messenger RNAs in the obese and hypertriglyceridemic rat with visceral fat accumulation. Hepatology 27:557–562
85. Bartels ED, Lauritsen M, Nielsen LB (2002) Hepatic expression of microsomal triglyceride transfer protein and in vivo secretion of triglyceride-rich lipoproteins are increased in obese diabetic mice. Diabetes 51:1233–1239
86. Phillips C, Owens D, Collins P, Tomkin GH (2002) Microsomal triglyceride transfer protein: does insulin resistance play a role in the regulation of chylomicron assembly? Atherosclerosis 160:355–360
87. Phillips C, Bennett A, Anderton K, Owens D, Collins P, White D, Tomkin GH (2002) Intestinal rather than hepatic microsomal triglyceride transfer protein as a cause of postprandial dyslipidemia in diabetes. Metabolism 51:847–852
88. Gleeson A, Anderton K, Owens D, Bennett A, Collins P, Johnson A, White D, Tomkin GH (1999) The role of microsomal triglyceride transfer protein and dietary cholesterol in chylomicron production in diabetes. Diabetologia 42:944–948

89. Iqbal J, Li X, Chang BH, Chan L, Schwartz GJ, Chua SC Jr, Hussain MM (2010) An intrinsic gut leptin-melanocortin pathway modulates intestinal microsomal triglyceride transfer protein and lipid absorption. J Lipid Res 51:1929–1942
90. Terasawa Y, Cases SJ, Wong JS, Jamil H, Jothi S, Traber MG, Packer L, Gordon DA, Hamilton RL, Farese RV Jr (1999) Apolipoprotein B-related gene expression and ultrastructural characteristics of lipoprotein secretion in mouse yolk sac during embryonic development. J Lipid Res 40:1967–1977
91. Shelton JM, Lee MH, Richardson JA, Patel SB (2000) Microsomal triglyceride transfer protein expression during mouse development. J Lipid Res 41:532–537
92. Raabe M, Flynn LM, Zlot CH, Wong JS, Veniant MM, Hamilton RL, Young SG (1998) Knockout of the abetalipoproteinemia gene in mice: reduced lipoprotein secretion in heterozygotes and embryonic lethality in homozygotes. Proc Natl Acad Sci U S A 95:8686–8691
93. Farese RV Jr, Ruland SL, Flynn LM, Stokowski RP, Young SG (1995) Knockout of the mouse apolipoprotein B gene results in embryonic lethality in homozygotes and protection against diet-induced hypercholesterolemia in heterozygotes. Proc Natl Acad Sci U S A 92:1774–1778
94. Farese RV Jr, Cases S, Ruland SL, Kayden HJ, Wong JS, Young SG, Hamilton RL (1996) A novel function for apolipoprotein B: lipoprotein synthesis in the yolk sac is critical for maternal-fetal lipid transport in mice. J Lipid Res 37:347–360
95. Marza E, Barthe C, Andre M, Villeneuve L, Helou C, Babin PJ (2005) Developmental expression and nutritional regulation of a zebrafish gene homologous to mammalian microsomal triglyceride transfer protein large subunit. Dev Dyn 232:506–518
96. Schlegel A, Stainier DY (2006) Microsomal triglyceride transfer protein is required for yolk lipid utilization and absorption of dietary lipids in zebrafish larvae. Biochemistry 45:15179–15187
97. Scoville DH, Sato T, He XC, Li L (2008) Current view: intestinal stem cells and signaling. Gastroenterology 134:849–864
98. Dai K, Khatun I, Hussain MM (2010) NR2F1 and IRE1beta suppress microsomal triglyceride transfer protein expression and lipoprotein assembly in undifferentiated intestinal epithelial cells. Arterioscler Thromb Vasc Biol 30:568–574
99. Hussain MM, Kancha RK, Zhou Z, Luchoomun J, Zu H, Bakillah A (1996) Chylomicron assembly and catabolism: role of apolipoproteins and receptors. Biochim Biophys Acta 1300:151–170
100. Sun H, Chow EC, Liu S, Du Y, Pang KS (2008) The Caco-2 cell monolayer: usefulness and limitations. Expert Opin Drug Metab Toxicol 4:395–411
101. Hughes TE, Sasak WV, Ordovas JM, Forte TM, Lamon-Fava S, Schaefer EJ (1987) A novel cell line (Caco-2) for the study of intestinal lipoprotein synthesis. J Biol Chem 262:3762–3767
102. Luchoomun J, Hussain MM (1999) Assembly and secretion of chylomicrons by differentiated Caco-2 cells. Nascent triglycerides and preformed phospholipids are preferentially used for lipoprotein assembly. J Biol Chem 274:19565–19572
103. Dashti N, Smith EA, Alaupovic P (1990) Increased production of apolipoprotein B and its lipoproteins by oleic acid in Caco-2 cells. J Lipid Res 31:113–123
104. Bertolotti A, Wang X, Novoa I, Jungreis R, Schlessinger K, Cho JH, West AB, Ron D (2001) Increased sensitivity to dextran sodium sulfate colitis in IRE1beta-deficient mice. J Clin Invest 107:585–593
105. Qiu W, Federico L, Naples M, Avramoglu RK, Meshkani R, Zhang J, Tsai J, Hussain M, Dai K, Iqbal J et al (2008) Phosphatase and tensin homolog (PTEN) regulates hepatic lipogenesis, microsomal triglyceride transfer protein, and the secretion of apolipoprotein B-containing lipoproteins. Hepatology 48:1799–1809
106. Pan X, Hussain FN, Iqbal J, Feuerman MH, Hussain MM (2007) Inhibiting proteasomal degradation of microsomal triglyceride transfer protein prevents CCl4-induced steatosis. J Biol Chem 282:17078–17089
107. Hussain MM, Pan X (2009) Clock genes, intestinal transport and plasma lipid homeostasis. Trends Endocrinol Metab 20:177–185
108. Maillot F, Garrigue MA, Pinault M, Objois M, Theret V, Lamisse F, Hoinard C, Antoine JM, Lairon D, Couet C (2005) Changes in plasma triacylglycerol concentrations after sequential lunch and dinner in healthy subjects. Diabetes Metab 31:69–77
109. Mondola P, Gambardella P, Santangelo F, Santillo M, Greco AM (1995) Circadian rhythms of lipid and apolipoprotein pattern in adult fasted rats. Physiol Behav 58:175–180
110. Balasubramaniam S, Szanto A, Roach PD (1994) Circadian rhythm in hepatic low-density-lipoprotein (LDL)-receptor expression and plasma LDL levels. Biochem J 298(Pt 1):39–43
111. Bruckdorfer KR, Kang SS, Khan IH, Bourne AR, Yudkin J (1974) Diurnal changes in the concentrations of plasma lipids, sugars, insulin and corticosterone in rats fed diets containing various carbohydrates. Horm Metab Res 6:99–106
112. Pan X, Hussain MM (2007) Diurnal regulation of microsomal triglyceride transfer protein and plasma lipid levels. J Biol Chem 282:24707–24719
113. Pan X, Zhang Y, Wang L, Hussain MM (2010) Diurnal regulation of MTP and plasma triglyceride by CLOCK is mediated by SHP. Cell Metab 12:174–186
114. Hirokane H, Nakahara M, Tachibana S, Shimizu M, Sato R (2004) Bile acid reduces the secretion of very low density lipoprotein by repressing microsomal

triglyceride transfer protein gene expression mediated by hepatocyte nuclear factor-4. J Biol Chem 279:45685–45692
115. Soh J, Iqbal J, Queiroz J, Fernandez-Hernando C, Hussain MM (2013) MicroRNA-30c reduces hyperlipidemia and atherosclerosis in mice by decreasing lipid synthesis and lipoprotein secretion. Nat Med 19:892–900
116. Avramoglu RK, Basciano H, Adeli K (2006) Lipid and lipoprotein dysregulation in insulin resistant states. Clin Chim Acta 368:1–19
117. Christian P, Su Q (2014) MicroRNA regulation of mitochondrial and ER stress signaling pathways: implications for lipoprotein metabolism in metabolic syndrome. Am J Physiol Endocrinol Metab 307: E729–E737
118. Su Q, Baker C, Christian P, Naples M, Tong X, Zhang K, Santha M, Adeli K (2014) Hepatic mitochondrial and ER stress induced by defective PPARalpha signaling in the pathogenesis of hepatic steatosis. Am J Physiol Endocrinol Metab 306: E1264–E1273
119. Soran H, Schofield JD, Adam S, Durrington PN (2016) Diabetic dyslipidaemia. Curr Opin Lipidol 27:313–322
120. Klop B, Elte JW, Cabezas MC (2013) Dyslipidemia in obesity: mechanisms and potential targets. Nutrients 5:1218–1240
121. Zimmet P, Alberti KG, Shaw J (2001) Global and societal implications of the diabetes epidemic. Nature 414:782–787
122. Iqbal J, Al QA, Hawwari A, Alghanem AF, Ahmed G (2018) Metabolic syndrome, dyslipidemia and regulation of lipoprotein metabolism. Curr Diabetes Rev 14:427–433
123. Ginsberg HN, Zhang YL, Hernandez-Ono A (2006) Metabolic syndrome: focus on dyslipidemia. Obesity (Silver Spring) 14(Suppl 1):41S–49S
124. Taskinen MR (2003) Diabetic dyslipidaemia: from basic research to clinical practice. Diabetologia 46:733–749
125. Adiels M, Olofsson SO, Taskinen MR, Boren J (2006) Diabetic dyslipidaemia. Curr Opin Lipidol 17:238–246
126. Veilleux A, Grenier E, Marceau P, Carpentier AC, Richard D, Levy E (2014) Intestinal lipid handling: evidence and implication of insulin signaling abnormalities in human obese subjects. Arterioscler Thromb Vasc Biol 34:644–653
127. Verges B (2005) New insight into the pathophysiology of lipid abnormalities in type 2 diabetes. Diabetes Metab 31:429–439
128. Lamarche B, Tchernof A, Moorjani S, Cantin B, Dagenais GR, Lupien PJ, Despres JP (1997) Small, dense low-density lipoprotein particles as a predictor of the risk of ischemic heart disease in men. Prospective results from the Quebec Cardiovascular Study. Circulation 95:69–75
129. Gardner CD, Fortmann SP, Krauss RM (1996) Association of small low-density lipoprotein particles with the incidence of coronary artery disease in men and women. JAMA 276:875–881
130. Wang Y, Viscarra J, Kim SJ, Sul HS (2015) Transcriptional regulation of hepatic lipogenesis. Nat Rev Mol Cell Biol 16:678–689
131. Barrows BR, Parks EJ (2006) Contributions of different fatty acid sources to very low-density lipoprotein-triacylglycerol in the fasted and fed states. J Clin Endocrinol Metab 91:1446–1452
132. Tamura S, Shimomura I (2005) Contribution of adipose tissue and de novo lipogenesis to nonalcoholic fatty liver disease. J Clin Invest 115:1139–1142
133. Goldberg IJ, Ginsberg HN (2006) Ins and outs modulating hepatic triglyceride and development of nonalcoholic fatty liver disease. Gastroenterology 130:1343–1346
134. Adiels M, Olofsson SO, Taskinen MR, Boren J (2008) Overproduction of very low-density lipoproteins is the hallmark of the dyslipidemia in the metabolic syndrome. Arterioscler Thromb Vasc Biol 28:1225–1236
135. Rivellese AA, De NC, Di ML, Patti L, Iovine C, Coppola S, Del PS, Riccardi G, Annuzzi G (2004) Exogenous and endogenous postprandial lipid abnormalities in type 2 diabetic patients with optimal blood glucose control and optimal fasting triglyceride levels. J Clin Endocrinol Metab 89:2153–2159
136. Kolovou GD, Anagnostopoulou KK, Pavlidis AN, Salpea KD, Iraklianou SA, Tsarpalis K, Damaskos DS, Manolis A, Cokkinos DV (2005) Postprandial lipemia in men with metabolic syndrome, hypertensives and healthy subjects. Lipids Health Dis 4:21
137. Masuda D, Yamashita S (2017) Postprandial hyperlipidemia and remnant lipoproteins. J Atheroscler Thromb 24:95–109
138. Jung UJ, Choi MS (2014) Obesity and its metabolic complications: the role of adipokines and the relationship between obesity, inflammation, insulin resistance, dyslipidemia and nonalcoholic fatty liver disease. Int J Mol Sci 15:6184–6223
139. Ristic-Medic D, Vucic V (2013) Dietary fats and metabolic syndrome. J Nutr Health Food Sci 1:1–8
140. Holland WL, Knotts TA, Chavez JA, Wang LP, Hoehn KL, Summers SA (2007) Lipid mediators of insulin resistance. Nutr Rev 65:S39–S46
141. Morino K, Petersen KF, Shulman GI (2006) Molecular mechanisms of insulin resistance in humans and their potential links with mitochondrial dysfunction. Diabetes 55(Suppl 2):S9–S15
142. Marchesini G, Brizi M, Bianchi G, Tomassetti S, Bugianesi E, Lenzi M, McCullough AJ, Natale S, Forlani G, Melchionda N (2001) Nonalcoholic fatty liver disease: a feature of the metabolic syndrome. Diabetes 50:1844–1850
143. Chitturi S, Abeygunasekera S, Farrell GC, Holmes-Walker J, Hui JM, Fung C, Karim R, Lin R,

Samarasinghe D, Liddle C et al (2002) NASH and insulin resistance: insulin hypersecretion and specific association with the insulin resistance syndrome. Hepatology 35:373–379
144. Arkan MC, Hevener AL, Greten FR, Maeda S, Li ZW, Long JM, Wynshaw-Boris A, Poli G, Olefsky J, Karin M (2005) IKK-beta links inflammation to obesity-induced insulin resistance. Nat Med 11:191–198
145. Chen L, Chen R, Wang H, Liang F (2015) Mechanisms linking inflammation to insulin resistance. Int J Endocrinol 2015:508409
146. Daniele G, Guardado MR, Winnier D, Fiorentino TV, Pengou Z, Cornell J, Andreozzi F, Jenkinson C, Cersosimo E, Federici M et al (2014) The inflammatory status score including IL-6, TNF-alpha, osteopontin, fractalkine, MCP-1 and adiponectin underlies whole-body insulin resistance and hyperglycemia in type 2 diabetes mellitus. Acta Diabetol 51:123–131
147. Shoelson SE, Lee J, Goldfine AB (2006) Inflammation and insulin resistance. J Clin Invest 116:1793–1801
148. Bergman RN, Kim SP, Hsu IR, Catalano KJ, Chiu JD, Kabir M, Richey JM, Ader M (2007) Abdominal obesity: role in the pathophysiology of metabolic disease and cardiovascular risk. Am J Med 120:S3–S8
149. Robins SJ, Lyass A, Zachariah JP, Massaro JM, Vasan RS (2011) Insulin resistance and the relationship of a dyslipidemia to coronary heart disease: the Framingham Heart Study. Arterioscler Thromb Vasc Biol 31:1208–1214
150. Sadur CN, Yost TJ, Eckel RH (1984) Insulin responsiveness of adipose tissue lipoprotein lipase is delayed but preserved in obesity. J Clin Endocrinol Metab 59:1176–1182
151. Xiao C, Lewis GF (2012) Regulation of chylomicron production in humans. Biochim Biophys Acta 1821:736–746
152. Duez H, Lamarche B, Uffelman KD, Valero R, Cohn JS, Lewis GF (2006) Hyperinsulinemia is associated with increased production rate of intestinal apolipoprotein B-48-containing lipoproteins in humans. Arterioscler Thromb Vasc Biol 26:1357–1363
153. Federico LM, Naples M, Taylor D, Adeli K (2006) Intestinal insulin resistance and aberrant production of apolipoprotein B48 lipoproteins in an animal model of insulin resistance and metabolic dyslipidemia: evidence for activation of protein tyrosine phosphatase-1B, extracellular signal-related kinase, and sterol regulatory element-binding protein-1c in the fructose-fed hamster intestine. Diabetes 55:1316–1326
154. Hussain MM, Bakillah A (2008) New approaches to target microsomal triglyceride transfer protein. Curr Opin Lipidol 19:572–578
155. Wierzbicki AS, Hardman T, Prince WT (2009) Future challenges for microsomal transport protein inhibitors. Curr Vasc Pharmacol 7:277–286
156. Rizzo M, Wierzbicki AS (2011) New lipid modulating drugs: the role of microsomal transport protein inhibitors. Curr Pharm Des 17:943–949
157. Burnett JR, Watts GF (2007) MTP inhibition as a treatment for dyslipidaemias: time to deliver or empty promises? Expert Opin Ther Targets 11:181–189
158. Chang G, Ruggeri RB, Harwood HJ Jr (2002) Microsomal triglyceride transfer protein (MTP) inhibitors: discovery of clinically active inhibitors using high-throughput screening and parallel synthesis paradigms. Curr Opin Drug Discov Devel 5:562–570
159. Paras C, Hussain MM, Rosenson RS (2010) Emerging drugs for hyperlipidemia. Expert Opin Emerg Drugs 15:433–451
160. Chandler CE, Wilder DE, Pettini JL, Savoy YE, Petras SF, Chang G, Vincent J, Harwood HJ Jr (2003) CP-346086: an MTP inhibitor that lowers plasma cholesterol and triglycerides in experimental animals and in humans. J Lipid Res 44:1887–1901
161. Cuchel M, Bloedon LT, Szapary PO, Kolansky DM, Wolfe ML, Sarkis A, Millar JS, Ikewaki K, Siegelman ES, Gregg RE et al (2007) Inhibition of microsomal triglyceride transfer protein in familial hypercholesterolemia. N Engl J Med 356:148–156
162. Samaha FF, McKenney J, Bloedon LT, Sasiela WJ, Rader DJ (2008) Inhibition of microsomal triglyceride transfer protein alone or with ezetimibe in patients with moderate hypercholesterolemia. Nat Clin Pract Cardiovasc Med 5:497–505
163. Cuchel M, Meagher EA, du Toit TH, Blom DJ, Marais AD, Hegele RA, Averna MR, Sirtori CR, Shah PK, Gaudet D et al (2013) Efficacy and safety of a microsomal triglyceride transfer protein inhibitor in patients with homozygous familial hypercholesterolaemia: a single-arm, open-label, phase 3 study. Lancet 381:40–46
164. Blom DJ, Averna MR, Meagher EA, du Toit TH, Sirtori CR, Hegele RA, Shah PK, Gaudet D, Stefanutti C, Vigna GB et al (2017) Long-term efficacy and safety of the microsomal triglyceride transfer protein inhibitor lomitapide in patients with homozygous familial hypercholesterolemia. Circulation 136:332–335
165. Wang N, Zhang J, Wu Y, Liu J, Liu L, Guo X (2016) Metformin improves lipid metabolism disorders through reducing the expression of microsomal triglyceride transfer protein in OLETF rats. Diabetes Res Clin Pract 122:170–178
166. Aggarwal D, West KL, Zern TL, Shrestha S, Vergara-Jimenez M, Fernandez ML (2005) JTT-130, a microsomal triglyceride transfer protein (MTP) inhibitor lowers plasma triglycerides and LDL cholesterol concentrations without increasing hepatic triglycerides in guinea pigs. BMC Cardiovasc Disord 5:30

167. Borradaile NM, de Dreu LE, Barrett PH, Behrsin CD, Huff MW (2003) Hepatocyte apoB-containing lipoprotein secretion is decreased by the grapefruit flavonoid, naringenin, via inhibition of MTP-mediated microsomal triglyceride accumulation. Biochemistry 42:1283–1291
168. Borradaile NM, de Dreu LE, Wilcox LJ, Edwards JY, Huff MW (2002) Soya phytoestrogens, genistein and daidzein, decrease apolipoprotein B secretion from HepG2 cells through multiple mechanisms. Biochem J 366:531–539
169. Lin MC, Wang EJ, Lee C, Chin KT, Liu D, Chiu JF, Kung HF (2002) Garlic inhibits microsomal triglyceride transfer protein gene expression in human liver and intestinal cell lines and in rat intestine. J Nutr 132:1165–1168
170. Jahangir Z, Bakillah A, Iqbal J (2018) Regulation of sphingolipid metabolism by microRNAs: a potential approach to alleviate atherosclerosis. Diseases 6:82
171. Horie T, Baba O, Kuwabara Y, Yokode M, Kita T, Kimura T, Ono K (2014) MicroRNAs and lipoprotein metabolism. J Atheroscler Thromb 21:17–22
172. Aryal B, Singh AK, Rotllan N, Price N, Fernandez-Hernando C (2017) MicroRNAs and lipid metabolism. Curr Opin Lipidol 28:273–280
173. Irani S, Iqbal J, Antoni WJ, Ijaz L, Hussain MM (2018) microRNA-30c reduces plasma cholesterol in homozygous familial hypercholesterolemic and type 2 diabetic mouse models. J Lipid Res 59:144–154
174. Zhou L, Hussain MM (2017) Human microRNA-548p decreases hepatic apolipoprotein B secretion and lipid synthesis. Arterioscler Thromb Vasc Biol 37:786–793

Circadian Clock Regulation on Lipid Metabolism and Metabolic Diseases

5

Xiaoyue Pan, Samantha Mota, and Boyang Zhang

Abstract

The basic helix-loop-helix-PAS transcription factor (CLOCK, Circadian locomotor output cycles protein kaput) was discovered in 1994 as a circadian clock. Soon after its discovery, the circadian clock, Aryl hydrocarbon receptor nuclear translocator-like protein 1 (ARNTL, also call BMAL1), was shown to regulate adiposity and body weight by controlling on the brain hypothalamic suprachiasmatic nucleus (SCN). Farther, circadian clock genes were determined to exert several of lipid metabolic and diabetes effects, overall indicating that CLOCK and BMAL1 act as a central master circadian clock. A master circadian clock acts through the neurons and hormones, with expression in the intestine, liver, kidney, lung, heart, SCN of brain, and other various cell types of the organization. Among circadian clock genes, numerous metabolic syndromes are the most important in the regulation of food intake (via regulation of circadian clock genes or clock-controlled genes in peripheral tissue), which lead to a variation in plasma phospholipids and tissue phospholipids. Circadian clock genes affect the regulation of transporters and proteins included in the regulation of phospholipid metabolism. These genes have recently received increasing recognition because a pharmacological target of circadian clock genes may be of therapeutic worth to make better resistance against insulin, diabetes, obesity, metabolism syndrome, atherosclerosis, and brain diseases. In this book chapter, we focus on the regulation of circadian clock and summarize its phospholipid effect as well as discuss the chemical, physiology, and molecular value of circadian clock pathway regulation for the treatment of plasma lipids and atherosclerosis.

Keywords

Circadian clock · Phospholipids metabolism · Lipid metabolism · Atherosclerosis · Diet-induced obesity

Abbreviations

ABCA-1	ATP-binding cassette transporter 1
ABCG5/8	ATP binding cassette subfamily G member 5/8
ApoE$^{-/-}$	Apolipoprotein E knockout

X. Pan (✉) · S. Mota · B. Zhang
Department of Foundations of Medicine, New York University Long Island School of Medicine, Mineola, NY, USA

Diabetes and Obesity Research Center, New York University Winthrop Hospital, Mineola, NY, USA
e-mail: Xiaoyue.Pan@nyulangone.org

X.-C. Jiang (ed.), *Lipid Transfer in Lipoprotein Metabolism and Cardiovascular Disease*, Advances in Experimental Medicine and Biology 1276, https://doi.org/10.1007/978-981-15-6082-8_5

aSMase	Acidic SMase
BMAL1	Aryl hydrocarbon receptor nuclear translocator-like protein 1
CerS	Ceramide synthase
Chkα	Choline kinase alpha
CLOCK	Circadian locomotor output cycles protein kaput
Clock$^{\Delta 19/\Delta 19}$ or *Clk*$^{mt/mt}$	Dominant-negative Clock mutant protein mice
CRY1/2	Cryptochrome 1,2
FA	Fatty acid
GSK-3β	Glycogen synthase kinase 3 beta
HAT	Histone acetyltransferase
HMG-CoA reductase	3-hydroxy-3-methyl-glutaryl-CoA reductase
Ldlr$^{-/-}$	Low-density lipoprotein receptor knockout
LDL	Low-density lipoprotein
MTTP	Microsomal triglyceride transfer protein
NPAS2	Neuronal PAS-containing protein 2
NPC1L1	NPC1 intracellular cholesterol transporter 1
nSMase	Neutral SMase
PC	Phosphatidylcholine
PCSK9	Proprotein convertase subtilisin/kexin type 9
PE	Phosphatidylethanolamine
Per1/2/3	Period genes 1, 2, 3 (Period1/2/3)
PPARδ	Peroxisome proliferator-activated receptor delta
Rev-erbα	Nuclear receptor subfamily 1, group D, member 1
RGCs	Retinal ganglion cells
Rorα	RAR-related orphan receptor A
SCN	Suprachiasmatic nucleus
SHP	Small heterodimer partner
SIRT1	Sirtuin 1
SMase	Sphingomyelinase
UBE3A	HECT-type E3 ligase
USF2	Upstream Transcription Factor 2
VLDL	Very low-density lipoprotein

5.1 Introduction

Sleep disorders are now a major health threat to our life, this may result in more than 22 million Americans suffering from sleep disorder annually [1–15]. At least 38,000 people die from heart disease directly complicated by a sleep disorder [1–15]. A sleep disorder now affects almost every ethnicity and cultural society, setting an enormous load on the modern healthcare system in the United States and worldwide. The numerous complications associated with sleep disorders are major metabolic syndrome, atherosclerosis, hypertension, dyslipidemia, obesity, diabetes mellitus, cardiovascular diseases, several cancers, and certain types of brain diseases such as Alzheimer's disease [1–15]. Emphasizing the consequences of sufficient lipid buffering, atherosclerosis represents, to date, the very high common cause of lipid-related diseases [16–20]. According to sleep being a major risk factor for development of atherosclerosis, ample sleep achieved by either dieting, lifestyle, pharmacology, changes in circadian rhythms, proinflammatory responses, and metabolic effects improving sleep quality and pattern have shown, in numerous preclinical studies, many promising effects. For example, the ATP-binding cassette subfamily G member 5/8 (ABCG5/8), N-terminal Niemann-Pick C1 (NPC1) intracellular cholesterol transporter 1 (NPC1L1), and Microsomal triglyceride transfer protein (MTTP) inhibitor are adequate to indicate significant improvements in systemic lipid metabolism and atherosclerosis-linked comorbidities [4, 21–

29]. Further emphasizing, the direct relationship between atherosclerosis and lipid regulation. Plasma cholesterols reduced by ABCG5/8, NPC1L1, and MTP inhibitors, which are regulated by circadian clock genes, most often result in whole resolution of atherosclerosis, an opinion that encouraged the American Heart Association and National Institutes of Health to recommend such inhibitors under assured conditions for the treatment of atherosclerosis. Since the correlation between food intake and phospholipid regulation is highly confirmed by several basic research studies, inhibitors to inhibit food intake intuitively appear promising to improve phospholipid metabolism [30–34]. Under this reason, remarkable examples of such strategies are the administration of MTP inhibitor, which not only reduces plasma cholesterol through their MTTP inhibition but also decreases phospholipid metabolism through their ability to reduce lipid absorption via circadian rhythm regulation of food intake [35, 36]. We have shown that the circadian clock genes can regulate plasma triglycerides and cholesterol and regulate cholesterol and triglyceride absorption and metabolism [28, 29, 37–40]. A prominent example of circadian clock gene regulation is the circadian clock with a mutant clock gene, which improves body fat mass and body weight through regulation of intestinal lipid absorption and adipose lipid metabolism [38, 40–43]. However, whether circadian clock genes regulate phospholipid metabolism is not commonly known.

Phospholipids are polar, ionic compounds composed of an alcohol that is attached by a phosphodiester bridge to either diacylglycerol or sphingosine [35]. There are two classes of phospholipids: those that have glycerol (from glucose) as a backbone are called glycerophospholipids and those that have a sphingosine (from serine and palmitate) are called sphingophospholipids. Most phospholipids are synthesized in the smooth endoplasmic reticulum [35]. From there, they are transported to the Golgi apparatus and then to membranes of organelles or the plasma membrane of organelles [44]. They could also be secreted from the cell by exocytosis. Phosphatidylcholine (also called PC) and phosphatidylethanolamine (also called PE) are the most abundant phospholipids in most eukaryotic cells [35, 45]. The primary route of their synthesis uses choline and ethanolamine obtained either from food intake or from the turnover of the body's phospholipids [44]. Sphingomyelin is one of the principal structural lipids of the membranes of nerve tissues. It is synthesized from ceramide (an acyl sphingosine) and phosphatidylcholine. Sphingomyelin is also hydrolyzed into ceramide and phosphorylcholine [46]. Ceramide is further degraded to sphingosine and free fatty acid (FA) [46]. In this book chapter, we summarized the regulation of circadian clock genes with a special focus on their role to control lipid metabolism and metabolic diseases. A key central field will thereby be the topic of whether disordering the circadian clock genes will regulate transcription factors, and will the function of a protein pathway be of chronotherapeutic value to progress phospholipid metabolism?

5.2 Origins of the Mammalian Clock

Several biological, physiological, and behavioral activities show characteristic recurrence with 24-h intervals related to sunrise and sunset. Light entrains the central clock present in two lateral SCNs in the hypothalamus via the retinohypothalamic tract. The master circadian clock arises from autoregulatory transcriptional, translational, and posttranslational feedback loops of few transcription factors encoded by "clock" genes, including circadian locomotor output cycles protein kaput (CLOCK), brain and muscle aryl hydrocarbon receptor nuclear translocator-like 1 (BMAL1), neuronal PAS-containing protein 2 (NPAS2), period genes (Period1/2/3, *PER1/2/3)*, and cryptochrome genes (*CRY1/2*) [24, 47–50]. The BMAL1:CLOCK and BMAL1:NPAS2 heterodimers bind to *cis*-acting E-box sequences present in the promoter regions of *PER1/2/3* and *CRY1/2* and enhance their expression, constituting a positive feed-forward loop. Unlike

CLOCK and BMAL1, PER1-3 and CRY1-2-protein can dimerize and translocate to the nucleus and then dimerize the PER1-3:CRY1-2 complex, inhibiting the activity of CLOCK: BMAL1 or NPAS2-BMAL1. In the center of the hypothalamus, circadian clock genes are localized in the SCN and express neuronally and hormonally [16, 25, 51]. Zhang et al. have reported that the liver had the most circadian genes and then kidney as the 2nd, whereas the hypothalamus had the fewest (Fig. 5.1) [52]. SCN is responsible for controlling circadian rhythms, these circadian clock proteins' neuronal and hormonal activities regulate different body functions in the 24-hour cycle, such as body temperature, wake up/sleep, and food intake. To activate CLOCK:BMAL1 or NPAS2:BMAL1, there are 450 unique protein modifications [52]. Clock genes also need multiple posttranslational modifications, including phosphorylation, ubiquitination, acetylation, and SUMOylation to regulate various physiological functions [53–55]. This posttranslational modification of BMAL1 is regulated by ubiquitin-specific protease 2 (USP2) [54, 55]. USP2 is essential to deubiquitinating PER1, CRY1, and CRY2 in vivo [54–57]. This mechanism was demonstrated by the absence of deubiquitinated Per1, Bmal1, Cry1, and Cry2 in mice deficient in USP2 [55].

BMAL1:CLOCK is a heterodimer formed via CLOCK with 361 amino acids and BMAL1 with 387 amino acids. CLOCK and BMAL1 have the same one, basic, helix-loop-helix (bHLH) binding of protein to DNA via recognized E-box sites, through hydrogen bonding, between serine residues and DNA. As we know, E-box sites are about 20 base pairs upstream of genes with a major 5′-CACGTG-3″ canonical motif. CLOCK and BMAL1 or NPAS2 (a paralog of clock) can not only recruit transcription factors to the E-box site but also can upregulate transcription of the target genes such as nuclear receptor subfamily 1, group D, member 1 (NR1D1, Rev-erbα), D-Box binding PAR BZIP transcription factor (DBP), peroxisome proliferator-activated receptor alpha (PPARα), small heterodimer partner (SHP), GATA binding protein 4 (GATA4), and paired box protein 4 (PAX4) [16, 24, 28, 29, 40, 47–50, 58] (Fig. 5.2). These genes can also, through recruitment of histone acetyltransferases, decondense the nucleosome into heterochromatin allowing transcriptional machinery access to the DNA, such as mutant CLOCK, which downregulates upstream transcription factor 2 (USF2). In addition, USF1 serves as a suppressor of the circadian clock mutant, revealing the nature of the DNA-binding of the Clock:Bmal1 complex in mice [59]. This data also suggests that USF1 and USF2 are important modulators of molecular and behavioral circadian rhythms in mammals. In addition, it is possible that CLOCK regulates USF2 through the histone acetyltransferase pathway. However, more experiments are required to understand this mechanism.

The reported levels of CLOCK and BMAL protein do not show dramatic circadian oscillations in mammalian brain; however, reflecting the species-related phosphorylation in circadian clock protein shows clear circadian oscillations with time-dependent, posttranslational regulation [53]. The degradation of CLOCK and BMAL1 is more important for transcription activation of clock-controlled genes through E-boxes in their promoters [60]. For example, estrogen receptors are regulated by CLOCK [61]. CLOCK:BMAL1 proteins can bioaccumulate by proteasome inhibitor MG132 by preventing their protein degradation. MG132 is an inhibitor that decreases E-Box-mediated transcription by interfering with CLOCK: BMAL1 regulation cycles in humans. Whereas in rodents, CLOCK19 protein is hypophosphorylated to a higher extent than those of wild-type CLOCK [62]. In vitro studies have also shown several enzymes (such as casein kinase I/II, glycogen synthase kinase 3 beta (GSK-3β), and cyclin-dependent kinase 5) are responsible for CLOCK:BMAL1 degradation [63–67]. For example, GSK-3β-catalyzed phosphorylation can phosphorylate Ser431 of CLOCK dependent site Ser427 and Thr21 of BMAL1 dependent site with Ser17, to induce higher activity of CLOCK and BMAL1 under unstable conditions. Similarly, protein kinase Ck2 can

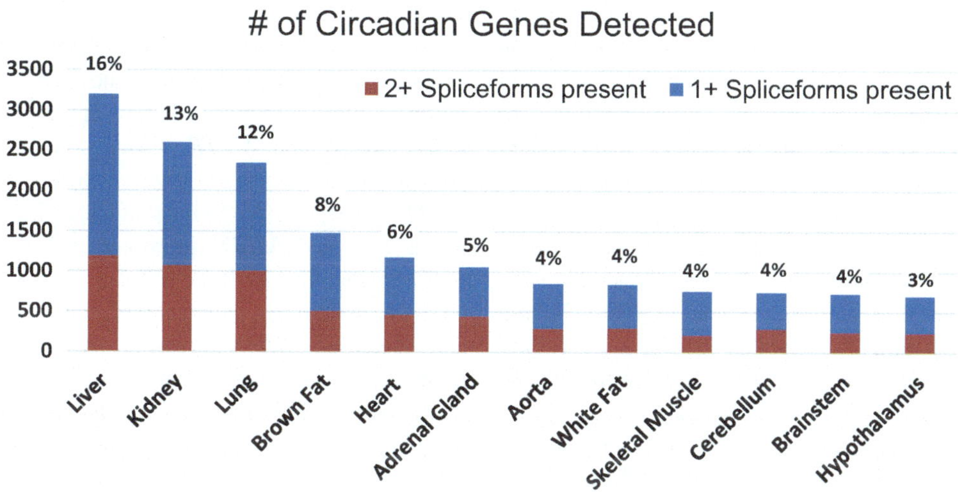

Fig. 5.1 *Number of circadian clock genes detected in each organ.* Circadian expression of protein-coding genes in different tissues. Blue marks indicate the number of genes with at least one spliceform detected by RNA-seq. Orange marks indicate the number of genes with at least two spliceforms detected by RNA-seq. Blue numbers to the top of each bar state the percentage of protein-coding genes with rhythmic expression in each organ of Zhang et al. publication. Figure modified according to Zhang et al. publication in PNAS [52]

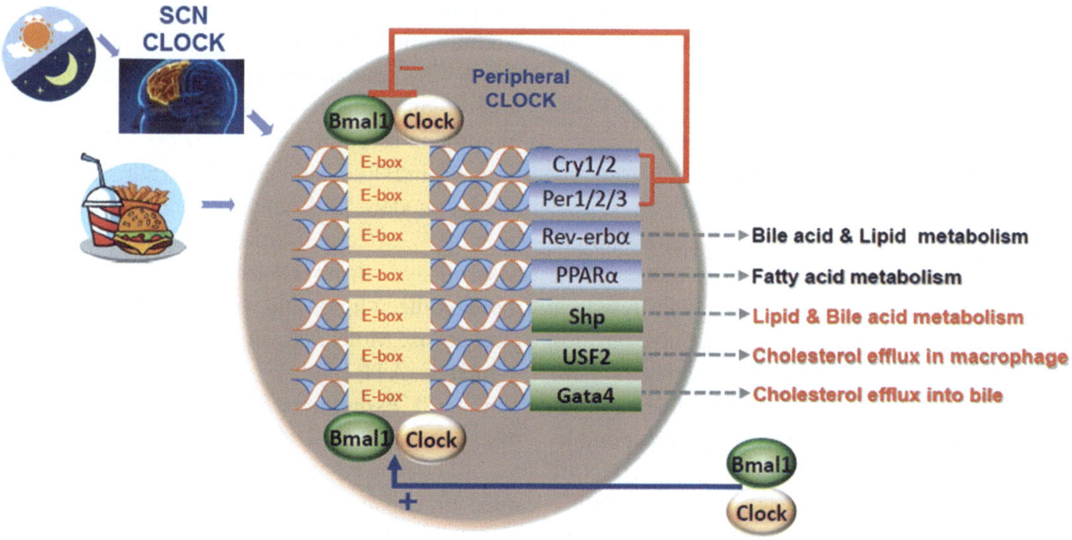

Fig. 5.2 *Clock and clock-collected genes regulate metabolic function.* Both light- and food-entrained oscillators appear to affect the expression of circadian clock genes and clock-collected genes in the peripheral tissue. In SCN and peripheral, Clock:Bmal1 heterodimerize to activate transcription of circadian target genes including the genes of *Per1/2/3 and Cry1/2*. *Per1/2/3 and Cry1/2* interact and inhibit Bmal1 and Clock. We have shown that Clock and Bmal1 regulate several transcription factors such as Shp, Usf2, and Gata4 regulating the expression of several genes involved in lipid metabolism as well as other pathways that affect metabolism

phosphorylate and stabilize BMAL1 by eliminating BMAL1 polyubiquitination [66, 67]. Regulation of CLOCK:BMAL1 phosphorylation affects transcription through alterations in DNA-binding [68]. In addition, CLOCK:BMAL1 activity is affected not only by phosphorylation but also by ubiquitination to induce its transactivation and degradation [69]. For example, SUMOylation and O-GlcNAcylation induce ubiquitination of BMAL1 at Lys259 and at Ser418, respectively, to increase BMAL1 transactivation and degradation [69]. In addition, HECT-type E3 ligase can promote ubiquitination of BMAL1 and CLOCK [66, 70, 71]. Moreover, there are several studies showing that sirtuin 1 binds to CLOCK and BMAL1 and deacetylates BMAL1 at lys537 [71], thus preventing CRY1 recruitment and restarting the transactivation of the clock gene. Furthermore, HAT activity required site motif A of CLOCK and acetylation site of BMAL1 are required to rescue the cellular clock-controlled gene rhythm [72, 73]. Histone modifications were essential for normal clock function [71, 72, 74–77]. CLOCK:BMAL1 heterodimers shuttle between the nucleus and the cytosol, thus suggesting that the dimer-protein modulation is involved in several post-translation and transcription levels.

5.3 Physiological Functions of Circadian Clock

While major studies indicate that most metabolic functions of circadian clock require transcription and post-translation levels, there is research indicating that circadian clock genes have physiologically related functions on a body metabolism, potentially through several pathways that have yet to be identified. Circadian clock genes respond to external stimuli, and the one prominent effect of the circadian clock gene is its ability to diurnally control food intake. We have shown that circadian clock genes and lipid transport proteins are expressed in the small intestinal enterocytes and respond to food entrainment in wild-type mice [38]. Dominant-negative Clock mutant protein mice (*Clock*$^{\Delta19/\Delta19}$ or *Clk*$^{mt/mt}$) disrupt the circadian expression and food entrainment of the clock genes [38, 41]. In addition, the absorption of lipids was high in Clock mutant mice [38, 40]. Our data also suggests that Clock plays an important role in light and food entrainment of intestinal function. To understand the mechanism of clock genes regulating lipid absorption and metabolism, we studied the role of clock gene in the diurnal regulation of plasma triglyceride-rich apolipoprotein B-lipoprotein and *MTTP*. Clock mutant mice showed sustained hypertriglyceridemia and high *MTTP* expression. We found that CLOCK knockdown activated *MTTP* promoter and reduced SHP, in the human liver cell line Huh7 cells, CLOCK temporally interacts with the E-box site and increases *SHP* expression, whereas SHP reduces *MTTP* expression by differentially interacting with hepatocyte nuclear factor 4 alpha and the liver receptor homolog-1 [40]. In *Clock*$^{mt/mt}$ mice, however, the binding of Clock to *Shp* promoter did not show cyclic change, and Shp mRNA levels were relatively low and did not change [40]. This data shows that a decreased interaction of *SHP* with these transcription factors is associated with increased *MTTP* expression. Therefore, S*HP* is a clock-controlled gene that transmits information from *CLOCK* to *MTTP*. Additionally, we showed, for the first time, that *Clock*$^{\Delta19/\Delta19}$ mutant protein enhances plasma cholesterol and atherosclerosis in the low-density lipoprotein receptor knockout (*Ldlr*$^{-/-}$) and apolipoprotein E knockout (*ApoE*$^{-/-}$) atherosclerosis animal models [29]. In addition, Clock mutant protein affects macrophage function. Macrophages from *Clock*$^{\Delta19/\Delta19}$ mice took up more oxidized lipids and were defective in cholesterol efflux. Molecular studies showed that Clock regulates ATP-binding cassette subfamily A member 1 expression and cholesterol efflux in macrophages via Usf2 [29]. In addition, we recently showed that global Bmal1-deficient mice or hepatic-specific Bmal1 knockout mice also have an impaired cholesterol metabolism, display hepatic cholesterol efflux into bile, develop atherosclerosis when fed with an atherogenic diet, and potentiate the development of atherosclerotic lesions in the *Ldlr*$^{-/-}$ and *ApoE*$^{-/-}$ atherosclerosis animal models [28]. Liver-specific inactivation of Bmal1 led to elevated plasma low-density lipoprotein (LDL) and very low-density lipoprotein (VLDL) cholesterol levels as a consequence of the disruption of the Pcsk9/Ldl receptor regulatory axis [22, 28, 78].

Phosphatidylcholine is one of the phospholipids that occupy 70% of VLDL phospholipids. Phosphatidylcholine biosynthesis is known to be required for VLDL secretion [78]. This has also shown that diurnal variation of VLDL concentration is linked to the clock-controlled production of phosphatidylcholine. Furthermore, Ma et al. have identified two distinct groups exhibiting rhythmic and nonrhythmic patterns of gene expression during light-dark cycles, according to the database of the circadian regulation of lipid-associated genome-wide association studies (GWAS) candidate genes in mouse liver [79]. Liver-specific Bmal1 knockout mice increased plasma Ldl/Vldl cholesterol levels through disordered Pcsk9/Ldl receptor expression [79].

In line with this idea, circadian clock genes affect food intake, body weight, plasma glucose, and lipids, have protective effects on the adipose tissue, heart, liver, and intestine, and affect phospholipid metabolism via several pathways [4, 21, 80–82]. In other words, the circadian clock may, through several mechanisms control phosphatidylcholine (phospholipid that makes up 50% of total cellular phospholipid biosynthesis), as the phosphatidylcholine phenotype can be copied by different circadian clock gene mutations [83]. Wild-type mice in normal light and dark cycles display a rhythmic accumulation of hepatic phosphatidylcholine with a peak at Zeitgeber time (ZT) 22-0. Bmal1-deficient ($Bmal1^{-/-}$) mice show elevated phosphatidylcholine levels in the liver associated with an atherogenic lipoprotein profile [78]. To investigate whether the circadian variation of phosphatidylcholine levels is the result of a circadian regulation of phosphatidylcholine biosynthesis, Grechez-Cassiau et al. found that choline kinase alpha (Chkα) gene is a clock-controlled gene in the liver [78]. Chkα gene expression is regulated by the Rev-erbα and RAR-related orphan receptor A (Rorα) nuclear receptors [78]. Thus, hepatic phosphatidylcholine is regulated by the circadian clock gene through a Bmal1-Rev-erbα-Chkα axis and suggests that an intact circadian timing system is important for the temporal coordination of phospholipid metabolism. The Rev-erbα subtype appears to be a key circadian regulator of phosphatidylcholine metabolism in the liver through the rhythmic transcriptional repression of the Chkα gene. Thus, a likely mechanism by which hepatic phosphatidylcholine levels are increased in the $Bmal1^{-/-}$ mice is that Chkα upregulates by the high total of choline kinase activity [78]. In addition, there is a low Rev-erbα gene expression level in the Per1/Per2 double knockout mice [84]. Twenty-four out of 27 phosphatidylcholine species were arrhythmic by the lipidomic profiling, although 16% of lipid metabolites were still oscillating in the liver [84]. These studies suggest that a genetic disruption of the circadian clock system compromises phosphatidylcholine homeostasis.

5.4 Preclinical Studies on Diurnal Rhythm in Phospholipid Metabolism

Minami et al. found oscillatory peaks of phospholipids were detected by liquid chromatography-mass spectrometry (among these time-indicating metabolites) [85]. Fourteen oscillatory peaks were identified as various types of lysophosphatidylcholines with different unsaturated FAs [85]. As mentioned above, in mammalian cells, phosphatidylcholine is one of the phospholipids constituting 50% of total cellular phospholipids [86]; phosphatidylcholine is also the main circulating phospholipid in plasma, where it is critical for the assembly and secretion of lipoproteins by the liver. Hepatic phospholipids enter in bile salt-mediated micelle formation in the intestinal lumen, which facilitates the absorption of lipid-soluble nutrients from the diet [87]. Several studies have shown that serum phosphatidylcholine is shown to be subjected to temporal control that could be correlated with rest-activity cycles and feeding [84, 88]. Phosphatidylcholine plays an important role in mammalian cell signaling [89] as well as in oncogenic signaling pathways [78, 89–91]. Numerous studies have evaluated the circadian clock genes' effect on phospholipid metabolism. Diurnal rhythm of retinal phospholipid synthetic enzyme has been shown in the retina of rats [92]. Retinal phospholipid synthetic enzymes showed daily variations, in retinal ganglion cells (RGCs) of chicken when in constant darkness. [^{32}P]Phospholipids display circadian oscillations both in in vivo chicken kept in constant light and in cultures of immunopurified embryonic RGCs [92]. Several distinct enzymes, lysophospholipid acyltransferases, phosphatidate phosphohydrolase, and diacylglycerol lipase, in the pathway of phospholipid biosynthesis and degradation have shown diurnal variation [93]. These activities of these enzymes are high during the subjective day and low at night, as were the metabolic changes observed in the in vivo labeling of phospholipid in cultures of purified embryonic RGCs [93, 94]. In addition, glycerophospholipid synthesis has also shown diurnal rhythm in retinal inner nuclear layer cells [93, 94]. Biosynthesis of phospholipid has shown the circadian cycle by serum shock in cultured quiescent NIH3T3 cells; this cycle is abolished by knock down Per1 gene, suggesting that the biosynthesis of phospholipid circadian cycle in cultured fibroblasts depends on the endogenous circadian clock [94–97]. Ruggiero et al. showed that the diurnal rhythm of phospholipid phosphatidylserine demarcation of photoreceptor outer segments tip is not intrinsic to rod photoreceptors but requires activities of the retinal pigment epithelium as well [98]. In line with the circadian cycle of phospholipid or phospholipid biosynthesis in vivo and in vitro, levels of serum phospholipids such as phosphatidylcholines (18:0/18:1) or 1-stearoyl-2-oleoyl-sn-glycero-3-phosphocholine are typically regulated in mice lacking circadian clock-collected gene PPAR gamma (PPARδ) activity [98]. Serum phosphatidylcholine (18:0/18:1) can reduce postprandial lipid levels, and phosphatidylcholine can increase FA utilization through muscle PPAR alpha (PPARα) [98]. When mice were fed with a high-fat diet, the rhythm of phosphatidylcholine (18:0/18:1) was diminished. Phosphatidylcholine (1:0/18:1) administration in db/db mice (a model for diabetic dyslipidemia) can improve metabolic homeostasis, suggesting that alterations in diurnal hepatic PPARδ-phosphatidylcholine (18:0/18:1) signaling affect metabolic disorders, including obesity [99]. Obesity can alter circadian rhythms in multiple tissues. Diet-induced obesity altered the rhythm pattern of serum phosphatidylcholine [99, 100]. As a phospholipid outcome, ceramide, a class of sphingolipids, Jang et al. showed that the ceramide concentration in WT mice showed a strong peak at Zeitgeber time 9 (ZT9; 9 h after lights on time) and ZT21, but no rhythmicity in ceramide expression was seen in Per1/Per2 double KO mice [101]. To understand the mechanism of diurnal rhythm of ceramide, they also measure several gene expressions including via sphingomyelinase (SMase) or by ceramide synthase (CerS)-mediated synthesis; both are important for sphingomyelin hydrolysis to ceramide. Jang et al. found that CerS2 expression levels showed a biphasic pattern of expression in WT mice but no rhythmicity in Per1/Per2 double KO mice [102]. While the neutral SMase (nSMase) and acidic SMase (aSMase) mRNA in WT mice were expressed in a circadian variation, the correlation between the expression levels of these SMases with times of day was weak in Per1/Per2 double KO mice [102]. Collectively, this study suggests that both SMases and CerS2 mRNA expression are regulated by the presence of mPer1/mPer2 circadian clock genes in vivo and imply that ceramide may play a vital role in circadian rhythms and physiology [102]. However, the molecular mechanism of circadian clock genes regulating phospholipid metabolism is still unclear and limited.

5.5 Clinical Studies on Circadian Clocks' Role in Phospholipid Metabolism

Animal research shows a clear involvement of membrane-derived phospholipid in circadian rhythms. Additionally, 7–20% of metabolites in human blood have been observed showing circadian variation [85, 103–106]. Under a series of preclinical studies, the existence of both daily change and seasonal variations affects the composition of phospholipids in human cell membranes [12, 107, 108]. Over 1 year, in 20 healthy subjects, Ruf et al. found that 11 of 13 phospholipids' FA content showed significant daily rhythms and were largely synchronous among subjects [108]. This data is supported by several other studies, overall indicating that human physiology is still dominated by geophysical sunrise and sunset, resulting in a strong daily cycle [107, 109]. However, seasonal rhythms are less well defined. FAs derived from phospholipids also play a role as precursors of prostaglandins, thromboxanes, and leukotrienes. A much more likely candidate for such a function of rhythmicity is the interaction between membrane FAs and transmembrane proteins. It is a possible explanation for rhythmic alterations of membrane composition [108]. In particular, a link between sleep deprivation and phosphatidylcholine is also shown by the result that both the circadian system and plasma lipids display a reciprocal correlation over the day with a subset of phosphatidylcholine and triglyceride species in plasma being high in sleep-deprived 20 total subjects of young-aged healthy ethnic Chinese males [110].

Some epidemiological researches comfort the relationship between the circadian system and the regulation of diurnal rhythm of phospholipids. A marked circadian variation was recorded in plasma total cholesterol, high-density lipoprotein cholesterol, phospholipid, and total lipid concentration in healthy Indians of different age groups of 162 total subjects [111]. Plasma phospholipid concentrations were characterized by a circadian rhythm in all age groups. Females had numerically higher values than males. However, the rhythm peak was significantly changed by age, reaching a maximum in middle adulthood and decreasing in the older age group [111, 112]. This suggests that the diurnal rhythm of plasma phospholipids is associated with age, gender, diet, and smoking and affects circulating plasma lipid components in healthy Indians. In addition, a 24-hr time series of plasma metabolites has been simultaneously assessed in type 2 diabetes, compared with an age- and weight-matched control group during a controlled daily routine [113]. Similarly, a total of 100 of 663 metabolites, representing all metabolite categories, showed diurnal rhythmic concentrations that exceeded the Bonferroni threshold, showing that the peak times of all phospholipids were clustered during the afternoon to midnight [114, 115].

We previously showed that peptide-like drugs, H+-peptide cotransporter 1, Pept1, showing diurnal rhythm, could influence the pharmacokinetics of peptide-like drugs [116–118]. Drug statins, a HMG-CoA reductase inhibitor that is in clinical evaluation for the treatment of type 2 diabetes and atherosclerosis, show beneficial effects on plasma lipids [119–121]. Interestingly, statins were recommended to be administered in the evening [119, 121]. However opinions differ on the best time to take statins. Simvastatin was reportedly better in the evening too, but simvastatin taken in the evening was not better than when it was taken in the morning by a different study group [122–124]. It remains in clinical evaluation for treatment. Lipidomics can be used to examine differences in circadian responses to medications that target lipid pathways, such as statins, and to better characterize the mode of action of such drugs. So far, there are collective preclinical and clinical studies overall suggesting a beneficial effect of chronotherapeutics. Beyond circadian clock's direct phospholipid role, it has to be noticed that food intake and body weight change because of circadian clock pathway regulation might provide a particular potential for second improvement of phospholipid treatment.

5.6 Conclusion

The circadian clock system has, over the last 20 years, been researched for it's involvement in a number of metabolic functions that go well over their primary classification as a regulator affecting wake up/sleep and food intake. Along with circadian clock regulation in phospholipid metabolism, various studies evaluated the therapeutic effect of phospholipid modulation. The circadian clock correlating to phospholipids might offer a potential treatment for atherosclerosis and obesity associated with pathological atherosclerosis. Circadian clock altering of molecular time will be of chronotherapeutic value to reduce metabolic disorders, impaired immune function, and accelerated aging and to improve phospholipid metabolism and cardiovascular diseases. Importantly, while disordered circadian clock genes and sleep disorder are known to affect more than 50 million US residents (https://www.ncbi.nlm.nih.gov/books/NBK19961/), it is possible that other physiological functions of circadian clock are not yet understood.

Acknowledgments This work was supported in part by NIH National Heart, Lung, and Blood Institute Grant R56 HL137912-01 and American Heart Association Grant-In-Aid 16GRNT30960027 to X. Pan.

References

1. Fernando S, Biggs SN, Horne RSC, Vollenhoven B, Lolatgis N, Hope N et al (2017) The impact of melatonin on the sleep patterns of women undergoing IVF: a double blind RCT. Hum Reprod Open 2017: hox027
2. Grajewski B, Whelan EA, Lawson CC, Hein MJ, Waters MA, Anderson JL et al (2015) Miscarriage among flight attendants. Epidemiology 26:192–203
3. Grajewski B, Whelan EA, Nguyen MM, Kwan L, Cole RJ (2016) Sleep disturbance in female flight attendants and teachers. Aerosp Med Hum Perform 87:638–645
4. Huang W, Ramsey KM, Marcheva B, Bass J (2011) Circadian rhythms, sleep, and metabolism. J Clin Invest 121:2133–2141
5. Jackson CL, Redline S, Emmons KM (2015) Sleep as a potential fundamental contributor to disparities in cardiovascular health. Annu Rev Public Health 36:417–440
6. Kim TW, Jeong JH, Hong SC (2015) The impact of sleep and circadian disturbance on hormones and metabolism. Int J Endocrinol 2015:591729
7. Knutson KL, Lauderdale DS (2007) Sleep duration and overweight in adolescents: self-reported sleep hours versus time diaries. Pediatrics 119:e1056–e1062
8. Labyak S, Lava S, Turek F, Zee P (2002) Effects of shiftwork on sleep and menstrual function in nurses. Health Care Women Int 23:703–714
9. Mahoney MM (2010) Shift work, jet lag, and female reproduction. Int J Endocrinol 2010:813764
10. Maung SC, El SA, Chapman C, Cohen D, Cukor D (2016) Sleep disorders and chronic kidney disease. World J Nephrol 5:224–232
11. McNeil J, Doucet E, Chaput JP (2013) Inadequate sleep as a contributor to obesity and type 2 diabetes. Can J Diabetes 37:103–108
12. Potter GD, Skene DJ, Arendt J, Cade JE, Grant PJ, Hardie LJ (2016) Circadian rhythm and sleep disruption: causes, metabolic consequences, and countermeasures. Endocr Rev 37:584–608
13. Spiegel K, Knutson K, Leproult R, Tasali E, Van CE (2005) Sleep loss: a novel risk factor for insulin resistance and Type 2 diabetes. J Appl Physiol (1985) 99:2008–2019
14. Srinivasan V, Singh J, Pandi-Perumal SR, Brown GM, Spence DW, Cardinali DP (2010) Jet lag, circadian rhythm sleep disturbances, and depression: the role of melatonin and its analogs. Adv Ther 27:796–813
15. Ting L, Malhotra A (2005) Disorders of sleep: an overview. Prim Care 32:305–318. v
16. Hussain MM, Pan X (2009) Clock genes, intestinal transport and plasma lipid homeostasis. Trends Endocrinol Metab 20:177–185
17. Hussain MM, Fatma S, Pan X, Iqbal J (2005) Intestinal lipoprotein assembly. Curr Opin Lipidol 16:281–285
18. Hussain MM, Rava P, Pan X, Dai K, Dougan SK, Iqbal J et al (2008) Microsomal triglyceride transfer protein in plasma and cellular lipid metabolism. Curr Opin Lipidol 19:277–284
19. Bobulescu IA (2010) Renal lipid metabolism and lipotoxicity. Curr Opin Nephrol Hypertens 19:393–402
20. Rui L (2017) Brown and Beige adipose tissues in health and disease. Compr Physiol 7:1281–1306
21. Bass J, Lazar MA (2016) Circadian time signatures of fitness and disease. Science 354:994–999
22. Gooley JJ (2016) Circadian regulation of lipid metabolism. Proc Nutr Soc 75:440–450
23. Hussain MM, Pan X (2015) Circadian regulation of macronutrient absorption. J Biol Rhythm 30:459–469
24. Hussain MM, Pan X (2015) Circadian regulators of intestinal lipid absorption. J Lipid Res 56:761–770
25. Hussain MM, Pan X (2012) Clock regulation of dietary lipid absorption. Curr Opin Clin Nutr Metab Care 15:336–341
26. Molusky MM, Hsieh J, Lee SX, Ramakrishnan R, Tascau L, Haeusler RA et al (2018) Metformin and AMP kinase activation increase expression of the sterol transporters ABCG5/8 (ATP-Binding Cassette Transporter G5/G8) with potential antiatherogenic consequences. Arterioscler Thromb Vasc Biol 38:1493–1503

27. Douris N, Kojima S, Pan X, Lerch-Gaggl AF, Duong SQ, Hussain MM et al (2011) Nocturnin regulates circadian trafficking of dietary lipid in intestinal enterocytes. Curr Biol 21:1347–1355
28. Pan X, Bradfield CA, Hussain MM (2016) Global and hepatocyte-specific ablation of Bmal1 induces hyperlipidaemia and enhances atherosclerosis. Nat Commun 7:13011
29. Pan X, Jiang XC, Hussain MM (2013) Impaired cholesterol metabolism and enhanced atherosclerosis in clock mutant mice. Circulation 128:1758–1769
30. Zambo V, Simon-Szabo L, Szelenyi P, Kereszturi E, Banhegyi G, Csala M (2013) Lipotoxicity in the liver. World J Hepatol 5:550–557
31. Wiedeman AM, Barr SI, Green TJ, Xu Z, Innis SM, Kitts DD (2018) Dietary choline intake: current state of knowledge across the life cycle. Nutrients 10:1513
32. Lordan R, Tsoupras A, Zabetakis I (2017) Phospholipids of animal and marine origin: structure, function, and anti-inflammatory properties. Molecules 22:1964
33. Parchem K, Bartoszek A (2016) Phospholipids and products of their hydrolysis as dietary preventive factors for civilization diseases. Postepy Hig Med Dosw (Online) 70:1343–1361
34. Palamiuc L, Ravi A, Emerling BM (2019) Phosphoinositides in autophagy: current roles and future insights. FEBS J 287(2):222–238
35. Jiang XC (2018) Phospholipid transfer protein: its impact on lipoprotein homeostasis and atherosclerosis. J Lipid Res 59:764–771
36. Li Z, Park TS, Li Y, Pan X, Iqbal J, Lu D et al (2009) Serine palmitoyltransferase (SPT) deficient mice absorb less cholesterol. Biochim Biophys Acta 1791:297–306
37. Pan X, Munshi MK, Iqbal J, Queiroz J, Sirwi AA, Shah S et al (2013) Circadian regulation of intestinal lipid absorption by apolipoprotein AIV involves forkhead transcription factors A2 and O1 and microsomal triglyceride transfer protein. J Biol Chem 288:20464–20476
38. Pan X, Hussain MM (2009) Clock is important for food and circadian regulation of macronutrient absorption in mice. J Lipid Res 50:1800–1813
39. Pan X, Hussain MM (2007) Diurnal regulation of microsomal triglyceride transfer protein and plasma lipid levels. J Biol Chem 282:24707–24719
40. Pan X, Zhang Y, Wang L, Hussain MM (2010) Diurnal regulation of MTP and plasma triglyceride by CLOCK is mediated by SHP. Cell Metab 12:174–186
41. Turek FW, Joshu C, Kohsaka A, Lin E, Ivanova G, McDearmon E et al (2005) Obesity and metabolic syndrome in circadian clock mutant mice. Science 308:1043 1045
42. Gekakis N, Staknis D, Nguyen HB, Davis FC, Wilsbacher LD, King DP et al (1998) Role of the CLOCK protein in the mammalian circadian mechanism. Science 280:1564–1569
43. Oishi K, Shirai H, Ishida N (2005) CLOCK is involved in the circadian transactivation of peroxisome-proliferator-activated receptor alpha (PPARalpha) in mice. Biochem J 386:575–581
44. Kent C (1995) Eukaryotic phospholipid biosynthesis. Annu Rev Biochem 64:315–343
45. Jiang XC, Goldberg IJ, Park TS (2011) Sphingolipids and cardiovascular diseases: lipoprotein metabolism, atherosclerosis and cardiomyopathy. Adv Exp Med Biol 721:19–39
46. Wymann MP, Schneiter R (2008) Lipid signalling in disease. Nat Rev Mol Cell Biol 9:162–176
47. Asher G, Schibler U (2011) Crosstalk between components of circadian and metabolic cycles in mammals. Cell Metab 13:125–137
48. Brown SA (2016) Circadian metabolism: from mechanisms to metabolomics and medicine. Trends Endocrinol Metab 27:415–426
49. Eckel-Mahan K, Sassone-Corsi P (2013) Metabolism and the circadian clock converge. Physiol Rev 93:107–135
50. Takahashi JS (2015) Molecular components of the circadian clock in mammals. Diabetes Obes Metab 17 (Suppl 1):6–11
51. Takahashi JS, Shimomura K, Kumar V (2008) Searching for genes underlying behavior: lessons from circadian rhythms. Science 322:909–912
52. Zhang R, Lahens NF, Ballance HI, Hughes ME, Hogenesch JB (2014) A circadian gene expression atlas in mammals: implications for biology and medicine. Proc Natl Acad Sci U S A 111:16219–16224
53. Hirano A, Fu YH, Ptacek LJ (2016) The intricate dance of post-translational modifications in the rhythm of life. Nat Struct Mol Biol 23:1053–1060
54. Yang Y, Duguay D, Bedard N, Rachalski A, Baquiran G, Na CH et al (2012) Regulation of behavioral circadian rhythms and clock protein PER1 by the deubiquitinating enzyme USP2. Biol Open 1:789–801
55. Yang Y, Duguay D, Fahrenkrug J, Cermakian N, Wing SS (2014) USP2 regulates the intracellular localization of PER1 and circadian gene expression. J Biol Rhythm 29:243–256
56. Scoma HD, Humby M, Yadav G, Zhang Q, Fogerty J, Besharse JC (2011) The de-ubiquitinylating enzyme, USP2, is associated with the circadian clockwork and regulates its sensitivity to light. PLoS One 6:e25382
57. Tong X, Buelow K, Guha A, Rausch R, Yin L (2012) USP2a protein deubiquitinases and stabilizes the circadian protein CRY1 in response to inflammatory signals. J Biol Chem 287:25280–25291
58. Sussman W, Stevenson M, Mowdawalla C, Mota S, Ragolia L, Pan X (2019) BMAL1 controls glucose uptake through paired-homeodomain transcription factor 4 in differentiated Caco-2 cells. Am J Physiol Cell Physiol 317:C492
59. Shimomura K, Kumar V, Koike N, Kim TK, Chong J, Buhr ED et al (2013) Usf1, a suppressor

of the circadian clock mutant, reveals the nature of the DNA-binding of the CLOCK:BMAL1 complex in mice. elife 2:e00426
60. Stratmann M, Suter DM, Molina N, Naef F, Schibler U (2012) Circadian Dbp transcription relies on highly dynamic BMAL1-CLOCK interaction with E boxes and requires the proteasome. Mol Cell 48:277–287
61. Reid G, Hubner MR, Metivier R, Brand H, Denger S, Manu D et al (2003) Cyclic, proteasome-mediated turnover of unliganded and liganded ERalpha on responsive promoters is an integral feature of estrogen signaling. Mol Cell 11:695–707
62. Yoshitane H, Takao T, Satomi Y, Du NH, Okano T, Fukada Y (2009) Roles of CLOCK phosphorylation in suppression of E-box-dependent transcription. Mol Cell Biol 29:3675–3686
63. Spengler ML, Kuropatwinski KK, Schumer M, Antoch MP (2009) A serine cluster mediates BMAL1-dependent CLOCK phosphorylation and degradation. Cell Cycle 8:4138–4146
64. Kwak Y, Jeong J, Lee S, Park YU, Lee SA, Han DH et al (2013) Cyclin-dependent kinase 5 (Cdk5) regulates the function of CLOCK protein by direct phosphorylation. J Biol Chem 288:36878–36889
65. Sahar S, Zocchi L, Kinoshita C, Borrelli E, Sassone-Corsi P (2010) Regulation of BMAL1 protein stability and circadian function by GSK3beta-mediated phosphorylation. PLoS One 5:e8561
66. Tamaru T, Hattori M, Honda K, Nakahata Y, Sassone-Corsi P, van der Horst GT et al (2015) CRY drives cyclic CK2-mediated BMAL1 phosphorylation to control the mammalian circadian clock. PLoS Biol 13:e1002293
67. Tamaru T, Hirayama J, Isojima Y, Nagai K, Norioka S, Takamatsu K et al (2009) CK2alpha phosphorylates BMAL1 to regulate the mammalian clock. Nat Struct Mol Biol 16:446–448
68. Eide EJ, Vielhaber EL, Hinz WA, Virshup DM (2002) The circadian regulatory proteins BMAL1 and cryptochromes are substrates of casein kinase Iepsilon. J Biol Chem 277:17248–17254
69. Lee J, Lee Y, Lee MJ, Park E, Kang SH, Chung CH et al (2008) Dual modification of BMAL1 by SUMO2/3 and ubiquitin promotes circadian activation of the CLOCK/BMAL1 complex. Mol Cell Biol 28:6056–6065
70. Gossan NC, Zhang F, Guo B, Jin D, Yoshitane H, Yao A et al (2014) The E3 ubiquitin ligase UBE3A is an integral component of the molecular circadian clock through regulating the BMAL1 transcription factor. Nucleic Acids Res 42:5765–5775
71. Nakahata Y, Kaluzova M, Grimaldi B, Sahar S, Hirayama J, Chen D et al (2008) The NAD+-dependent deacetylase SIRT1 modulates CLOCK-mediated chromatin remodeling and circadian control. Cell 134:329–340
72. Doi M, Hirayama J, Sassone-Corsi P (2006) Circadian regulator CLOCK is a histone acetyltransferase. Cell 125:497–508
73. Hirayama J, Sahar S, Grimaldi B, Tamaru T, Takamatsu K, Nakahata Y et al (2007) CLOCK-mediated acetylation of BMAL1 controls circadian function. Nature 450:1086–1090
74. Koike N, Yoo SH, Huang HC, Kumar V, Lee C, Kim TK et al (2012) Transcriptional architecture and chromatin landscape of the core circadian clock in mammals. Science 338:349–354
75. Katada S, Sassone-Corsi P (2010) The histone methyltransferase MLL1 permits the oscillation of circadian gene expression. Nat Struct Mol Biol 17:1414–1421
76. DiTacchio L, Le HD, Vollmers C, Hatori M, Witcher M, Secombe J et al (2011) Histone lysine demethylase JARID1a activates CLOCK-BMAL1 and influences the circadian clock. Science 333:1881–1885
77. Duong HA, Weitz CJ (2014) Temporal orchestration of repressive chromatin modifiers by circadian clock Period complexes. Nat Struct Mol Biol 21:126–132
78. Grechez-Cassiau A, Feillet C, Guerin S, Delaunay F (2015) The hepatic circadian clock regulates the choline kinase alpha gene through the BMAL1-REV-ERBalpha axis. Chronobiol Int 32:774–784
79. Ma D, Liu T, Chang L, Rui C, Xiao Y, Li S et al (2015) The liver clock controls cholesterol homeostasis through Trib1 protein-mediated regulation of PCSK9/low density lipoprotein receptor (LDLR) axis. J Biol Chem 290:31003–31012
80. Marcheva B, Ramsey KM, Peek CB, Affinati A, Maury E, Bass J (2013) Circadian clocks and metabolism. Handb Exp Pharmacol 217:127–155
81. Maury E, Hong HK, Bass J (2014) Circadian disruption in the pathogenesis of metabolic syndrome. Diabetes Metab 40:338–346
82. Peek CB, Ramsey KM, Marcheva B, Bass J (2012) Nutrient sensing and the circadian clock. Trends Endocrinol Metab 23:312–318
83. Vollmers C, Gill S, DiTacchio L, Pulivarthy SR, Le HD, Panda S (2009) Time of feeding and the intrinsic circadian clock drive rhythms in hepatic gene expression. Proc Natl Acad Sci U S A 106:21453–21458
84. Adamovich Y, Rousso-Noori L, Zwighaft Z, Neufeld-Cohen A, Golik M, Kraut-Cohen J et al (2014) Circadian clocks and feeding time regulate the oscillations and levels of hepatic triglycerides. Cell Metab 19:319–330
85. Minami Y, Kasukawa T, Kakazu Y, Iigo M, Sugimoto M, Ikeda S et al (2009) Measurement of internal body time by blood metabolomics. Proc Natl Acad Sci U S A 106:9890–9895
86. Yao ZM, Vance DE (1990) Reduction in VLDL, but not HDL, in plasma of rats deficient in choline. Biochem Cell Biol 68:552–558
87. Li Z, Vance DE (2008) Phosphatidylcholine and choline homeostasis. J Lipid Res 49:1187–1194
88. Gorne LD, Costa-Rodriguez VA, Pasquare SJ, Salvador GA, Giusto NM, Guido ME (2015) The mouse liver displays daily rhythms in the metabolism

of phospholipids and in the activity of lipid synthesizing enzymes. Chronobiol Int 32:11–26
89. van MG, Voelker DR, Feigenson GW (2008) Membrane lipids: where they are and how they behave. Nat Rev Mol Cell Biol 9:112–124
90. Glunde K, Bhujwalla ZM, Ronen SM (2011) Choline metabolism in malignant transformation. Nat Rev Cancer 11:835–848
91. de Ramirez MA, Gallego-Ortega D, Sarmentero-Estrada J, Lagares D, Del Gomez PT, Bandres E et al (2008) Choline kinase as a link connecting phospholipid metabolism and cell cycle regulation: implications in cancer therapy. Int J Biochem Cell Biol 40:1753–1763
92. Ikemoto A, Fukuma A, Fujii Y, Okuyama H (2000) Diurnal rhythms of retinal phospholipid synthetic enzymes are retained but their activities are decreased in rats under alpha-linolenic acid deficiency. Arch Biochem Biophys 383:108–113
93. Garbarino-Pico E, Carpentieri AR, Castagnet PI, Pasquare SJ, Giusto NM, Caputto BL et al (2004) Synthesis of retinal ganglion cell phospholipids is under control of an endogenous circadian clock: daily variations in phospholipid-synthesizing enzyme activities. J Neurosci Res 76:642–652
94. Garbarino-Pico E, Valdez DJ, Contin MA, Pasquare SJ, Castagnet PI, Giusto NM et al (2005) Rhythms of glycerophospholipid synthesis in retinal inner nuclear layer cells. Neurochem Int 47:260–270
95. Costa-Rodriguez VA, Marquez S, Salvador GA, Pasquare SJ, Gorne LD, Garbarino-Pico E et al (2013) Daily rhythms of glycerophospholipid synthesis in fibroblast cultures involve differential enzyme contributions. J Lipid Res 54:1798–1811
96. de Arriba Zerpa GA, Guido ME, Bussolino DF, Pasquare SJ, Castagnet PI, Giusto NM et al (1999) Light exposure activates retina ganglion cell lysophosphatidic acid acyl transferase and phosphatidic acid phosphatase by a c-fos-dependent mechanism. J Neurochem 73:1228–1235
97. Marquez S, Crespo P, Carlini V, Garbarino-Pico E, Baler R, Caputto BL et al (2004) The metabolism of phospholipids oscillates rhythmically in cultures of fibroblasts and is regulated by the clock protein PERIOD 1. FASEB J 18:519–521
98. Ruggiero L, Connor MP, Chen J, Langen R, Finnemann SC (2012) Diurnal, localized exposure of phosphatidylserine by rod outer segment tips in wild-type but not Itgb5−/− or Mfge8−/− mouse retina. Proc Natl Acad Sci U S A 109:8145–8148
99. Liu S, Brown JD, Stanya KJ, Homan E, Leidl M, Inouye K et al (2013) A diurnal serum lipid integrates hepatic lipogenesis and peripheral fatty acid use. Nature 502:550–554
100. Kohsaka A, Laposky AD, Ramsey KM, Estrada C, Joshu C, Kobayashi Y et al (2007) High-fat diet disrupts behavioral and molecular circadian rhythms in mice. Cell Metab 6:414–421
101. Guan D, Xiong Y, Borck PC, Jang C, Doulias PT, Papazyan R et al (2018) Diet-induced circadian enhancer remodeling synchronizes opposing hepatic lipid metabolic processes. Cell 174:831–842
102. Jang YS, Kang YJ, Kim TJ, Bae K (2012) Temporal expression profiles of ceramide and ceramide-related genes in wild-type and mPer1/mPer2 double knockout mice. Mol Biol Rep 39:4215–4221
103. Gooley JJ, Chua EC (2014) Diurnal regulation of lipid metabolism and applications of circadian lipidomics. J Genet Genomics 41:231–250
104. Pan X, Hussain MM (2012) Gut triglyceride production. Biochim Biophys Acta 1821:727–735
105. Dallmann R, Viola AU, Tarokh L, Cajochen C, Brown SA (2012) The human circadian metabolome. Proc Natl Acad Sci U S A 109:2625–2629
106. Kumar JP, Challet E, Kalsbeek A (2015) Circadian rhythms in glucose and lipid metabolism in nocturnal and diurnal mammals. Mol Cell Endocrinol 418 (Pt 1):74–88
107. Navara KJ, Nelson RJ (2007) The dark side of light at night: physiological, epidemiological, and ecological consequences. J Pineal Res 43:215–224
108. Ruf T, Arnold W (2015) Daily and seasonal rhythms in human mucosa phospholipid fatty acid composition. J Biol Rhythm 30:331–341
109. Nelson RJ (2004) Seasonal immune function and sickness responses. Trends Immunol 25:187–192
110. Chua EC, Shui G, Cazenave-Gassiot A, Wenk MR, Gooley JJ (2015) Changes in plasma lipids during exposure to total sleep deprivation. Sleep 38:1683–1691
111. Singh R, Sharma S, Singh RK, Mahdi AA, Singh RK, Lee GC et al (2016) Effect of gender, age, diet and smoking status on chronomics of circulating plasma lipid components in healthy Indians. Clin Chim Acta 459:10–18
112. Singh R, Sharma S, Singh RK, Cornelissen G (2016) Circadian time structure of circulating plasma lipid components in healthy Indians of different age groups. Indian J Clin Biochem 31:215–223
113. Isherwood CM, van der Veen, Johnston JD, Skene DJ (2017) Twenty-four-hour rhythmicity of circulating metabolites: effect of body mass and type 2 diabetes. FASEB J 31:5557–5567
114. Gu F, Klerman EB, Kim S, Moore S, Yu K, Albert PS et al (2019) Diurnal variation of metabolites in three individual participants. Chronobiol Int 36:332–342
115. Gu F, Xu S, Devesa SS, Zhang F, Klerman EB, Graubard BI et al (2017) Longitude position in a time zone and cancer risk in the United States. Cancer Epidemiol Biomark Prev 26:1306–1311
116. Pan X, Terada T, Okuda M, Inui K (2003) Altered diurnal rhythm of intestinal peptide transporter by fasting and its effects on the pharmacokinetics of ceftibuten. J Pharmacol Exp Ther 307:626–632
117. Pan X, Terada T, Irie M, Saito H, Inui K (2002) Diurnal rhythm of H+-peptide cotransporter in rat

small intestine. Am J Physiol Gastrointest Liver Physiol 283:G57–G64
118. Pan X, Terada T, Okuda M, Inui K (2004) The diurnal rhythm of the intestinal transporters SGLT1 and PEPT1 is regulated by the feeding conditions in rats. J Nutr 134:2211–2215
119. Miettinen TA (1980) Diurnal variation of LDL and HDL cholesterol. Ann Clin Res 12:295–298
120. Saito Y, Yoshida S, Nakaya N, Hata Y, Goto Y (1991) Comparison between morning and evening doses of simvastatin in hyperlipidemic subjects. A double-blind comparative study. Arterioscler Thromb 11:816–826
121. De GA, Mallozzi MA, Fabbian F, Portaluppi F, Manfredini R (2013) Circadian rhythms and medical diseases: does it matter when drugs are taken? Eur J Intern Med 24:698–706
122. Cilla DD Jr, Gibson DM, Whitfield LR, Sedman AJ (1996) Pharmacodynamic effects and pharmacokinetics of atorvastatin after administration to normocholesterolemic subjects in the morning and evening. J Clin Pharmacol 36:604–609
123. Wallace A, Chinn D, Rubin G (2003) Taking simvastatin in the morning compared with in the evening: randomised controlled trial. BMJ 327:788
124. Wright DF, Pavan KV, Al-Sallami HS, Duffull SB (2011) The influence of dosing time, variable compliance and circadian low-density lipoprotein production on the effect of simvastatin: simulations from a pharmacokinetic-pharmacodynamic model. Basic Clin Pharmacol Toxicol 109:494–498

ABC Transporters, Cholesterol Efflux, and Implications for Cardiovascular Diseases

6

Nan Wang and Marit Westerterp

Abstract

Most types of cells in the body have no or very limited capacity of catabolizing cholesterol, so cholesterol efflux is essential for cholesterol homeostasis. There are multiple mechanisms responsible for cellular cholesterol efflux. Among them, the active efflux pathways are mediated primarily by the ATP-binding cassette (ABC) transporters ABCA1 and ABCG1. ABCA1 is essential for cholesterol and phospholipid efflux to apolipoprotein A-I and high density lipoprotein (HDL) biogenesis. ABCG1 promotes cholesterol efflux primarily to HDL particles. Atherosclerotic cardiovascular disease is a chronic inflammatory disease characterized by marked macrophage foam cell accumulation in atherosclerotic plaques and associated pro-inflammatory responses in lesional cells. Findings from both animal and human studies indicate a critical role of disturbed cholesterol homeostasis in pro-inflammatory responses in these cells, particularly in lesional macrophages. ABCA1 and ABCG1 are highly expressed in macrophages, particularly in response to cholesterol accumulation, and are essential in maintenance of cholesterol homeostasis. Functional deficiency of ABCA1 and ABCG1 in macrophage markedly increases atherogenesis, with exacerbated inflammatory responses. ABCA1 and ABCG1 also play a critical role in control of hematopoietic stem and progenitor cell (HSPC) proliferation and extramedullary hematopoiesis. Hematopoietic ABCA1 and ABCG1 deficiencies cause marked HSPC expansion and extramedullary hematopoiesis, particularly in hypercholesterolemia, and lead to marked monocytosis and neutrophilia with exacerbated pro-inflammatory responses. All these contribute to atherosclerosis. In this chapter, we describe these findings and discuss the current understanding of the underlying mechanisms. We also discuss other ABC transporters such as ABCG4, which also promotes cholesterol efflux to HDL and controls megakaryocyte proliferation and platelet biogenesis. By this pathway, ABCG4 also modulates atherogenesis. Therapeutic approaches targeting the pathways and mechanisms described also are discussed.

Keywords

ABC transporters · ABCA1 · ABCG1 · Cholesterol efflux · HDL · Cardiovascular diseases

N. Wang (✉)
Department of Medicine, Columbia University, New York, NY, USA
e-mail: nw30@cumc.columbia.edu

M. Westerterp
Department of Pediatrics, University Medical Center Groningen, University of Groningen, Groningen, The Netherlands

X.-C. Jiang (ed.), *Lipid Transfer in Lipoprotein Metabolism and Cardiovascular Disease*, Advances in Experimental Medicine and Biology 1276, https://doi.org/10.1007/978-981-15-6082-8_6

6.1 Introduction

Cholesterol homeostasis is essential for cell structure and function, explaining why cells devote a relatively large portion of transcriptome and energy to regulation of cellular cholesterol homeostasis [1]. Cholesterol is an important lipid component of the cell membrane and modulates membrane fluidity and permeability [2, 3]. Thus, cholesterol plays a key role in regulation of membrane protein spatial distribution and function as well as transmembrane signaling processes involved in various fundamental biological activities such as immunity and cell proliferation [2, 4]. Cholesterol also can modulate cell signaling by covalent modification of signaling molecules. Cholesterol modification of Hedgehog and Smoothened is essential for their spatial distribution and function [5, 6]. In addition, cholesterol is the precursor of all steroid hormones, vitamin D and bile acids.

Due to the importance of cholesterol in cell biology, cells have developed complex mechanisms to regulate cellular cholesterol homeostasis. Animal cells acquire cholesterol by de novo cholesterol biosynthesis and from the circulation in the form of lipoproteins [1, 7]. In addition, phagocytes could obtain cholesterol by phagocytosis of apoptotic cells, damaged cells, and senescent cells, or via phagocytosis of cells that have died via other regulated cell death processes such as necroptosis, pyroptosis, and ferroptosis. Cells that rely primarily on endogenous cholesterol synthesis generally do not accumulate excess endogenous cholesterol due to the homeostatic feedback regulation of cholesterol biosynthesis at multiple steps. Cells that internalize exogenous cholesterol, from lipoproteins or phagocytosis of dead cells, repress de novo cholesterol synthesis and low-density lipoprotein (LDL) receptor expression in response to cholesterol loading [7]. However, downregulation of LDL receptor (LDLr) expression alone may not be sufficient to prevent excess cholesterol accumulation. Most types of cells in the body have no or very limited capacity of catabolizing cholesterol. Therefore, cells need to evolve other mechanisms to prevent accumulation of excess cholesterol. One mechanism is cholesterol esterification, which is mediated by the microsomal enzyme acyl-coenzyme A:cholesterol acyltransferase (ACAT) [8]. This reaction converts a potentially highly toxic "free" cholesterol to cholesterol ester which can be stored in lipid droplets [8]. However, even this mechanism appears to be insufficient to prevent excess free cholesterol accumulation in cells under certain context. To deal with the pool of accumulated cholesterol, cells also develop other mechanisms to remove excess cholesterol from the cell. These mechanisms involve conversion of cholesterol to oxysterols and passive or active processes to remove cholesterol and oxysterols from the cell [9]. Dysregulation of these processes to maintain cholesterol homeostasis could lead to excess free cholesterol accumulation in cells, disrupting membrane fluidity and transmembrane cell signaling, and could eventually lead to cell death. It is not surprising that dysregulated cellular cholesterol homeostasis has been linked to various chronic metabolic diseases including atherosclerotic cardiovascular disease (ACD), diabetes, and non-alcoholic hepatic disease [7].

In this chapter, we will discuss the mechanism of how cellular cholesterol is actively transported out of the cell, a process belonging to the overall mechanism of cholesterol efflux, and how this mechanism regulates cell activity and functions. We will also discuss how these regulated cell activity and functions are involved in important systemic physiology and pathology, particularly the ones related to ACD. The key molecules involved in active cellular cholesterol efflux are ATP-binding cassette transporters (ABC transporters). Therefore, our discussion will focus on the role of ABC transporters, while mechanisms involving other molecules relevant in regulation of ABC transporter-mediated cholesterol efflux will be discussed as well.

6.2 ABC Transporters and Cholesterol Efflux

The first ABC transporter identified to mediate cholesterol efflux is ABC transporter A1 (ABCA1). Since the hypothesis of HDL as the major lipoprotein fraction responsible for transport of cholesterol from peripheral tissues to the liver was proposed [10], studies were started to characterize the role of HDL particles in promotion of cholesterol efflux from cultured cells. This led to identification of high density, small HDL particles as lipoprotein subclasses that potently promote cholesterol efflux [11]. Follow-up studies identified specific high-affinity binding of these HDL particles to the surface of the cell [12]. Interestingly, these HDL binding sites are upregulated upon cholesterol loading of the cells [13]. In addition, the binding of HDL is associated with removal of cholesterol from an intracellular pool [14] and apolipoprotein A-I (apoA-I), the major apolipoprotein component of HDL, small HDL particles, or apoA-I like mimetic peptides promote cholesterol efflux. Importantly, cholesterol efflux promoted by apoA-I is impaired in fibroblasts derived from Tangier Disease patients [15]. Tangier Disease is a rare genetic disorder characterized by extremely low plasma levels of HDL and apoA-I and deposition of cholesteryl esters in tissues [16]. These studies link apoA-I binding and cholesterol efflux to low HDL and tissue cholesterol accumulation in Tangier Disease. The identity of the specific cell surface apoA-I binding site remains elusive until human genetic studies have identified ABCA1 as the mutated gene in Tangier Disease [17–19]. This breakthrough links ABCA1 with potential apoA-I binding and cholesterol efflux. Follow-up studies provided direct evidence that overexpression of ABCA1 in cultured cells as a gain of function model markedly increases apoA-I binding and cholesterol efflux from the cell [20]. Direct binding of apoA-I to ABCA1 is suggested by chemical crosslinking of labeled apoA-I to cell surface ABCA1 with chemical crosslinkers with a spacer arm of ~12 angstroms [20]. Furthermore, single-molecule imaging using total internal reflection fluorescence microscopy reveals that a direct interaction forms between apoA-I and ABCA1 on the plasma membrane during the initial step of HDL formation [21]. Together, these studies define ABCA1 as the authentic transporter that mediates cellular cholesterol efflux to apoA-I and initiates HDL biogenesis. These studies also led to efforts to assess whether other ABC transporters are involved in cholesterol transport and whether mutations of these ABC transporters cause genetic disease with disturbed sterol metabolism and homeostasis. These efforts led to the discovery of ABCG5 and ABCG8 as the mutated genes in sitosterolemia, a genetic metabolic disease with marked elevation of plasma plant sterols and increased risk of ACD [22]. Detailed discussion of ABCG5 and ABCG8 and their role in sterol metabolism is presented in another chapter of the book.

ABCA1 promotes cellular cholesterol efflux to lipid-poor apoA-I or small HDL particles but not to large, lipid-rich HDL particles [20, 23]. However, large HDL particles also potently promote cellular cholesterol efflux, particularly in combination with HDL-associated lecithin cholesterol acyltransferase (LCAT) activity. LCAT converts cholesterol to cholesterol esters, expanding the cores of HDL particles and facilitating cholesterol efflux to HDL. Further studies have identified ABCG1 and ABCG4 as ABC transporters that facilitate cellular cholesterol efflux to HDL particles but not to lipid-poor apoA-I [24]. Like ABCG5 and ABCG8 but unlike ABCA1, ABCG1, and ABCG4 are half transporters. While ABCG5 and ABCG8 form heterodimers to function, ABCG1 and ABCG4 function mainly as homodimers to promote cholesterol efflux. Unlike ABCA1, ABCG1 and ABCG4 do not bind apoA-I or HDL to facilitate cholesterol efflux [24].

6.3 Mechanisms Underlying ABC Transporter-Mediated Cholesterol Efflux

Point mutation analysis demonstrates that an active ATPase activity of ABCA1 is required for apoA-I-mediated cholesterol efflux, indicating an active cholesterol efflux process [23]. In addition to cholesterol, ABCA1 also promotes phospholipid efflux to apoA-I [20], including phosphatidylcholine, phosphatidylserine, and sphingomyelin with a preference for phosphatidylcholine [25], a finding consistent with defective phospholipid efflux from cells of Tangier Disease patients [26]. While apoA-I binding to cell surface and to ABCA1 enhanced by ABCA1 activity has been well established, controversy persists regarding the mechanism by which apoA-I bound to the surface of cell acquires membrane phospholipids and cholesterol. In the "direct loading model," apoA-I acquires lipids directly from ABCA1, while it is bound to the transporter [27, 28]. By contrast, apoA-I is proposed to acquire lipids from the specific membrane domains created by ABCA1 in the "indirect model" [29, 30]. The structure of human ABCA1 revealed by cryo-EM with nominal resolutions of 4.1 angstrom has been described and confirms ABCA1 as a phospholipid translocase. A "lateral access" mechanism for ABCA1-mediated lipid export where the lipid substrate on the inner leaflet of the membrane may bind to the transporter from the lateral membrane has been speculated on [31]. However, the molecular mechanism for the lipid substrate delivery from ABCA1 to apoA-I for nascent HDL formation remains elusive [31], and the structural information from this study has been used to support the "direct loading model" or "indirect model" accordingly [27, 32]. Also, additional controversy exists over ABCA1-mediated apoA-I binding to cell surface. While chemical crosslinking, single-molecule imaging and mutant ABCA1 studies in some earlier reports support direct binding of apoA-I to ABCA1, a more recent study suggests that ABCA1 has floppase activity for phosphatidylinositol (4, 5) bisphosphate (PIP2) [33]. PIP2 directly binds apoA-I, and the increased cell surface PIP2 functions as a bridge for ABCA1-mediated apoA-I binding and to promote cholesterol efflux [33]. Disruption of membrane PIP2 formation led to reduced apoA-I binding and cholesterol efflux, indicating a critical role of PIP2 in ABCA1-mediated apoA-I binding and lipid efflux [33]. While the structural and biochemical findings strongly suggest ABCA1 as a phospholipid transporter, it is not clear whether ABCA1 directly transports cholesterol. Some clues may be gained from structural analysis of ABCG5/ABCG8 dimers. Electron density map of human ABCG5/ABCG8 using crystallization in lipid bilayers has revealed features that may represent cholesterol. Mutagenesis studies of amino acid residues that may represent binding surfaces or entryway for sterols to access the core of the heterodimer interface confirm the essential role of these residues in cholesterol transport. Structural comparison of the transmembrane domains (TMDs) of ABCA1 and ABCG5/G8 reveals similarity in the general structural organization of TMDs, despite ABCA1 and ABCG5/G8 belonging to different ABC subfamilies and ABCA1 comprising one single polypeptide chain, whereas ABCG5/G8 is a heterodimer. The TMDs of ABCA1 and ABCG5/G8 share 14% sequence identity and 35%–40% similarity, suggesting evolutionary relevance. Earlier studies showed that treatment of cells with ABCA1 overexpression by cyclodextrin, a chemical compound that potently promotes cholesterol efflux, could dissociate phospholipid efflux from cholesterol efflux to apoA-I [23]. However, recent studies indicated that trypsin treatment could release extracellular domains of ABCA1 from cell surface into media in parallel to a rapid release of phospholipid and cholesterol [27]. This release of membrane lipids was dependent on the ATPase activity of ABCA1. Based on these findings, it has been proposed that phospholipids and cholesterol transported by ABCA1 are temporarily sequestered within the extracellular domains of ABCA1 during lipid efflux and nascent HDL formation [27]. This model also is consistent, to some extent, with a concurrent process of phospholipid and cholesterol efflux to apoA-I for nascent HDL assembly in murine macrophages with high ABCA1 expression [34] and with the finding that ABCA7, a member of ABCA family, promotes phospholipid but not cholesterol efflux to lipid-poor apoA-I [35]. Together, these findings suggest ABCA1 as a direct transporter for both phospholipid and cholesterol.

The protein ligand for ABCA1 is not highly specific. In addition to apoA-I, other apolipoproteins such as apolipoprotein E (apoE) or even amphipathic α-helical apoA-I mimetic peptides can serve as ABCA1 ligands for lipid acceptance [36]. While a major portion of cellular ABCA1 is localized on the plasma membrane and promotes cholesterol efflux from the plasma membrane, a preferred source for ABCA1-mediated cholesterol efflux is the pool of cholesterol in late endosomes and lysosomes. The functional importance of this pathway for intracellular cholesterol efflux is demonstrated by the findings that macrophages isolated from the mice modeled for Niemann-Pick type C1 disease, a genetic disorder with accumulation of cholesterol and other lipids in late endosomes and lysosomes, show marked defect in ABCA1-mediated cholesterol efflux from late endosomes and lysosomes [37]. Mechanistically, ABCA1 on the plasma membrane can be internalized and traffic to late endosomes and lysosomes [38]. A PEST sequence mutant ABCA1 shows impaired internalization and defective cholesterol efflux from late endosomes, while cholesterol efflux from cell surface mediated by the mutant ABCA1 is unaffected [38]. These studies indicate the functional importance of ABCA1 internalization and trafficking in mediating cholesterol efflux from intracellular cholesterol pools.

Like ABCA1, ABCG1 has been identified on plasma membrane and in intracellular organelles. Earlier studies report localization of ABCG1 to plasma membrane, Golgi, and recycling endosomes in transfected HEK293 cells [39]. In macrophage-like cells, activation of liver X receptor (LXR) increases ABCG1 expression and presentation to the cell surface, in association with increased cellular cholesterol efflux to HDL [39, 40]. Follow-up studies indicate that ABCG1 is primarily localized intracellularly and acts to promote cholesterol transport and traffic from intracellular sites to plasma membrane for efflux [41]. Thus, it has been proposed that ABCG1 at the plasma membrane mobilizes plasma membrane cholesterol and ABCG1 in late endosomes and lysosomes generates mobile pools of cholesterol that can traffic by both vesicular and non-vesicular pathways to the plasma membrane where it can also be transferred to extracellular acceptors with a lipid surface [41]. Conceptually, these two mechanisms are not mutually exclusive, and both mechanisms may act in vivo. Macrophage deficiency of ABCG1 leads to suppression of *Ldlr* and *Hmgcr* expression relative to wild-type cells and increased cholesteryl ester formation by ACAT, even in the absence of acceptors in the media to promote cholesterol efflux [39]. This suggests redistribution of cholesterol from plasma membrane to the ER, leading to suppression of cholesterol biosynthetic genes, independent of cholesterol efflux. This is consistent with the finding that overexpression of ABCG1 leads to an increase in the mature form of SREBP-2 and its target gene expression. While ABCG1 activity promotes mostly cholesterol efflux, ABCG1 also facilitates a low magnitude of phospholipid efflux [42]. However, the precise structure, substrate, and function of ABCG1 are still largely unknown.

ABCG4 is the only other ABCG family member that has been shown to promote cholesterol efflux to HDL when overexpressed in cultured mammalian cells [24]. While even less well studied, ABCG4 is the most homologous and closest to ABCG1 in structure and function in ABC transporters [24]. Thus, it is likely that ABCG4 promotes cholesterol efflux to HDL via a mechanism that is similar to ABCG1.

6.4 Regulation of ABC Transporters

ABCA1 and ABCG1 are essential in cellular cholesterol homeostasis in multiple tissues and cell types. Therefore, ABCA1 and ABCG1 expression and function are tightly regulated. The primary regulation of ABCA1 and ABCG1 expression at a transcriptional level is mediated by nuclear receptors liver X receptor (LXR)α and LXRβ, the master transcription factors regulating cholesterol homeostasis by regulation of expression of multiple effectors in cholesterol transport and metabolism, such as ABCA1, ABCG1, apolipoprotein E (apoE), cholesteryl ester transfer protein (CETP), and inducible degrader of the LDL receptor (IDOL) [43]. In phagocytes such as macrophages, the primary function of LXR in regulation of cholesterol homeostasis is to prevent excessive cholesterol accumulation. When unesterified cholesterol accumulates in macrophages due to uptake of cholesterol-rich apoB containing lipoprotein particles or phagocytosis of apoptotic cells, production of oxysterols such as 27-hydroxycholesterol, 25-hydroxycholesterol, 22-hydroxycholesterol, and 24(S),25-epoxycholesterol is increased. These oxysterols act as LXR ligands, activate LXR, and upregulate ABCA1, ABCG1, and apoE transcription [43]. As a result, cholesterol efflux from the cells is increased, and this will help to remove excess cellular cholesterol and maintain cholesterol homeostasis.

LXRs form obligate heterodimer with retinoic X receptors (RXRs) to regulate ABCA1 and ABCG1 expression and RXR agonists such as retinoic acid increase ABCA1 and ABCG1 expression [44, 45]. Activation of peroxisome proliferator-activated receptor (PPAR)α and PPARγ, nuclear receptors with free fatty acids and eicosanoids as endogenous ligands, also increases ABCA1 expression [46, 47]. In addition to the nuclear receptors, ABCA1 and ABCG1 expressions are also regulated by classic signaling pathways such as cyclic AMP-mediated signaling pathways. Cyclic AMP analogs are known to upregulate ABCA1 expression in macrophage or macrophage-like cells [48]. Conversely, ABCA1 and ABCG1 expression are reported to be downregulated by multiple signaling pathways involved in inflammation, particularly pathways involving NF-kappaB activation [49–51].

Cells also develop mechanisms to regulate ABCA1 and ABCG1 expression beyond direct regulation of transcription. As discussed above, many tissues and cell types develop mechanisms to upregulate ABCA1 or ABCG1 expression as a way to prevent excessive cholesterol accumulation. Sterol regulatory element-binding protein (SREBP)-2 and SREBP-1 are master transcription factors in regulation of cholesterol and fatty acid biosynthesis and homeostasis, as demonstrated by the seminal discoveries by Brown and Goldstein [7]. miR-33a or miR-33b, intronic microRNAs (miRNA) located within the gene encoding SREBP-2 or SREBP-1, respectively, inhibit the expression of ABCA1 and ABCG1 [52]. miR-33 antagomirs increase ABCA1 and ABCG1 expression, plasma HDL levels, and reverse cholesterol transport in vivo [52].

6.5 ABC Transporters in HDL Metabolism and ACD

As the mutated gene in Tangier Disease, ABCA1 is the primary gene product that is essential for HDL biogenesis, which explains the extremely low plasma HDL cholesterol (HDL-C) levels in Tangier Disease patients with homozygous ABCA1 deficiency [53]. This is recapitulated in whole body ABCA1 deficient mice. The liver is the primary organ for HDL biogenesis and hepatocyte-specific ABCA1 deficiency causes ~70–80% lower plasma HDL-C in rodents [54]. The second major organ for ABCA1-mediated HDL biosynthesis is intestine, and ABCA1 deficiency in enterocytes causes 20–30% decrease in plasma HDL-C levels [55]. Other tissues also contribute to HDL generation in an ABCA1-dependent fashion. Adipocyte-specific ABCA1 deficiency leads to 15% reduction of plasma HDL [56]. While ABCA1 has an important role in mediating cholesterol efflux from macrophages to lipid-poor apolipoproteins and maintenance of cellular cholesterol homeostasis, transplantation of bone marrow from *Abca1*$^{-/-}$ mice into wild-type mice or from wild-type mice into *Abca1*$^{-/-}$ mice has little effect on plasma HDL concentrations in the recipient [57, 58], suggesting that myeloid ABCA1 expression has minimal impact on plasma HDL levels. Since ABCA1 is essential for nascent HDL biosynthesis, ABCA1 has profound impact on HDL metabolism and its function. ABCA1 not only mediates HDL biogenesis but also modulates apoA-I turnover. Lipidation of apoA-I for nascent HDL formation and the subsequent HDL maturation by acquiring additional lipids via ABCA1 and other pathways such as ABCG1- or scavenger receptor class B, type 1 (SR-BI)-mediated cholesterol efflux, CETP, or PLTP-mediated lipid exchange or LCAT-facilitated HDL packaging profoundly modulate apoA-I turnover. In ABCA1 deficiency, lipid-poor apoA-I due to defective lipidation has increased clearance rate from plasma, causing marked decrease in plasma apoA-I levels [54]. Infusion of reconstituted human HDL into liver-specific but not whole body ABCA1 deficient mice can restore plasma HDL-C and apoA-I levels [54]. Together, these studies indicate the important role of hepatic ABCA1 in generation of early HDL particles and the essential role of extrahepatic ABCA1 in further lipidation and maturation of the early HDL particles [54].

The prominent phenotypes of Tangier Disease patients include extremely low plasma HDL, enlarged tonsils with a yellow and orange appearance, splenomegaly, hepatomegaly, and peripheral neuropathy, indicating the essential role of ABCA1 in HDL biogenesis and regulation of cellular cholesterol efflux and homeostasis, particularly for cells that accumulate massive amounts of lipids in the absence of ABCA1, such as macrophage, other reticuloendothelial cells, and Schwann cells [59].

Reverse cholesterol transport (RCT), originally proposed by Glomset [10], is the process by which cholesterol in peripheral tissues is transported by HDL to the liver for excretion into the bile and feces. In this process, the initial step is cholesterol efflux from peripheral cells to HDL. Like ABCA1, ABCG1 also is essential in promoting cellular cholesterol efflux to HDL in multiple tissues and cell types. Mice that are deficient in ABCG1 have lipid accumulation in macrophages within multiple tissues when they are fed a high-fat, high-cholesterol diet, particularly in the lung [60]. However, ABCG1 deficiency does not affect plasma lipoprotein levels, indicating a minor role of ABCG1-mediated cholesterol efflux in HDL metabolism. The lack of impact of ABCG1 deficiency on plasma lipoprotein levels may reflect the fact that its expression in hepatocytes is low, and the low hepatic expression of ABCG1 may reflect primarily its expression in Kupffer and endothelial cells [61]. Nevertheless, ABCA1 and ABCG1 show additive or synergistic activity in facilitating cholesterol efflux to HDL in macrophages, consistent with an important role of ABCA1 and ABCG1 in RCT initiated from macrophages in vivo [62, 63].

As expected, activity of the gene products that regulate ABCA1 and ABCG1 expression also regulate ABCA1- and ABCG1-mediated RCT. LXR agonists promote RCT in vivo in mouse models and human cells [64, 65]. While LXR agonists induce hepatic and intestinal expression of ABCG5 and ABCG8, which likely contributes to cholesterol excretion into the bile and feces, enhanced RCT from macrophages in response to LXR agonists in vivo is likely attributed at least partly to induced macrophage ABCA1 and ABCG1 expression as indicated by the increased cholesterol tracer in the plasma without significant change of plasma lipoprotein levels [64, 65]. Farnesoid X receptor (FXR) also is a nuclear receptor that regulates lipid metabolism. FXR activation in liver increases hepatic miR-144 levels, which in turn lowers hepatic ABCA1 and plasma HDL levels [66]. This implies that bile acids regulate plasma HDL levels via a FXR-miR-144-ABCA1 pathway in hepatocytes. Interestingly, it has been reported that selective hepatic ABCA1 deficiency increases RCT [67], as hepatic ABCA1 promotes efflux of hepatic cholesterol back to plasma but not excretion into the bile. Together, these findings suggest the possibility that bile salts promote RCT in the postprandial state by downregulation of hepatic ABCA1 and upregulation of SR-BI [68] via FXR activation.

It has been well established that plasma HDL-C levels correlate inversely with the incidence of ACD, suggesting a protective role of HDL. Efforts have been made to understand the mechanisms underlying the anti-atherogenic properties of HDL. ACD is a nonresolving chronic inflammatory disease, and the atherogenic process is thought to be triggered by the subendothelial retention of apoB-containing, cholesterol-rich lipoprotein particles at sites of arterial walls susceptible to blood flow disturbance [69]. In response, tissue macrophages engulf these lipoprotein particles via pattern recognition receptors selected in evolution for handling components of microbial pathogens and also mediating internalization of modified lipoproteins. Unlike the LDL receptor, the expression and activity of these scavenger receptors are not suppressed by increased cellular content of cholesterol, leading to continued uptake of lipoprotein particles [69]. Eventually, the continued cholesterol loading will overwhelm the cellular mechanisms that act to prevent cholesterol overloading and cause excessive cholesterol accumulation, leading to increased free cholesterol content in cell membranes and even cholesterol microcrystal formation [70]. These events result in activation of cellular signaling pathways that promote pro-inflammation responses such as Toll-like receptor-mediated response or inflammasome activation with increased production of pro-inflammatory cytokines [70], and the molecular mechanisms will be discussed in detail below. Free cholesterol accumulation is also potently cytotoxic, leading to cell death via apoptosis, pyroptosis, or secondary necroptosis [69]. Tissue repair and inflammation resolution require efficient clearance of dead or damaged cells and phenotypic conversion of pro-inflammatory macrophages into macrophages that suppress inflammation and promote healing [71]. Cholesterol overloading of macrophages not only induces pro-inflammatory responses but also leads to defective efferocytosis of dead or damaged lesional cells and impaired resolution of inflammation [71]. These maladaptive responses act in a vicious cycle to promote atherosclerosis progression, exacerbate plaque necrotic core formation, thinning of the fibrous cap, and other features of plaque instability, and eventually lead to plaque rupture and atherothrombosis [72]. Since ABCA1 and ABCG1 are essential in prevention of cholesterol overloading in macrophages, it is expected that hematopoietic- or macrophage-specific ABCA1 and ABCG1 deficiency have profound impact on macrophage cholesterol homeostasis, inflammatory response, efferocytosis, and, possibly, the resolution of inflammation.

6.6 ABCA1- and ABCG1-Mediated Cholesterol Efflux Pathways in Inflammation and Atherogenesis

ABCA1- and ABCG1-mediated cholesterol efflux suppress the secretion of pro-inflammatory cytokines from macrophages and endothelial cells [73, 74], which may contribute to their atheroprotective effects. Several pro-inflammatory cytokines and adhesion molecules, including monocyte chemoattractant protein-1 (MCP-1) [75], vascular cell adhesion molecule-1 (VCAM-1) [76], intracellular adhesion molecule-1 (ICAM-1) [77], type I interferons (IFNs) [78], interleukin-1β (IL-1β) [79], and IL-18 [80–82], accelerate atherosclerosis in mice. These pro-inflammatory factors promote monocyte adhesion to endothelial cells and monocyte infiltration into the intima, as such contributing to the initiation and progression of atherosclerosis [75–78]. The CANTOS (canakinumab atherothrombosis anti-inflammatory outcome study) trial has shown that antibodies to IL-1β suppress recurrent cardiovascular events, demonstrating for the first time that inflammation accelerates atherosclerosis in humans [83]. Extensive evidence thus shows a role for inflammation in atherogenesis. Mechanisms of pro-inflammatory cytokine secretion and the role of ABCA1 and ABCG1 therein are described below.

Macrophage cytokine secretion is to a large extent regulated by pathways downstream of Toll-like receptors (TLRs) [84]. These TLRs are mostly activated during acute-septic shock, infections [84], and also in atherosclerosis [85]. The TLR family consists of at least ten different TLRs [86]. Of all TLRs, TLR4 has been described most extensively. One of the main ligands for TLR4 is lipopolysaccharide (LPS). LPS is secreted by Gram-negative bacteria when they multiply or lyse during infection [86]. Additional ligands for TLR4, especially of importance during atherosclerosis, include minimally modified forms of LDL (mmLDL) [87]. Upon activation, TLR4 dimerizes in lipid rafts, cholesterol-enriched domains of the plasma membrane, where it forms a complex together with its effector protein MD2 [88]. Downstream of the TLR4-MD2 complex, both myeloid differentiation primary response 88 (MyD88) and TIR-domain-containing adapter-inducing interferon-β (TRIF) signaling, is activated, eventually leading to the activation of nuclear factor (NF)-kB [86]. NF-kB then induces transcription of several pro-inflammatory genes, including tumor necrosis factor a (TNFa), IL-6, MCP-1, and macrophage inflammatory protein-2 (MIP-2) [86]. Transcription of type I IFNs, including IFN-α and IFN-β, occurs downstream of TRIF signaling [86]. Endothelial cells also express TLRs. Activation of TLR4 in endothelial cells increases the expression of several adhesion molecules and chemokines that promote monocyte adhesion, including VCAM-1, ICAM-1, and MCP-1 [89].

Mechanistic studies have shown that mouse macrophages deficient in both *Abca1* and *Abcg1* show increased surface expression of the TLR4-MD2 complex at the plasma membrane compared to wild-type macrophages [73]. In line, macrophages deficient in *Abca1* and/or *Abcg1* show increased secretion of pro-inflammatory cytokines compared to wild-type macrophages upon LPS-induced TLR4 activation [73]. Membrane cholesterol depletion with methyl-β-cyclodextrin suppresses the LPS-induced pro-inflammatory gene expression in macrophages deficient in *Abca1* and/or *Abcg1* [73]. Collectively, these data show that deficiency of *Abca1* and *Abcg1* enhances membrane cholesterol accumulation, which stabilizes the TLR4-MD2 complex in lipid rafts, as such increasing TLR4 surface expression, and downstream secretion of pro-inflammatory cytokines [73]. Hyperlipidemic mice with deficiency of *Abca1* and *Abcg1* in macrophages show increased plasma levels of pro-inflammatory cytokines and, importantly, increased inflammatory gene expression in atherosclerotic plaques [61], suggesting in vivo relevance. These effects likely contribute to the increased atherosclerosis observed in hyperlipidemic mice with macrophage *Abca1* and *Abcg1* deficiency [61].

Deficiency of *Abca1* and *Abcg1* in endothelial cells also enhances LPS-induced inflammation in vitro, as well as monocyte adhesion [74]. Similar to observations in macrophages, *Abca1* and *Abcg1* deficiency likely stabilizes TLR4 surface expression in cholesterol-enriched domains in endothelial cells. In addition, endothelial *Abca1* and *Abcg1* deficiency suppress the activity of endothelial nitric oxide synthase (eNOS) [74], which produces NO. NO production suppresses NF-kB activation and expression of adhesion molecules in endothelial cells, as well as monocyte adhesion [90]. Hence, endothelial *Abca1* and *Abcg1* deficiency enhance endothelial cell inflammation and monocyte adhesion by stabilizing TLR4 surface expression at the plasma membrane and decreasing NO production. As a consequence, hyperlipidemic mice with endothelial *Abca1* and *Abcg1* deficiency show increased atherosclerosis [74].

HDL suppresses surface expression of the TLR4-MD2 complex in wild-type macrophages and LPS-induced pro-inflammatory cytokine expression [91]. HDL suppresses this LPS-response to a significantly larger extent in wild-type macrophages than macrophages deficient in *Abca1* and *Abcg1*, suggesting that cholesterol efflux to HDL is required for its anti-inflammatory effects [92]. In addition, studies in humans have shown that *ABCA1* heterozygous mutation carriers, with a ~ 50% decrease in plasma HDL levels as well as a ~ 50% decrease in *ABCA1* expression in all cells, show increased plasma levels of pro-inflammatory cytokines and increased inflammation in the vessel wall [93]. The latter was shown by a PET-CT scan that monitors the metabolic activity of cells in the vessel wall, where high metabolic activity reflects a high level of inflammation [93]. These data indicate translational potential of the findings in animal models and show that HDL and cholesterol efflux pathways suppress vascular inflammation in humans.

6.7 Abca1- and Abcg1-Mediated Cholesterol Efflux Suppress Monocytosis and Neutrophilia

Elevated levels of monocytes and neutrophils in blood are associated with increased cardiovascular disease (CVD) events in humans [94]. Monocytes and neutrophils mainly originate from their progenitor cells in the bone marrow [95]. Macrophage *Abca1* and *Abcg1* deficiency increases pro-inflammatory cytokines and, especially, in hyperlipidemic mice, secretion of macrophage colony-stimulating factor (M-CSF), MCP-1, and granulocyte colony-stimulating factor (G-CSF) from monocytes and macrophages [61]. MCP-1 is the ligand for C-C chemokine receptor 2 (CCR2) and stimulates egress of monocytes from the bone marrow [96], while M-CSF and G-CSF instruct granulocyte-macrophage progenitors (GMPs) in the bone marrow to produce monocytes/macrophages and neutrophils, respectively [95]. Hyperlipidemic mice with macrophage *Abca1* and *Abcg1* deficiency show a twofold increase in monocytes and neutrophils in blood [61], presumably as a consequence of increased levels of MCP-1, M-CSF, and G-CSF in plasma.

Production of GMPs is driven by hematopoietic stem cells (HSCs), including $Lin^-Sca1^+cKit^+$ (LSK) cells. Abca1 and Abcg1 are highly expressed in LSKs [97]. LSKs are also referred to as hematopoietic stem and multipotential progenitor cells (HSPCs). Mice with deficiency of the cholesterol transporters *Abca1* and *Abcg1* exhibit a dramatic ~five-fold increase in blood monocyte and neutrophil counts, reflected by a ~ five-fold increase in the HSPC population in the bone marrow [97]. Mechanistic studies have shown that *Abca1* and *Abcg1* deficiency enhance HSPC proliferation due to an increased surface expression of the common β subunit of the receptor for granulocyte-macrophage colony-stimulating factor (GM-CSF) and interleukin-3 (IL-3) on HSPCs [97]. This common β subunit is critical for the expansion of HSPCs and the downstream generation of progenitor cells and leukocytes. The common β subunit is localized in lipid rafts [97] and, similar to the TLR4-MD2 complex in $Abca1^{-/-}Abcg1^{-/-}$ macrophages, stabilized by membrane cholesterol accumulation [97]. Extramedullary hematopoiesis involves mobilization of HSPCs from the bone marrow via the blood into the spleen and other organs [98]. $Abca1^{-/-}Abcg1^{-/-}$ mice also exhibited splenomegaly and extramedullary hematopoiesis; $Abca1^{-/-}Abcg1^{-/-}$ mice displayed increased HSPC mobilization from the bone marrow to the spleen [99]. The mechanism underlying this phenomenon involves increased IL-23 secretion from $Abca1^{-/-}Abcg1^{-/-}$ macrophages and dendritic cells as a result of upregulation of the TLR4 and TLR3 signaling in these cells [99]. IL-23 is known to initiate a signaling cascade leading to enhanced production of IL-17 by T helper 17 cells and G-CSF by bone marrow stromal cells [100], thus directing GMPs in the bone marrow toward neutrophil production [95]. This subsequently decreases the abundance of osteoblasts and $nestin^+$ mesenchymal stem cells that express C-X-C motif chemokine 12 (CXCL12), which is a key retention ligand for C-X-C chemokine receptor type 4 (CXCR4) on HSPCs [101, 102]. Thus, the bone marrow niche is altered, decreasing its ability to retain HSPCs, and HSPCs are mobilized to organs, including the spleen. Extramedullary hematopoiesis likely contributes to monocytosis and neutrophilia in $Abca1^{-/-}Abcg1^{-/-}$ mice as well [103]. In summary, cholesterol efflux pathways mediated by Abca1 and Abcg1 suppress HSPC proliferation in the bone marrow, extramedullary hematopoiesis, and inflammatory cytokine secretion by macrophages, as such suppressing monocytosis and neutrophilia.

Atherosclerosis studies in mice show that deficiency of *Abca1* and *Abcg1* in hematopoietic cells leads to a more dramatic increase in atherosclerosis (2.7-fold) than deficiency of these transporters in macrophages alone (~73%) [61]. While inflammation and macrophage cholesterol accumulation contributed to atherosclerosis in mice with hematopoietic or macrophage *Abca1/Abcg1* deficiency, these data show that monocytosis and neutrophilia, which is more pronounced in mice with hematopoietic than macrophage deficiency of *Abca1* and *Abcg1*, are clearly pro-atherogenic [61]. Indeed monocytosis and neutrophilia are associated with increased CVD in humans [94].

6.8 Role of ABCA1- and ABCG1-Mediated Cholesterol Efflux Pathways in Inflammasome Activation

The CANTOS (canakinumab atherothrombosis anti-inflammatory outcome study) trial has shown that antibodies to IL-1β suppress recurrent cardiovascular events, thus proving for the first time that inflammation accelerates atherosclerosis in humans [83]. IL-1β is a main regulator of inflammation and is secreted by several cell types including macrophages and dendritic cells. Its secretion is controlled by a multimeric protein complex called the inflammasome [104]. Especially the NLRP3 inflammasome plays a role in atherosclerosis [79] and controls both IL-1β and IL-18 secretion. The NLRP3 inflammasome mandates two signals for activation: a priming signal and a so-called second signal. The priming signal leads to activation of NF-κB and occurs downstream of several receptors including TLR4. As a consequence, transcriptions of the several subunits of the NLRP3 inflammasome complex, including NLRP3, ASC, and pro-caspase-1, are increased, as is expression of pro-IL-1β. The second signal leads to cleavage of pro-caspase-1 into its active form, which subsequently cleaves pro-IL-1β and pro-IL-18, required for their secretion [104]. Initial atherosclerosis studies have shown that accumulation of cholesterol crystals or free cholesterol in lysosomes leads to lysosomal damage [79, 105], which itself serves as a second signal. While the presence of cholesterol crystals in atherosclerotic plaques has been called into question [106], these data [79, 105] do clearly demonstrate a link between cholesterol accumulation and inflammasome activation.

Hyperlipidemic mice with deficiency of *Abca1* and *Abcg1* in macrophages show accumulation of free cholesterol in lysosomes and activation of the NLRP3 inflammasomes as evident from increased IL-1β and IL-18 secretion, as well as caspase-1 cleavage, which is key to inflammasome activation [107]. Deficiency of *NLRP3* or *caspase-1* suppresses atherosclerosis in hyperlipidemic mice with *Abca1* and *Abcg1* deficiency in macrophages [107]. These findings thus show that macrophage cholesterol efflux pathways suppress atherosclerosis by decreasing inflammasome activation [107]. Tangier Disease patients, who carry a homozygous loss-of-function for the *ABCA1* gene, show increased plasma levels of IL-1β and IL-18, the products of inflammasome activation [107], suggesting human relevance. Similarly, decreased cholesterol efflux to HDL due to reduced expression of *ABCA1/ABCG1* in blood monocytes as observed in patients with poorly controlled diabetes mellitus [108, 109], chronic kidney diseases [110], or rheumatoid arthritis [111] may contribute to inflammasome activation and the increased inflammation in these diseases.

6.9 Abca1- and Abcg1-Mediated Cholesterol Efflux Suppress Lupus-like Autoimmunity

Normolipidemic mice deficient in *Abca1* and *Abcg1* show enlarged lymph nodes and at 40 weeks of age develop autoimmune glomerulonephritis, suggestive of systemic lupus erythematosus (SLE). Several immune cell types play a role in SLE, including B-cells, T-cells, macrophages, and dendritic cells (DCs). Mice with deficiency of *Abca1* and *Abcg1* in DCs, but not in T-cells or macrophages, show a similar autoimmune phenotype compared to *Abca1*$^{-/-}$*Abcg1*$^{-/-}$ mice [112]. DCs present antigens to T-cells, leading to their activation. However, DC *Abca1/Abcg1* deficiency in vitro or in vivo does not affect antigen presentation [112]. Instead, *Abca1/Abcg1* deficiency enhances secretion of pro-inflammatory cytokines from DCs, including IL-1β and IL-18 [112], accompanied by cleavage of caspase-1, a hallmark of inflammasome activation [112]. Presumably, TLR4 signaling and cholesterol accumulation act as signals for NLRP3 inflammasome activation in DC *Abca1/Abcg1* deficiency. Downstream of NLRP3 inflammasome activation in *Abca1/Abcg1* deficient DCs, T helper 1 cells, and T helper 17 cells is expanded [112], contributing to the autoimmune phenotype. Interestingly, *NLRP3* or *IL-18* polymorphisms have been associated with increased SLE risk in humans [113]. Deficiency of the *NLRP3* inflammasome diminishes the autoimmune phenotype in mice with DC *Abca1/Abcg1* deficiency and decreases the expansion of the T helper 1 cell population [112], suggesting a major role of inflammasome activation in autoimmunity when cholesterol efflux from dendritic cells is impaired.

Together, cholesterol efflux pathways exert anti-inflammatory effects by suppressing the activation of the NLRP3 inflammasome, with downstream effects on atherosclerosis and a lupus-like autoimmune phenotype.

6.9.1 ABCG4 in Platelet Biogenesis and Atherothrombosis

While ABCA1 and ABCG1 are involved in regulation of HSPC proliferation and HSPC mobilization from the bone marrow and extramedullary hematopoiesis in the spleen, ABCG4 is highly expressed in megakaryocyte progenitors (MkP), a type of progenitor cell in megakaryocyte/platelet lineage, but not in mature platelets [114]. Interestingly, little ABCA1 or ABCG1 is expressed in MkPs. ABCG4 promotes cholesterol efflux to HDL, and ABCG4-deficient MkPs are defective in cholesterol efflux to HDL, in association with free cholesterol accumulation, particularly in the plasma membrane [114]. Hematopoietic ABCG4 deficiency promotes atherosclerosis and accelerates arterial thrombosis in hypercholesterolemic *Ldlr*$^{-/-}$ mice [114]. Mechanistically, hematopoietic ABCG4 deficiency increased platelet counts, reticulated platelets, platelet/leukocyte complexes, and platelet-derived microparticles [114], all with proven pro-atherosclerotic and pro-thrombotic properties. *Abcg4*$^{-/-}$ MkPs show increased proliferation in response to thrombopoietin (TPO), the most important growth factor-regulating megakaryocyte/platelet lineage development in vivo, in association with increased cell surface levels of c-MPL, the TPO receptor [114]. Hematopoietic ABCG4 deficient mice display increased numbers of megakaryocytes in the bone marrow and spleen. The increased cell surface c-MPL levels in *Abcg4*$^{-/-}$ MkPs result from disruption of the negative feedback regulation of c-MPL in response to TPO and involve a defective activation of Lyn kinase and c-CBL E3 ligase. Lyn kinase is a palmitoylated membrane protein, and membrane association regulates Lyn kinase activation. Lyn kinase seems to act as a membrane cholesterol sensor. Increased plasma membrane cholesterol content in *Abcg4*$^{-/-}$ MkPs may increase Lyn kinase association with the membrane and decrease its tyrosine kinase activity in response to TPO, causing defective phosphorylation of c-CBL. This disrupts the negative feedback regulation of c-MPL, decreasing its turnover, increasing its expression at cell surface, increasing TPO/c-MPL signaling, and leading to increased megakaryocyte and platelet production [114]. These studies link increased platelet production, initiated from its lineage progenitor cells, to accelerated atherosclerosis and arterial thrombosis.

6.10 Summary

Studies in the past two decades have identified ABCA1, ABCG1, and ABCG4 as key transporters that regulate cholesterol efflux from various types of cells in vivo. Functional disruption of these transporters leads to dysregulated cholesterol homeostasis in cells and in the whole body. Genetic studies, mostly from animal models, indicate a role of these transporters in modulation of atherogenesis, in part via regulation of the production and function of monocytes, macrophages, dendritic cells, neutrophils, and platelets, and suggest translational relevance. HDL and apoA-I are the major lipoprotein components that promote cholesterol efflux from these transporters. Large cohort studies reproducibly show a strong and inverse relationship between HDL-C levels and the risk of incident

CVD independent of other lipids [115, 116]. In contrast, human genome-wide association studies have called into question the causal relationship between the SNPs of genes associated with HDL metabolism and CVD risk [117], and the clinical trials so far largely provide disappointing results in efforts to raise HDL-C as therapeutic intervention for CVD [118]. These results have contributed to the emerging concept that HDL functionality rather than HDL-C is the more relevant determinant in reducing the risk of CVD. Consistently, using cholesterol efflux function assays several studies demonstrate that cholesterol efflux capacity of human plasma or serum devoid of apoB containing lipoproteins is inversely associated with incident CVD events, often independent of circulating HDL-C or apoA-I levels [119–122]. In this regard, cholesterol efflux pathways could be therapeutically targeted, and one approach is to upregulate ABCA1 and ABCG1 expression by LXR activators, provided that their adverse effects on liver triglyceride metabolism could be circumvented. The other approach could include infusion of reconstituted HDL particles. This approach has shown anti-atherogenic effects in animal models [123, 124] and, in several small scale human trials, demonstrated acceptable safety and some promising effects. More definitive answers may come from ongoing large-scale clinical trials such as ApoA-I Event Reducing in Ischemic Syndromes II (AEGIS-II) trial.

References

1. Bloch K (1965) The biological synthesis of cholesterol. Science 150(3692):19–28
2. Lingwood D, Simons K (2010) Lipid rafts as a membrane-organizing principle. Science 327 (5961):46–50
3. Ikonen E (2008) Cellular cholesterol trafficking and compartmentalization. Nat Rev Mol Cell Biol 9 (2):125–138
4. Riscal R, Skuli N, Simon MC (2019) Even cancer cells watch their cholesterol! Mol Cell 76(2):220–231
5. Xiao X, Tang JJ, Peng C et al (2017) Cholesterol modification of smoothened is required for hedgehog signaling. Mol Cell 66(1):154–62 e10
6. Porter JA, Young KE, Beachy PA (1996) Cholesterol modification of hedgehog signaling proteins in animal development. Science 274(5285):255–259
7. Goldstein JL, Brown MS (2015) A century of cholesterol and coronaries: from plaques to genes to statins. Cell 161(1):161–172
8. Chang TY, Chang CC, Ohgami N, Yamauchi Y (2006) Cholesterol sensing, trafficking, and esterification. Annu Rev Cell Dev Biol 22:129–157
9. Westerterp M, Bochem AE, Yvan-Charvet L, Murphy AJ, Wang N, Tall AR (2014) ATP-binding cassette transporters, atherosclerosis, and inflammation. Circ Res 114(1):157–170
10. Glomset JA (1968) The plasma lecithins: cholesterol acyltransferase reaction. J Lipid Res 9(2):155–167
11. Oram JF, Albers JJ, Cheung MC, Bierman EL (1981) The effects of subfractions of high density lipoprotein on cholesterol efflux from cultured fibroblasts. Regulation of low density lipoprotein receptor activity. J Biol Chem 256(16):8348–8356
12. Biesbroeck R, Oram JF, Albers JJ, Bierman EL (1983) Specific high-affinity binding of high density lipoproteins to cultured human skin fibroblasts and arterial smooth muscle cells. J Clin Invest 71 (3):525–539
13. Oram JF, Brinton EA, Bierman EL (1983) Regulation of high density lipoprotein receptor activity in cultured human skin fibroblasts and human arterial smooth muscle cells. J Clin Invest 72(5):1611–1621
14. Slotte JP, Oram JF, Bierman EL (1987) Binding of high density lipoproteins to cell receptors promotes translocation of cholesterol from intracellular membranes to the cell surface. J Biol Chem 262 (27):12904–12907
15. Francis GA, Knopp RH, Oram JF (1995) Defective removal of cellular cholesterol and phospholipids by apolipoprotein A-I in Tangier Disease. J Clin Invest 96(1):78–87
16. Hooper AJ, Hegele RA, Burnett JR (2020) Tangier disease: update for 2020. Curr Opin Lipidol 31 (2):80–84
17. Bodzioch M, Orso E, Klucken J et al (1999) The gene encoding ATP-binding cassette transporter 1 is mutated in Tangier disease. Nat Genet 22(4):347–351
18. Rust S, Rosier M, Funke H et al (1999) Tangier disease is caused by mutations in the gene encoding ATP-binding cassette transporter 1. Nat Genet 22 (4):352–355
19. Brooks-Wilson A, Marcil M, Clee SM et al (1999) Mutations in ABC1 in Tangier disease and familial high-density lipoprotein deficiency. Nat Genet 22 (4):336–345
20. Wang N, Silver DL, Costet P, Tall AR (2000) Specific binding of ApoA-I, enhanced cholesterol efflux, and altered plasma membrane morphology in cells expressing ABC1. J Biol Chem 275 (42):33053–33058
21. Nagata KO, Nakada C, Kasai RS, Kusumi A, Ueda K (2013) ABCA1 dimer-monomer interconversion

during HDL generation revealed by single-molecule imaging. Proc Natl Acad Sci U S A 110 (13):5034–5039
22. Berge KE, Tian H, Graf GA et al (2000) Accumulation of dietary cholesterol in sitosterolemia caused by mutations in adjacent ABC transporters. Science 290 (5497):1771–1775
23. Wang N, Silver DL, Thiele C, Tall AR (2001) ATP-binding cassette transporter A1 (ABCA1) functions as a cholesterol efflux regulatory protein. J Biol Chem 276(26):23742–23747
24. Wang N, Lan D, Chen W, Matsuura F, Tall AR (2004) ATP-binding cassette transporters G1 and G4 mediate cellular cholesterol efflux to high-density lipoproteins. Proc Natl Acad Sci U S A 101 (26):9774–9779
25. Quazi F, Molday RS (2013) Differential phospholipid substrates and directional transport by ATP-binding cassette proteins ABCA1, ABCA7, and ABCA4 and disease-causing mutants. J Biol Chem 288(48):34414–34426
26. Remaley AT, Schumacher UK, Stonik JA, Farsi BD, Nazih H, Brewer HB Jr (1997) Decreased reverse cholesterol transport from Tangier disease fibroblasts. Acceptor specificity and effect of brefeldin on lipid efflux. Arterioscler Thromb Vasc Biol 17 (9):1813–1821
27. Ishigami M, Ogasawara F, Nagao K et al (2018) Temporary sequestration of cholesterol and phosphatidylcholine within extracellular domains of ABCA1 during nascent HDL generation. Sci Rep 8(1):6170
28. Kawanobe T, Shiranaga N, Kioka N, Kimura Y, Ueda K (2019) Apolipoprotein A-I directly interacts with extracellular domain 1 of human ABCA1. Biosci Biotechnol Biochem 83(3):490–497
29. Vedhachalam C, Ghering AB, Davidson WS, Lund-Katz S, Rothblat GH, Phillips MC (2007) ABCA1-induced cell surface binding sites for ApoA-I. Arterioscler Thromb Vasc Biol 27(7):1603–1609
30. Phillips MC (2014) Molecular mechanisms of cellular cholesterol efflux. J Biol Chem 289 (35):24020–24029
31. Qian H, Zhao X, Cao P, Lei J, Yan N, Gong X (2017) Structure of the human lipid exporter ABCA1. Cell 169(7):1228–39 e10
32. Phillips MC (2018) Is ABCA1 a lipid transfer protein? J Lipid Res 59(5):749–763
33. Gulshan K, Brubaker G, Conger H et al (2016) PI (4,5)P2 is translocated by ABCA1 to the cell surface where it mediates apolipoprotein A1 binding and nascent HDL assembly. Circ Res 119(7):827–838
34. Smith JD, Le Goff W, Settle M et al (2004) ABCA1 mediates concurrent cholesterol and phospholipid efflux to apolipoprotein A-I. J Lipid Res 45 (4):635–644
35. Wang N, Lan D, Gerbod-Giannone M et al (2003) ATP-binding cassette transporter A7 (ABCA7) binds apolipoprotein A-I and mediates cellular phospholipid but not cholesterol efflux. J Biol Chem 278(44):42906–42912
36. Fitzgerald ML, Morris AL, Chroni A, Mendez AJ, Zannis VI, Freeman MW (2004) ABCA1 and amphipathic apolipoproteins form high-affinity molecular complexes required for cholesterol efflux. J Lipid Res 45(2):287–294
37. Choi HY, Karten B, Chan T et al (2003) Impaired ABCA1-dependent lipid efflux and hypoalphalipoproteinemia in human Niemann-Pick type C disease. J Biol Chem 278(35):32569–32577
38. Chen W, Sun Y, Welch C et al (2001) Preferential ATP-binding cassette transporter A1-mediated cholesterol efflux from late endosomes/lysosomes. J Biol Chem 276(47):43564–43569
39. Wang N, Ranalletta M, Matsuura F, Peng F, Tall AR (2006) LXR-induced redistribution of ABCG1 to plasma membrane in macrophages enhances cholesterol mass efflux to HDL. Arterioscler Thromb Vasc Biol 26(6):1310–1316
40. Gu HM, Wang F, Alabi A, Deng S, Qin S, Zhang DW (2016) Identification of an amino acid residue critical for plasma membrane localization of ATP-binding cassette transporter G1 – brief report. Arterioscler Thromb Vasc Biol 36(2):253–255
41. Tarling EJ, Edwards PA (2011) ATP binding cassette transporter G1 (ABCG1) is an intracellular sterol transporter. Proc Natl Acad Sci U S A 108 (49):19719–19724
42. Kobayashi A, Takanezawa Y, Hirata T et al (2006) Efflux of sphingomyelin, cholesterol, and phosphatidylcholine by ABCG1. J Lipid Res 47(8):1791–1802
43. Wang B, Tontonoz P (2018) Liver X receptors in lipid signalling and membrane homeostasis. Nat Rev Endocrinol 14(8):452–463
44. Costet P, Lalanne F, Gerbod-Giannone MC et al (2003) Retinoic acid receptor-mediated induction of ABCA1 in macrophages. Mol Cell Biol 23 (21):7756–7766
45. Repa JJ, Turley SD, Lobaccaro JA et al (2000) Regulation of absorption and ABC1-mediated efflux of cholesterol by RXR heterodimers. Science 289 (5484):1524–1529
46. Chinetti G, Lestavel S, Bocher V et al (2001) PPAR-alpha and PPAR-gamma activators induce cholesterol removal from human macrophage foam cells through stimulation of the ABCA1 pathway. Nat Med 7(1):53–58
47. Chawla A, Boisvert WA, Lee CH et al (2001) A PPAR gamma-LXR-ABCA1 pathway in macrophages is involved in cholesterol efflux and atherogenesis. Mol Cell 7(1):161–171
48. Oram JF, Lawn RM, Garvin MR, Wade DP (2000) ABCA1 is the cAMP-inducible apolipoprotein receptor that mediates cholesterol secretion from macrophages. J Biol Chem 275(44):34508–34511
49. Chen M, Li W, Wang N, Zhu Y, Wang X (2007) ROS and NF-kappaB but not LXR mediate IL-1beta signaling for the downregulation of ATP-binding

cassette transporter A1. Am J Physiol Cell Physiol 292(4):C1493–C1501
50. Zhao GJ, Mo ZC, Tang SL et al (2014) Chlamydia pneumoniae negatively regulates ABCA1 expression via TLR2-nuclear factor-kappa B and miR-33 pathways in THP-1 macrophage-derived foam cells. Atherosclerosis 235(2):519–525
51. McGillicuddy FC, de la Llera MM, Hinkle CC et al (2009) Inflammation impairs reverse cholesterol transport in vivo. Circulation 119(8):1135–1145
52. Rayner KJ, Suarez Y, Davalos A et al (2010) MiR-33 contributes to the regulation of cholesterol homeostasis. Science 328(5985):1570–1573
53. Clifton-Bligh P, Nestel PJ, Whyte HM (1972) Tangier disease. Report of a case and studies of lipid metabolism. N Engl J Med 286(11):567–571
54. Timmins JM, Lee JY, Boudyguina E et al (2005) Targeted inactivation of hepatic Abca1 causes profound hypoalphalipoproteinemia and kidney hypercatabolism of apoA-I. J Clin Invest 115 (5):1333–1342
55. Brunham LR, Kruit JK, Iqbal J et al (2006) Intestinal ABCA1 directly contributes to HDL biogenesis in vivo. J Clin Invest 116(4):1052–1062
56. Chung S, Sawyer JK, Gebre AK, Maeda N, Parks JS (2011) Adipose tissue ATP binding cassette transporter A1 contributes to high-density lipoprotein biogenesis in vivo. Circulation 124(15):1663–1672
57. Haghpassand M, Bourassa PA, Francone OL, Aiello RJ (2001) Monocyte/macrophage expression of ABCA1 has minimal contribution to plasma HDL levels. J Clin Invest 108(9):1315–1320
58. van Eck M, Bos IS, Kaminski WE et al (2002) Leukocyte ABCA1 controls susceptibility to atherosclerosis and macrophage recruitment into tissues. Proc Natl Acad Sci U S A 99(9):6298–6303
59. Hobbs HH, Rader DJ (1999) ABC1: connecting yellow tonsils, neuropathy, and very low HDL. J Clin Invest 104(8):1015–1017
60. Kennedy MA, Barrera GC, Nakamura K et al (2005) ABCG1 has a critical role in mediating cholesterol efflux to HDL and preventing cellular lipid accumulation. Cell Metab 1(2):121–131
61. Westerterp M, Murphy AJ, Wang M et al (2013) Deficiency of ATP-binding cassette transporters A1 and G1 in macrophages increases inflammation and accelerates atherosclerosis in mice. Circ Res 112 (11):1456–1465
62. Out R, Jessup W, Le Goff W et al (2008) Coexistence of foam cells and hypocholesterolemia in mice lacking the ABC transporters A1 and G1. Circ Res 102(1):113–120
63. Wang X, Collins HL, Ranalletta M et al (2007) Macrophage ABCA1 and ABCG1, but not SR-BI, promote macrophage reverse cholesterol transport in vivo. J Clin Invest 117(8):2216–2224
64. Naik SU, Wang X, Da Silva JS et al (2006) Pharmacological activation of liver X receptors promotes reverse cholesterol transport in vivo. Circulation 113(1):90–97
65. Zanotti I, Poti F, Pedrelli M et al (2008) The LXR agonist T0901317 promotes the reverse cholesterol transport from macrophages by increasing plasma efflux potential. J Lipid Res 49(5):954–960
66. de Aguiar Vallim TQ, Tarling EJ, Kim T et al (2013) MicroRNA-144 regulates hepatic ATP binding cassette transporter A1 and plasma high-density lipoprotein after activation of the nuclear receptor farnesoid X receptor. Circ Res 112(12):1602–1612
67. Bashore AC, Liu M, Key CC et al (2019) Targeted deletion of hepatocyte Abca1 increases plasma HDL (high-density lipoprotein) reverse cholesterol transport via the LDL (low-density lipoprotein) receptor. Arterioscler Thromb Vasc Biol 39(9):1747–1761
68. Zhang Y, Yin L, Anderson J et al (2010) Identification of novel pathways that control farnesoid X receptor-mediated hypocholesterolemia. J Biol Chem 285(5):3035–3043
69. Tabas I, Glass CK (2013) Anti-inflammatory therapy in chronic disease: challenges and opportunities. Science 339(6116):166–172
70. Moore KJ, Sheedy FJ, Fisher EA (2013) Macrophages in atherosclerosis: a dynamic balance. Nat Rev Immunol 13(10):709–721
71. Back M, Yurdagul A Jr, Tabas I, Oorni K, Kovanen PT (2019) Inflammation and its resolution in atherosclerosis: mediators and therapeutic opportunities. Nat Rev Cardiol 16(7):389–406
72. Hansson GK, Libby P, Tabas I (2015) Inflammation and plaque vulnerability. J Intern Med 278 (5):483–493
73. Yvan-Charvet L, Welch C, Pagler TA et al (2008) Increased inflammatory gene expression in ABC transporter-deficient macrophages: free cholesterol accumulation, increased signaling via toll-like receptors, and neutrophil infiltration of atherosclerotic lesions. Circulation 118(18):1837–1847
74. Westerterp M, Tsuchiya K, Tattersall IW et al (2016) Deficiency of ATP-binding cassette transporters A1 and G1 in endothelial cells accelerates atherosclerosis in mice. Arterioscler Thromb Vasc Biol 36 (7):1328–1337
75. Combadiere C, Potteaux S, Rodero M et al (2008) Combined inhibition of CCL2, CX3CR1, and CCR5 abrogates Ly6C(hi) and Ly6C(lo) monocytosis and almost abolishes atherosclerosis in hypercholesterolemic mice. Circulation 117(13):1649–1657
76. Dansky HM, Barlow CB, Lominska C et al (2001) Adhesion of monocytes to arterial endothelium and initiation of atherosclerosis are critically dependent on vascular cell adhesion molecule-1 gene dosage. Arterioscler Thromb Vasc Biol 21(10):1662–1667
77. Collins RG, Velji R, Guevara NV, Hicks MJ, Chan L, Beaudet AL (2000) P-Selectin or intercellular adhesion molecule (ICAM)-1 deficiency substantially protects against atherosclerosis in apolipoprotein E-deficient mice. J Exp Med 191(1):189–194

78. Goossens P, Gijbels MJ, Zernecke A et al (2010) Myeloid type I interferon signaling promotes atherosclerosis by stimulating macrophage recruitment to lesions. Cell Metab 12(2):142–153
79. Duewell P, Kono H, Rayner KJ et al (2010) NLRP3 inflammasomes are required for atherogenesis and activated by cholesterol crystals. Nature 464 (7293):1357–1361
80. Elhage R, Jawien J, Rudling M et al (2003) Reduced atherosclerosis in interleukin-18 deficient apolipoprotein E-knockout mice. Cardiovasc Res 59 (1):234–240
81. Mallat Z, Corbaz A, Scoazec A et al (2001) Interleukin-18/interleukin-18 binding protein signaling modulates atherosclerotic lesion development and stability. Circ Res 89(7):E41–E45
82. Whitman SC, Ravisankar P, Daugherty A (2002) Interleukin-18 enhances atherosclerosis in apolipoprotein E(−/−) mice through release of interferon-gamma. Circ Res 90(2):E34–E38
83. Ridker PM, Everett BM, Thuren T et al (2017) Antiinflammatory therapy with canakinumab for atherosclerotic disease. N Engl J Med 377 (12):1119–1131
84. Takeuchi O, Hoshino K, Kawai T et al (1999) Differential roles of TLR2 and TLR4 in recognition of gram-negative and gram-positive bacterial cell wall components. Immunity 11(4):443–451
85. Michelsen KS, Wong MH, Shah PK et al (2004) Lack of Toll-like receptor 4 or myeloid differentiation factor 88 reduces atherosclerosis and alters plaque phenotype in mice deficient in apolipoprotein E. Proc Natl Acad Sci U S A 101(29):10679–10684
86. Akira S, Takeda K (2004) Toll-like receptor signalling. Nat Rev Immunol 4(7):499–511
87. Miller YI, Viriyakosol S, Worrall DS, Boullier A, Butler S, Witztum JL (2005) Toll-like receptor 4-dependent and -independent cytokine secretion induced by minimally oxidized low-density lipoprotein in macrophages. Arterioscler Thromb Vasc Biol 25(6):1213–1219
88. Triantafilou M, Miyake K, Golenbock DT, Triantafilou K (2002) Mediators of innate immune recognition of bacteria concentrate in lipid rafts and facilitate lipopolysaccharide-induced cell activation. J Cell Sci 115(Pt 12):2603–2611
89. Yumoto H, Chou HH, Takahashi Y, Davey M, Gibson FC 3rd, Genco CA (2005) Sensitization of human aortic endothelial cells to lipopolysaccharide via regulation of Toll-like receptor 4 by bacterial fimbria-dependent invasion. Infect Immun 73 (12):8050–8059
90. De Caterina R, Libby P, Peng HB et al (1995) Nitric oxide decreases cytokine-induced endothelial activation. Nitric oxide selectively reduces endothelial expression of adhesion molecules and proinflammatory cytokines. J Clin Invest 96 (1):60–68
91. Fotakis P, Kothari V, Thomas DG et al (2019) Anti-inflammatory effects of HDL (high-density lipoprotein) in macrophages predominate over proinflammatory effects in atherosclerotic plaques. Arterioscler Thromb Vasc Biol 39(12):e253–ee72
92. Yvan-Charvet L, Kling J, Pagler T et al (2010) Cholesterol efflux potential and antiinflammatory properties of high-density lipoprotein after treatment with niacin or anacetrapib. Arterioscler Thromb Vasc Biol 30(7):1430–1438
93. Bochem AE, van der Valk FM, Tolani S, Stroes ES, Westerterp M, Tall AR (2015) Increased systemic and plaque inflammation in ABCA1 mutation carriers with attenuation by statins. Arterioscler Thromb Vasc Biol 35(7):1663–1669
94. Coller BS (2005) Leukocytosis and ischemic vascular disease morbidity and mortality: is it time to intervene? Arterioscler Thromb Vasc Biol 25(4):658–670
95. Rieger MA, Hoppe PS, Smejkal BM, Eitelhuber AC, Schroeder T (2009) Hematopoietic cytokines can instruct lineage choice. Science 325(5937):217–218
96. Tsou CL, Peters W, Si Y et al (2007) Critical roles for CCR2 and MCP-3 in monocyte mobilization from bone marrow and recruitment to inflammatory sites. J Clin Invest 117(4):902–909
97. Yvan-Charvet L, Pagler T, Gautier EL et al (2010) ATP-binding cassette transporters and HDL suppress hematopoietic stem cell proliferation. Science 328 (5986):1689–1693
98. Wright DE, Wagers AJ, Gulati AP, Johnson FL, Weissman IL (2001) Physiological migration of hematopoietic stem and progenitor cells. Science 294(5548):1933–1936
99. Westerterp M, Gourion-Arsiquaud S, Murphy AJ et al (2012) Regulation of hematopoietic stem and progenitor cell mobilization by cholesterol efflux pathways. Cell Stem Cell 11(2):195–206
100. Stark MA, Huo Y, Burcin TL, Morris MA, Olson TS, Ley K (2005) Phagocytosis of apoptotic neutrophils regulates granulopoiesis via IL-23 and IL-17. Immunity 22(3):285–294
101. Mendez-Ferrer S, Michurina TV, Ferraro F et al (2010) Mesenchymal and haematopoietic stem cells form a unique bone marrow niche. Nature 466 (7308):829–834
102. Dar A, Goichberg P, Shinder V et al (2005) Chemokine receptor CXCR4-dependent internalization and resecretion of functional chemokine SDF-1 by bone marrow endothelial and stromal cells. Nat Immunol 6 (10):1038–1046
103. Robbins CS, Chudnovskiy A, Rauch PJ et al (2012) Extramedullary hematopoiesis generates Ly-6C (high) monocytes that infiltrate atherosclerotic lesions. Circulation 125(2):364–374
104. Martinon F, Tschopp J (2004) Inflammatory caspases: linking an intracellular innate immune system to autoinflammatory diseases. Cell 117 (5):561–574

105. Sheedy FJ, Grebe A, Rayner KJ et al (2013) CD36 coordinates NLRP3 inflammasome activation by facilitating intracellular nucleation of soluble ligands into particulate ligands in sterile inflammation. Nat Immunol 14(8):812–820
106. Randolph GJ (2014) Mechanisms that regulate macrophage burden in atherosclerosis. Circ Res 114 (11):1757–1771
107. Westerterp M, Fotakis P, Ouimet M et al (2018) Cholesterol efflux pathways suppress inflammasome activation, NETosis and atherogenesis. Circulation 138(9):898–912
108. Mauldin JP, Nagelin MH, Wojcik AJ et al (2008) Reduced expression of ATP-binding cassette transporter G1 increases cholesterol accumulation in macrophages of patients with type 2 diabetes mellitus. Circulation 117(21):2785–2792
109. Patel DC, Albrecht C, Pavitt D et al (2011) Type 2 diabetes is associated with reduced ATP-binding cassette transporter A1 gene expression, protein and function. PLoS One 6(7):e22142
110. Ganda A, Yvan-Charvet L, Zhang Y et al (2017) Plasma metabolite profiles, cellular cholesterol efflux, and non-traditional cardiovascular risk in patients with CKD. J Mol Cell Cardiol 112:114–122
111. Dragoljevic D, Kraakman MJ, Nagareddy PR et al (2018) Defective cholesterol metabolism in haematopoietic stem cells promotes monocyte-driven atherosclerosis in rheumatoid arthritis. Eur Heart J 39 (23):2158–2167
112. Westerterp M, Gautier EL, Ganda A et al (2017) Cholesterol accumulation in dendritic cells links the inflammasome to acquired immunity. Cell Metab 25 (6):1294–304 e6
113. Kahlenberg JM, Kaplan MJ (2014) The inflammasome and lupus: another innate immune mechanism contributing to disease pathogenesis? Curr Opin Rheumatol 26(5):475–481
114. Murphy AJ, Bijl N, Yvan-Charvet L et al (2013) Cholesterol efflux in megakaryocyte progenitors suppresses platelet production and thrombocytosis. Nat Med 19(5):586–594
115. Rosenson RS, Brewer HB Jr, Barter PJ et al (2018) HDL and atherosclerotic cardiovascular disease: genetic insights into complex biology. Nat Rev Cardiol 15(1):9–19
116. Rader DJ (2016) New therapeutic approaches to the treatment of dyslipidemia. Cell Metab 23(3):405–412
117. Voight BF, Peloso GM, Orho-Melander M et al (2012) Plasma HDL cholesterol and risk of myocardial infarction: a mendelian randomisation study. Lancet 380(9841):572–580
118. Libby P, Buring JE, Badimon L et al (2019) Atherosclerosis. Nat Rev Dis Primers 5(1):56
119. Khera AV, Cuchel M, de la Llera-Moya M et al (2011) Cholesterol efflux capacity, high-density lipoprotein function, and atherosclerosis. N Engl J Med 364(2):127–135
120. Soria-Florido MT, Castaner O, Lassale C et al (2020) Dysfunctional high-density lipoproteins are associated with a greater incidence of acute coronary syndrome in a population at high cardiovascular risk: a nested case-control study. Circulation 141 (6):444–453
121. Ebtehaj S, Gruppen EG, Bakker SJL, Dullaart RPF, Tietge UJF (2019) HDL (high-density lipoprotein) cholesterol efflux capacity is associated with incident cardiovascular disease in the general population. Arterioscler Thromb Vasc Biol 39(9):1874–1883
122. Shea S, Stein JH, Jorgensen NW et al (2019) Cholesterol mass efflux capacity, incident cardiovascular disease, and progression of carotid plaque. Arterioscler Thromb Vasc Biol 39(1):89–96
123. Murphy AJ, Akhtari M, Tolani S et al (2011) ApoE regulates hematopoietic stem cell proliferation, monocytosis, and monocyte accumulation in atherosclerotic lesions in mice. J Clin Invest 121 (10):4138–4149
124. Murphy AJ, Funt S, Gorman D, Tall AR, Wang N (2013) Pegylation of high-density lipoprotein decreases plasma clearance and enhances antiatherogenic activity. Circ Res 113(1):e1–e9

Apolipoprotein M: Research Progress and Clinical Perspective

7

Guanghua Luo and Ning Xu

Abstract

Apolipoprotein M (apoM) was first identified and characterized to the apolipoprotein family in 1999. Human apoM gene is located in a highly conserved segment in the major histocompatibility complex (MHC) class III locus on chromosome 6 and codes for an about 23 kDa protein that structurally belongs to the lipocalin superfamily. ApoM is selectively expressed in hepatocytes and in the tubular epithelium of kidney. In human plasma, apoM is mainly confined to the high-density lipoprotein (HDL) particles, but it may also occur in other lipoprotein classes, such as in the triglyceride-rich particles after fat intake. It has been demonstrated that apoM is critical for the formation of HDL, notably pre-beta HDL1. The antiatherogenic function of HDL is well established, and its ability to promote cholesterol efflux from foam cells in the atherosclerotic lesions is generally regarded as one of the key mechanisms behind this protective function. However, HDL could also display a variety of properties that may affect the complex atherosclerotic processes by other mechanisms, thus being involved in processes related to antioxidant defense, immune system, and systemic effects in septicemia, which may be partly contributed via its apolipoproteins and/or phospholipids. Moreover, it has been demonstrated that apoM functions as a natural carrier of sphingosin-1-phosphate (S1P) in vivo which may be related to its antiatherosclerotic and protective effects on endothelial cell barrier and anti-inflammatory properties. These may also provide a link between the diverse effects of HDL.

Keywords

Apolipoprotein M · High-density lipoprotein · Sphingosin-1-phosphate · Atherosclerosis · Endothelial cell barrier · Inflammatory

7.1 Introduction

The antiatherogenic function of high-density lipoproteins (HDL) is well established, and its ability to promote cholesterol efflux from foam cells in the atherosclerotic lesions is generally regarded as one of the key mechanisms behind this protective function [5, 63]. However, HDL could also display a variety of properties that may affect the complex atherosclerotic processes by other mechanisms, thus being involved in processes related to

G. Luo
Comprehensive Laboratory, The Third Affiliated Hospital of Soochow University, Changzhou, People's Republic of China

N. Xu (✉)
Division of Clinical Chemistry and Pharmacology, Department of Laboratory Medicine, Lunds University, Lund, Sweden
e-mail: ning.xu@med.lu.se

X.-C. Jiang (ed.), *Lipid Transfer in Lipoprotein Metabolism and Cardiovascular Disease*, Advances in Experimental Medicine and Biology 1276, https://doi.org/10.1007/978-981-15-6082-8_7

antioxidant defense, immune system, and systemic effects in septicemia, which may be partly contributed via its apolipoproteins and/or phospholipids [4, 11, 22, 36, 65, 83]. Human apolipoprotein M (apoM) is the latest HDL apolipoprotein [85] and functions as carrier of sphingosin-1-phosphate (S1P) in vivo [3, 17, 20, 30, 43, 62, 69] and has been demonstrated having antiatherosclerotic [3, 8, 9, 18, 44] and protective effects on endothelial cell barrier [79] and anti-inflammatory properties [6, 17, 29, 30, 43, 76, 78, 81, 103], which may provide a link between these diverse effects. ApoM was first identified and characterized to the apolipoprotein family in 1999 [85]. The human apoM gene is located in a highly conserved segment in the major histocompatibility complex (MHC) class III locus on chromosome 6 and codes for an about 23 kDa protein that structurally belongs to the lipocalin superfamily [24, 85, 86]. In human plasma, apoM is mainly confined to the HDL particles, but it may also occur in other lipoprotein classes, such as in the triglyceride-rich particles after fat intake [85]. ApoM is selectively expressed in hepatocytes and in the tubular epithelium of kidney [94]. It has been demonstrated that apoM is essential for the formation of HDL, especially pre-beta HDL1 [84]. Moreover, in transgenic mouse models, apoM has a strong protective effect against atherosclerosis [84]. And it has been demonstrated that the apoM-S1P axis could be related to lipid metabolism and remodeling endothelial function [20], which may act as the key physiological function of apoM in vivo. More recently, it has been reported that overexpression of apoM could reduce the degree of nephropathy in mice model with IgA nephropathy [48]. And hepatic apoM expression may involve in the non-alcoholic fatty liver diseases [61], and apoM deficiency could cause an autophagy dysregulation in the liver [101]. In this review we summarized research progress and clinical perspective of apoM.

7.2 Identification and Cloning of Human ApoM

Human apoM was first identified and isolated by Xu and Dahlback in 1999, from triglyceride-rich lipoproteins (TGRLP) of the postprandial plasma [85]. When they performed SDS-PAGE of delipidated human TGRLP and sequenced protein bands ranging from 6 kDa to 45 kDa, the N-terminal sequence of one of the sequences was characterized as MFHQIWAALLYFYGI. Except for human expressed sequence tags (EST) showing similarities to this N-terminal amino acid sequence, no homologous proteins were identified in public databases. Based on above sequences, a full-length cDNA of the novel protein containing 188 amino acids was obtained [85] (Fig. 7.1). According to the sequence of the protein, rabbit antibodies against five mixed synthetic peptides were raised, and the distribution of the protein among various lipoprotein subclasses was analyzed by the western blotting using the pooled antisera. Under reducing conditions, a 26 kDa band was particularly abundant in HDL, but minor amounts were also observed in low-density lipoprotein (LDL) and TGRLP (Figs. 7.2 and 7.3). In addition, a less pronounced band (approximately 23 kDa), corresponding in size to a nonglycosylated variant of the protein [85], was also observed. As the protein is extensively associated with lipoproteins in plasma, it fulfills the criteria for classification as an apolipoprotein. This novel protein was therefore named as apoM, following the last previously identified apolipoprotein, apoL [25]. The human apoM gene is located in the MHC class III locus (chromosome 6, p21.31) (Fig. 7.4) and contains six exons [85]. Both in mouse and human, the apoM gene is predicted to contain six exons enclosed in a 1.6 kb genomic region, which is consistent with the results of Southern blotting. Southern blot analysis of different species gave positive signals in all mammalian genomes, but not in DNA from chicken and yeast (Fig. 7.5) [85]. The human apoM cDNA (734 base pairs) encodes for a 188 amino acid residue protein. The 5'-untranslated region was 33 nucleotides and the 3'-untranslated region 120 nucleotides, excluding the poly(A) tail. The calculated molecular mass of the protein is 21 256. The amino acid sequences of human and mouse apoM are 79 % identical (human and rat apoM: 82 %) (Fig. 7.1a). In man, mouse, and rat, the apoM gene sequences predict the presence in the protein of a signal peptide sequence. Generally, such sequences are split

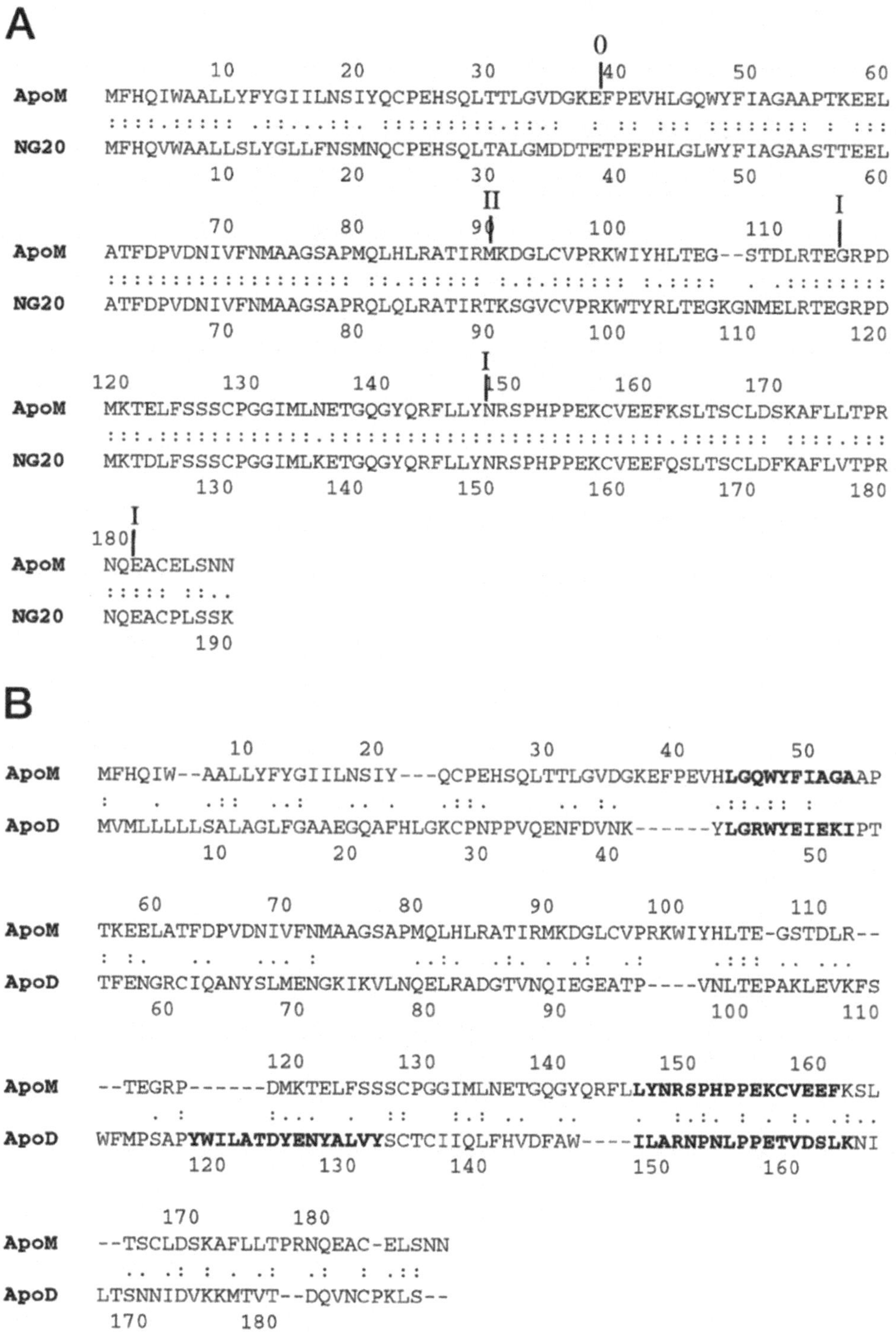

Fig. 7.1 *Amino acid sequence of human apoM and alignment with sequences of mouse apoM and human apoD.* (**a**) The amino acid sequence of human apoM compared to the mouse apoM (NG20). The positions of the five introns are indicated by *vertical lines*, and the intron types are given by *roman numbers*. (**b**) The human apoM sequence is aligned with that of human apoD using CLUSTAL W (1.74) multiple sequence alignment. The *bold letters* identify the two areas in apoM and the three in apoD that have typical lipocalin motifs

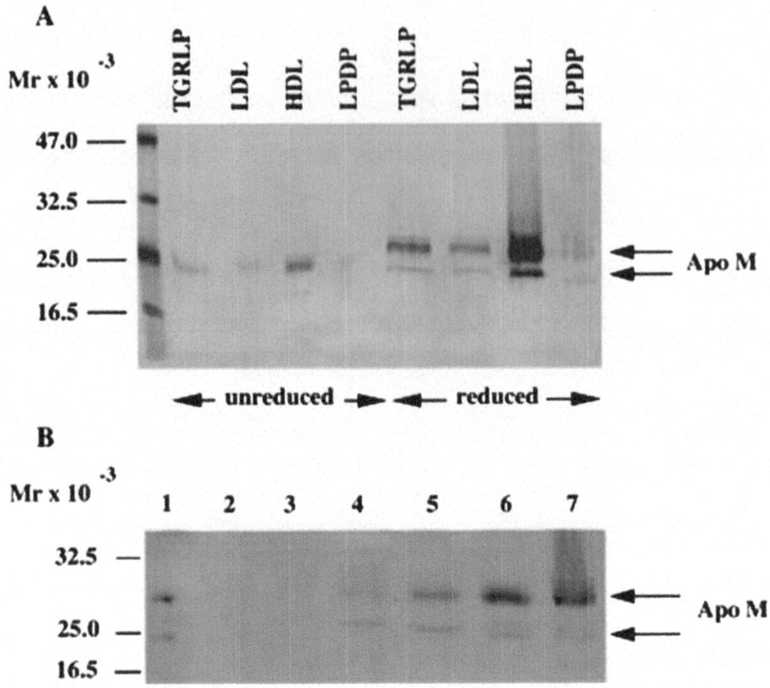

Fig. 7.2 *ApoM in different lipoprotein subclasses and in plasma.* (**a**) Apolipoproteins from TGRLP, LDL, HDL (5 mg in each *lane*), and LPDP (5 mg of plasma proteins) were applied to 4–15% gradient SDS-PAGE under reducing and nonreducing conditions and detected by Western blotting with pooled anti-peptide apoM antisera. (**b**) Increasing amounts of normal plasma proteins were applied to 8–18% gradient SDS-PAGE and analyzed by Western blotting using the pooled anti-peptide antisera. *Lanes 2–7* contain 0.75, 1.25, 2.5, 5, 10, and 20 mg of plasma proteins, respectively. In parallel, LPDP (10 mg of protein) was analyzed

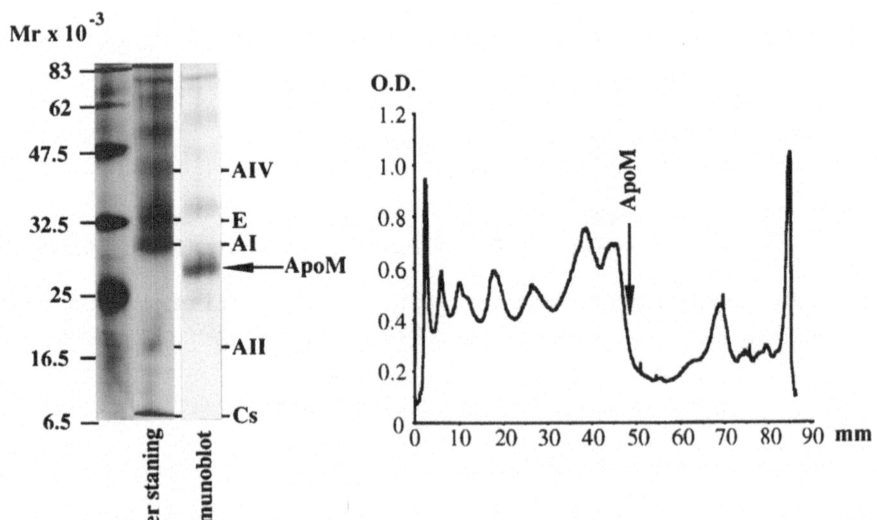

Fig. 7.3 *ApoM is a minor component of HDL.* To elucidate the relative amount of apoM in HDL as compared to the other lipoproteins, 10 mg of delipidated HDL was applied in duplicate to 10%-PAGE run in the presence of SDS under reducing conditions

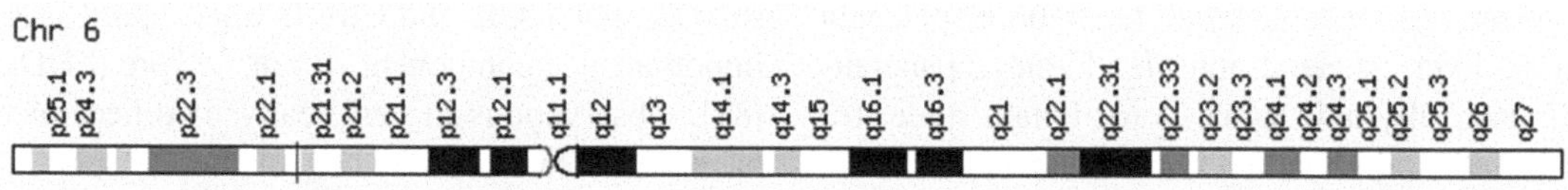

Fig. 7.4 *ApoM Gene in genomic location:* Bands according to Ensembl, locations according to GeneLoc (and/or Entrez Gene and/or Ensembl if different). https://www.genecards.org/cgi-bin/carddisp.pl?gene=APOM

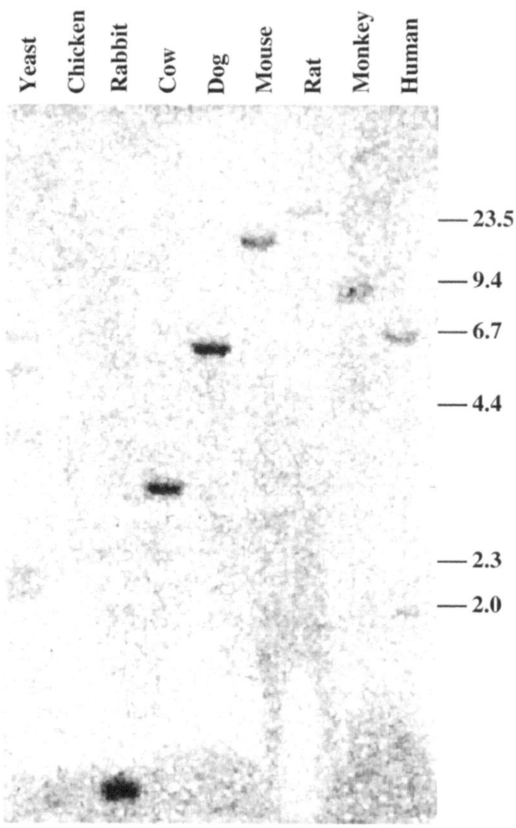

Fig. 7.5 *Southern blot analysis of DNA from various species. The Zoo blot contained nine different species.* The positions and sizes of marker DNA (*Hin*dIII-digested lDNA) are indicated at the right

from extracellular proteins prior to secretion from the cell of origin. ApoM, however, retains this signal peptide sequence (about 20 amino acids) in the mature protein, as the signal peptide sequence is not followed by a signal peptidase cleavage site [85]. The amino acid sequence of apoM contains six cysteines, which may be involved in the formation of three disulfide bridges. There is one potential site for N-linked glycosylation at Asn135 (Asn-Glu-Thr), whereas Asn148 (Asn-Arg-Ser-Pro) is less likely to be glycosylated because Pro-151 follows Ser-150. Through sensitive sequence searches, it was proposed that apoM is related, like apoD, to the lipocalin protein superfamily [85]. Lipocalins are involved in numerous biological functions: some are enzymatically active, and others bind signal substances such as pheromones, while still others are have regulatory functions in cellular metabolism and immunological responses [15]. Later, it has been demonstrated that apoM's hydrophobic binding pocket could predominantly carry sphingosine-1-phosphate (S1P) in plasma [20]. As mentioned above, apoM retains an uncleaved N-terminal signal peptide; this hydrophobic sequence most probably serves to anchor the molecule into the single layer of amphipathic lipids on the surface of the lipoprotein particle [85]. The predominant phospholipid in HDL is phosphatidylcholine, which has a positively charged choline group exposed on the surface of the lipoprotein particle. Several isoforms of apoM have been identified in plasma. These most probably represent various degrees of glycosylation (there is a glycosylation site at residue Asn135), sialylation, or phosphorylation. Karlsson et al., using two-dimensional gel electrophoresis and mass spectrometry, demonstrated that two isoforms of apoM are present in human HDL and three isoforms in LDL particles, probably due to differences in glycosylation or sialylation [40, 41]. However, there is only one form of apoM found in VLDL [56].

7.3 Cellular Expression and Plasma Distribution of ApoM

Northern blot analyses of multiple tissues (including spleen, thymus, prostate, testis, ovary, small intestine, colon, leukocytes, heart, brain, placenta, lung, liver, skeletal muscle, kidney, pancreas, stomach, thyroid, spinal cord, lymph node, trachea, adrenal gland, and bone marrow) indicated that apoM was expressed mainly in the liver and kidney [95]. Human tissue expression array studies showed that apoM is exclusively expressed in the liver and in the kidney (Fig. 7.6), while small amounts were also found in fetal liver and kidney [94]. To elucidate whether and when apoM is expressed, Zhang et al. investigated apoM expression patterns during mouse and human embryogenesis [95]. ApoM transcripts were detectable in mouse embryos from day 7.5 to day 18.5, and apoM was expressed at low levels at day 7.5 and then increased up to day 18.5 (i.e., almost to parturition; Fig. 7.7). ApoM-positive cells appeared mainly in the livers of day 12 embryos as detected by in situ hybridization. In day 15 embryos, apoM was expressed in both the liver and kidney. During human embryogenesis, apoM was strongly expressed in the livers of 3- to 5-month-old embryos and continued to be so throughout embryogenesis. In the kidney, apoM expression was highest in 5- to 9-month-old embryos (Fig. 7.8). The plasma concentration of apoM in man has been reported at 20–150 mg/L; although these estimates are uncertain for methodological reasons, it has been estimated that apoM constitutes a minor proportion of HDL apolipoproteins in man (less than 5% of the concentrations of apoAI) [34, 70]. Like most other small apolipoproteins, apoM can transfer between lipoprotein particles [85]. During the postprandial phase, for example, the concentrations of apoM in TGRLP increase, probably as a result of transfer from HDL particles [85]. Although apoM originally identified in TGRLP, human apoM is mainly transported with HDL. Using monoclonal antibodies and immunoaffinity chromatography, Christoffersen et al. [19] demonstrated that about 5% of all HDL particles in human plasma contain apoM; these were defined as HDL-apoM$^+$. Such apoM$^+$ particles contained more cholesterol than apoM$^-$ particles. ApoM$^+$ HDL is quite heterogeneous in protein composition; besides apoAI and AII, it also contains several other apolipoproteins such as apoCI, CII, and CIII. In mice, apoM is an important component of pre-β-HDL [84]. Observations in genetically modified mice have added to understanding of the transport of apoM in plasma. ApoM is thus associated with HDL-sized particles in wild-type and apoAI-deficient mice, whereas in LDL receptor-deficient (hypercholesterolemia) mice, it is found in HDL- and LDL-sized particles [27]. In apoE-deficient mice fed a high-fat, high-cholesterol diet, apoM is found mainly in VLDL-sized particles [27]. ApoM thus associates primarily with HDL under normal conditions, but it may also occur in pathologically increased lipoprotein fractions regardless of the nature of the lipoprotein particles. To investigate the impact on plasma lipoprotein metabolism of primary derangements in apoM processing, Wolfrum et al. [84] modulated hepatic apoM expression in mice through the use of apoM-silencing RNA or apoM adenovirus. Decreased apoM expression was accompanied by the accumulation of large HDL1 particles in plasma, while pre-β-HDL disappeared. In analogy, HNF-1-α knockout mice exhibited a lipoprotein pattern similar to that induced by apoM-silencing RNA; in this model, the aberrations in HDL fractions could be reversed by injection of apoM adenovirus [84]. Taken together, these observations demonstrate that apoM is critically involved in the formation of HDL, notably pre-β-HDL1.

7.4 Regulation of ApoM Expression

The hepatic expression of apoM and its concentration in plasma are dependent upon a number of nuclear transcription factors and also subject to hormonal and metabolic regulation. Several different regulatory pathways are involved in the regulation of apoM. Also, it appears that alterations in apoM metabolism are linked to clinically important entities such as inflammation, diabetes, and obesity. Hepatocyte nuclear factor 1α (HNF-1α) belongs to the helix-loop-helix homeodomain transcription factor family and was first identified by its interaction with regulatory sequences of liver-specific gene promoters.

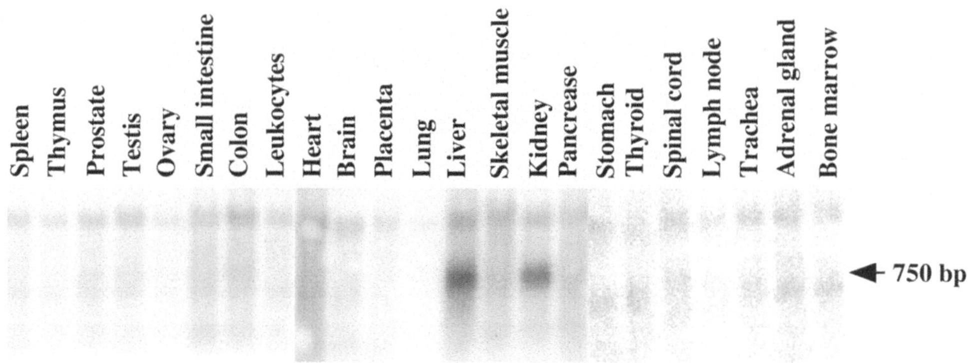

Fig. 7.6 *Northern blot analysis of apoM.* The human multiple tissue Northern blots were probed at high stringency with a radiolabeled full-length cDNA of apoM. The positions and sizes of apoM mRNA are indicated at the right

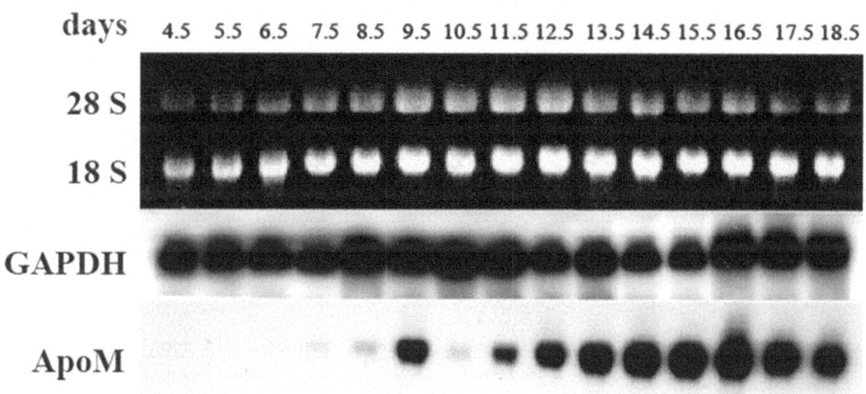

Fig. 7.7 *Onset of apoM expression during mouse embryogenesis.* Northern blots of total RNA of 4.5- to 18.5-day-old mouse embryos was hybridized to a randomly primed cDNA probe of apoM

It has important roles in development, cell differentiation, and metabolism, primarily in the liver, intestine, kidney, and exocrine pancreas [21, 37, 57]. HNF-1α protein can bind to the HNF-1 binding site of apoM promoter in vitro [67], while HNF-1α in vivo is a potent activator of apoM gene promoter [67]. Mutant HNF-1α$^{-/-}$ mice thus completely lack expression of apoM in the liver and kidney, and apoM is absent from plasma. In heterozygous HNF-1α+/- mice, serum levels of apoM are reduced by 50% in relation to wild-type animals. The HNF-binding site of the apoM promoter, which is highly preserved, has been identified, and specific mutations to this binding site abolished transcriptional activation of the apoM gene [67]. As described in more detail below, mutations in the HNF-1α gene are closely related to diabetes, notably to the MODY3 (maturity onset diabetes in the young) type. These patients have low plasma concentrations of apoM [67], and serum apoM levels can well distinguish HNF-1α-MODY and type 1 diabetes [59]. Peroxisome proliferator-activated receptors (PPARs) are nuclear transcription factors that regulate lipid and lipoprotein metabolism, glucose homeostasis, and the inflammatory response [31, 35, 73]. The PPAR family consists of three proteins – α, β/δ, and γ – that all display tissue-specific expression patterns reflecting their biological functions. PPARα is principally expressed in tissues exhibiting high rates of beta-oxidation, such as the liver, kidney, heart, and muscle, while PPARγ is expressed at high levels in adipose tissue. PPARβ/δ, however, is ubiquitously expressed [73]. It has been reported that the PPARβ/δ antagonist, GSK3787, could

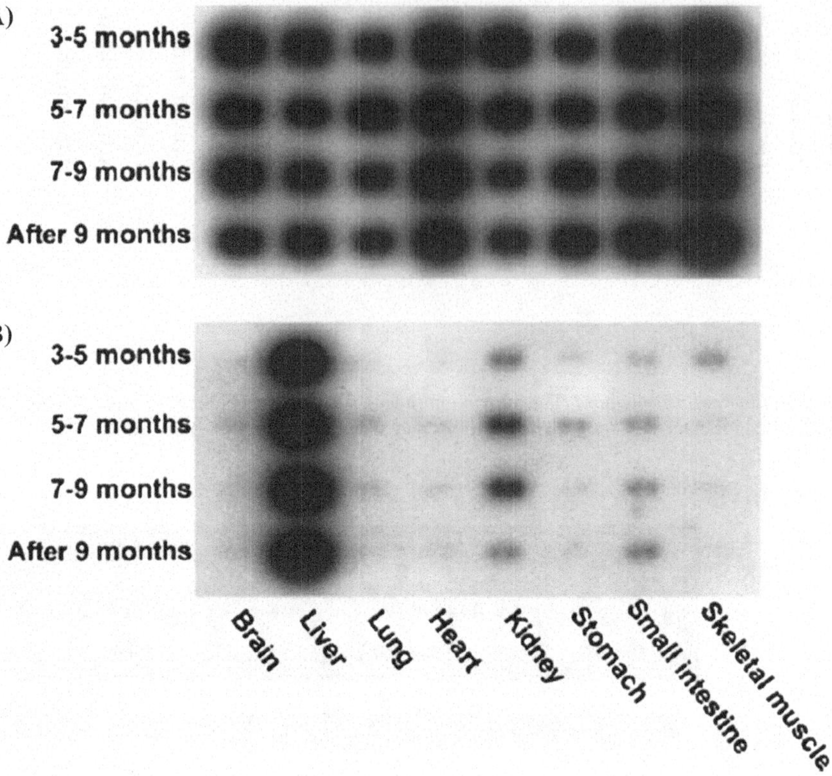

Fig. 7.8 *ApoM expression during human fetal tissues.* The mRNA arrays containing a series of fetal tissues from different embryonal stages were hybridized to a randomly primed cDNA apoM probe (panel B). Embryonal stages are indicated as months. Human GAPDH cDNA probe (panel A) was used as control

completely reverse the downregulation of apoM expression induced by palmitic acid, indicating that palmitic acid-induced downregulation of apoM expression is mediated by the PPARβ/δ pathway [55]. The molecular actions of fibrates and statins, two of the conventional hypolipidemic agents, involve the functions of hepatic PPAR(α) [33]. Exposure of HepG2 and Hep3B cells to the PPARα activator, gemfibrozil, resulted in a twofold induction of apoAI mRNA and a one-third reduction in apoB mRNA but had no significant effect on apoE mRNA levels [89]. Ciprofibrate treatment decreases hepatic apoB mRNA editing and alters the pattern of hepatic lipoprotein secretion [89]. Linden et al. reported that the PPARα agonist WY14643 could decrease the secretion of apoB-100 in VLDL, but not that of apoB-48, and decreased triglyceride biosynthesis and secretion from primary rat hepatocytes [51]. However, there is no data could be found that PPARα influences expression of apoM in vivo or in vitro. Xu et al. reported that PPARβ/δ could inhibit expressions of apoM and apoB in HepG2 cells, which were regulated by PI3-kinase pathways [89]. More recently, it has been demonstrated that treatment with pioglitazone, a PPARγ agonist, could decrease both the hepatic and plasma apoM and S1P levels in obese mice induced by diet [46]. In HepG2 cells, apoM overexpression could increase, whereas apoM knockdown could suppress PPARγ activities [46]. These results suggested that PPARγ regulates the S1P levels by modulating apoM. When PPARγ was lightly expressed, the level of apoM/S1P was highest, and that hepatic apoM/PPARγ axis might maintain the homeostasis of S1P metabolism [46]. Liver X receptors (LXR) are key regulators of cholesterol and bile acid metabolism in hepatocytes and also target genes involved in steroid hormone synthesis, growth hormone signaling, and inflammation. The retinoid X receptors (RXR) bind the biologically active vitamin A, 9-cis-retinoic acid, and are involved in a variety of cellular functions including cell differentiation and fatty acid metabolism. To integrate the cellular responses to various stimuli, there is excessive "cross-talk" not only between LXRs and RXR but also with the PPARs [49, 80]. As part of a microarray study on the interaction between these receptors, Calayir E et al. found that LXR agonists could inhibit apoM expression in vivo [13]. Moreover, HepG2 cells demonstrated that both LXR and RXR agonists could regulate apoM expression in vitro. Both T0-901317 (a LXR agonist) and 9-cis-retinoic acid (a RXR ligand) significantly inhibited apoM expression, but not apoB expression, in HepG2 cell cultures [97], indicating that apoM expression may also be modulated by the LXR-RXR pathway (Fig. 7.9). Several growth factors could influence the transcription and secretion of apolipoproteins in HepG2 cells. ApoB expression, for example, is markedly downregulated by transforming growth factor-β (TGF-β) [87]. In case of apoM, it has been reported that TGF-β was also able to downregulate apoM expression and secretion from HepG2 cells [87]. In addition, estrogen could also regulate hepatic apoM expression via the estrogen receptor α-specific binding motif [82]. Hepatic apoM overexpression could stimulate formation of large apoM-/S1P-enriched HDL in plasma [52]. The unique apoM-/S1P-enriched HDL may service to deliver S1P to extrahepatic tissues.

7.5 ApoM-S1P Axis

As mentioned above, the plasma apoM is one of the most important natural carriers of S1P in blood [20], and release of S1P from HDL-apoM probably requires the tight interaction with S1P receptors [98]. S1P is an important bioactive lysophospholipid mediator which plays a variety of physiological functions through S1P receptors on cell surfaces, such as antiapoptosis [44, 93], cell proliferation [44, 93], vasorelaxation [14, 44], and the maintenance of vascular permeability [12]. Plasma S1P is mainly derived from erythrocytes [74], the activated platelets [1], and endothelial cells [1]. In terms of the kinetics of S1P in the circulation, apoM plays a crucial role in the distribution of plasma S1P compared to other lysophospholipids. In blood, about two-thirds of S1P are carried by HDL, and only one-third by the albumin [1, 2]. The polar tail of S1P is orientated to the inside of the binding pocket of apoM, which can prevent the degradation of S1P by the phosphatase or S1P lyase [45]. Although S1P is a typical mediator of the sphingo-lysophospholipid, an analog of S1P, dihydrosphingosine 1-phosphate (DH-S1P) is another effective mediator of sphingo-lysophospholipid. DH-S1P lacks one double bond at the 4–5 carbon position of S1P and has a concentration of 20–30% S1P in plasma [58]. DH-S1P is reported to activate S1P receptors and has similar biological activity to S1P [58]. However, although S1P has an intracellular biological effect and agonist properties on S1P receptors [44], DH-S1P is not reported to act intracellularly [58]. It has been demonstrated that S1P receptor-1 (S1P1) signaling in endothelial cells modulates vascular responses to immune complex (IC) deposition [12]. S1P1 agonists and a fusion protein (apoM-constant domain of immunoglobulins, apoM-Fc) could enhance the endothelial barrier, limit leukocyte escape from capillaries, and provide protection against inflammatory injury. The S1P/S1P1 axis is thereafter identified as the target to attenuate tissue responses to the IC deposition and inhibits organ damage [12]. Moreover, Terao et al., reported that albumin-bound S1P could disrupt the barrier integrity of retinal pigment epithelial cells via the S1P2, whereas apoM-S1P strengthened this integrity [77]. Liu et al. [53] reported that apoM secretion is rate-limiting for hepatocyte S1P secretion and that its uncleaved signal peptide delays apoM trafficking out of the cell, promoting formation of larger nascent apoM- and S1P-enriched HDL particles that are probably precursors of larger apoM-/S1P-enriched plasma HDL.

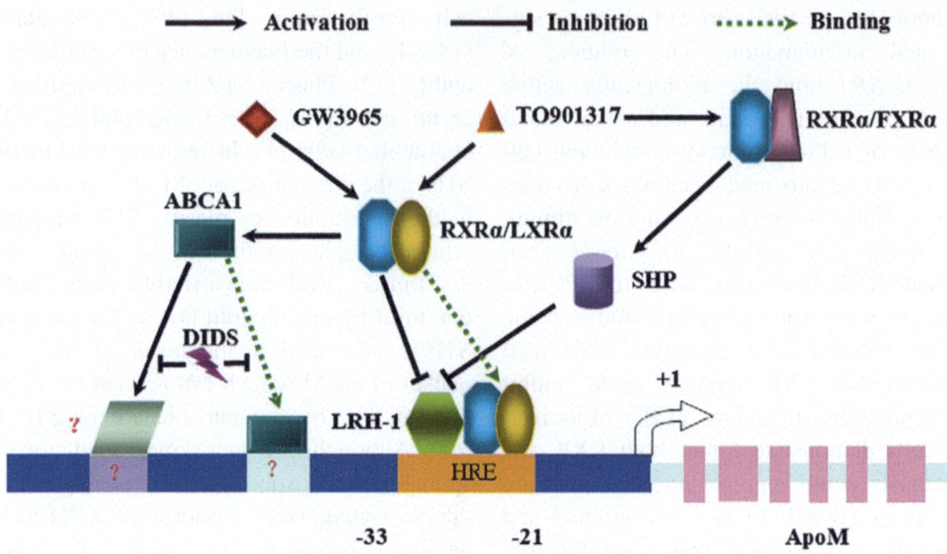

Fig. 7.9 *Possible mechanism of different effects of GW3965 and TO901317 on mRNA levels of apoM*

7.6 Importance of ApoM on Inflammation and Immune Response

Since the apoM gene is located in a highly conserved region (histocompatibility complex III (HMC-III) region on chromosome 6) in close proximity to genes related to the immune response (i.e., TNF, lymphotoxin B, and BAT3), it is reasonable to hypothesize that apoM may also be related to the immune response system or regulated by cytokines or other inflammatory factors. In addition, in HepG2 cells, platelet-activating factor (PAF) could upregulate apoM expression, whereas lexipafant (a PAF receptor antagonist) significantly suppressed mRNA levels and secretion of apoM in a dose-dependent manner [86]. However, neither tumor necrosis factor-a (TNF-a) nor interleukin-1a (IL-1a) influenced apoM expression or secretion in HepG2 cell cultures [86]. A couple of observations in more complex models are compatible with the idea that apoM may be involved in tissue defense mechanisms. During local ischemia-reperfusion injury of the livers in rats, hepatic apoM mRNA levels increased significantly during 1-h ischemia followed by 0.5–3 h reperfusion, which was similar to what has been observed for the heat shock protein HSP70 [90]. However, the plasma concentrations of apoM were not affected by ischemia-reperfusion injury. More recently, it has been demonstrated that apoM could regulate immune response via the apoM-S1P [75]. In septic patients, the reduction of plasma apoM and S1P reflects the severity of the disease [43]. A similar degree of disease severity was observed in baboons that were dependent on decreases in plasma S1P and apoM levels. S1P was reduced within 6–8 h of septic shock, whereas the apoM reduction was only occurred at 12–24 h, reflecting the almost complete loss of apoM and S1P in the HDL [43]. Perhaps, the decrease of S1P may be one of the reasons for the decrease of endothelial barrier function in sepsis, and apoM could be a new biomarker for the diagnosis of sepsis [17]. Increased endothelial glycocalyx shedding and vascular permeability are key characteristics of the sepsis pathophysiology and the organ failure progression. In a cohort of 184 septic patients or infected patients without organ failure, levels of glypican 1, 3, and 4 were significantly higher in septic patients than in those infected patients without organ failure [47]. The glypican 1, 3, and 4 were positively correlated with plasma levels of the syndecan 1 (glycocalyx degradation marker) and negatively correlated with plasma apoM and S1P levels. Some studies have investigated whether apoM has anti-inflammatory effects. As an example, apoM knockout mice exhibited more severe autoimmune encephalomyelitis [7]. The characteristics of this model included an increase in the number of lymphocytes in the brain parenchyma and a disruption of the blood-brain barrier. Furthermore, in a carrageenan-induced local inflammation model of the paw, apoM knockout mice had more vascular leakage than wild-type mice [16]. ApoM overexpression in apoM knockout mice could reverse this phenomenon. In another study, apoM knockout mice treated with lipopolysaccharide (LPS) could result in more severe acute lung injury than in wild-type mice [104]. And apoM overexpression could improve the survival rate of mice exposed to LPS, whereas the apoM gene knockout or knockdown decreased survival. In an in vitro study, tumor necrosis factor-α could reduce the levels of the vascular adhesion molecule-1 (VCAM-1) and E-selectin of primary endothelial cells in the presence of apoM-bound S1P [68]. And in mice, endothelial-specific S1P1 deletion resulted in increased ICAM expression of endothelial cells, whereas ICAM expression was reduced in those with endothelial-specific S1P1 overexpression [30]. Moreover, it has been demonstrated that apoM-induced inhibitory effects against the inflammatory response probably be mediated via the S1P1 and 3β-hydroxysterol Δ-24-reductase (DHCR24) pathways [81]. These existence studies strongly suggest the apoM/S1P/S1P1 axis may be a target for attenuating tissue inflammatory responses.

7.7 ApoM in Relation to Diabetes and Obesity

As mentioned above, there is a strong relationship between mutations in the HNF-1α gene and specific types of maturity onset diabetes in the young (MODY3) [67]. Mutations in the HNF-1α gene lead to impaired pancreatic β-cell function and impaired insulin secretion. Because of the multiple actions of HNF-1α, however, it is not surprising that such mutations also affect other critical metabolic functions. HNF-1α is a potent activator of the apoM promoter. Richter et al. [67], following up observations in HNF-1-α-deficient rats with low apoM levels, measured apoM concentrations in the sera of HNF-1α/maturity-onset diabetes of MODY3 patients compared to the normal matched control subjects (HNF-1a$^{+/+}$) and HNF-4α/MODY1 subjects (carrying a mutation in HNF-4α). Serum levels of apoM were significantly decreased in the HNF-1α/MODY3 subjects, in relation both to control subjects and to HNF-4α/MODY1 subjects, which may be partly related to the hyperglycemia [39, 96]. Serum levels of apoM may therefore be a useful serum marker for the identification of MODY3 patients [67]. Moreover, it was reported that a single-nucleotide polymorphism of the apoM proximal promoter region of the apoM gene (SNP T-778C) is associated with type 2 diabetes in a Chinese population [60]. Although it is well established that such patients develop atherogenic disturbances in lipoprotein metabolism, including low HDL concentrations, hypertriglyceridemia, and small dense LDL, it was not possible to evaluate the impact of this polymorphism on plasma lipoprotein concentrations since the patients were all under treatment. In another study, the relationships between plasma apoM, insulin and leptin levels, and lipoprotein concentrations were studied in normal and overweight females [88]. ApoM concentrations were positively correlated to leptin, BMI, and fasting insulin and negatively correlated to total cholesterol and LDL-cholesterol. The correlations between apoM and cholesterol and between apoM and leptin remained significant after adjustment for the influence of BMI. Forward stepwise multiple regressions showed that cholesterol and leptin were independent predictors of circulating apoM. Together, these two parameters explained about 30% of the variance in apoM. Hence, apoM is positively correlated to leptin and negatively correlated to cholesterol levels in humans [88]. In a mouse obese model, hepatic mRNA level of Foxa2 and protein mass of apoM were significantly decreased, which could be inverted by the administration of adiponectin [92]. Lee et al. reported that apoM T-855C and T-778C polymorphisms were associated with the obesity by regulating HDL metabolism [50], and Zhang et al. reported that the polymorphism C-724del in the promoter region of the apoM gene could confer the risk of type 2 diabetes among eastern Han Chinese [99]. More recently, Liu et al. reported that plasma S1P and apoM concentrations are inversely and independently associated with mortality, but not coronary artery calcium (CAC), in African Americans with type 2 diabetes after accounting for conventional risk factors [54].

7.8 Modulation of ApoM Levels Affects the Development of Atherosclerosis

The circulating lipoproteins are, in turn, closely bound up with cardiovascular status and the development of atherosclerotic lesions in the arterial vessel wall. HDL, which carries the predominant portion of apoM in plasma, is generally regarded as antiatherogenic, an attribute mainly ascribed to its role in "reverse cholesterol transport" [5, 63]. In mice, apoM is essential for the formation of HDL in the liver and its metabolism in the circulation. Wolfrum et al. [84] demonstrated that treatment with apoM-silencing RNA led to the accumulation of large HDL1 particles in plasma at the expense of normal pre-β-HDL particles, while overexpression of apoM in HNF-1α knockout mice by treatment with apoM adenovirus increased the formation of such pre-β-HDL particles. Moreover, Wolfrum et al. [84] used LDL receptor knockout mice fed a cholesterol-rich diet for 12 weeks and then administered apoM adenovirus, which increased apoM levels about twofold. After 3 weeks, atherosclerotic lesions were reduced by about 50% in animals with elevated apoM and pre-β-HDL levels. The remarkable antiatherogenic effect of elevated apoM levels may reflect several mechanisms. As demonstrated by in vitro experiments, HDL without pre-β-HDL were less efficient in promoting the efflux of cholesterol from cultured cells [84], so the concomitant rise in pre-β-HDL after administration of apoM adenovirus may conceivably increase the efficacy of "reverse cholesterol transport." To test whether this mechanism was also relevant, Christoffersen et al. [19] compared the properties of human HDL particles that contained apoM with those that did not. ApoM$^+$ particles were significantly more efficient in promoting the efflux of cholesterol from prelabeled THP-1 cells, lending support to the notion that one mechanism behind the antiatherogenic effect of apoM reflects a role in reverse cholesterol transport. However, it is also possible that apoM interacts with other steps in the complex formation of atherosclerotic lesions. Moreover, data [103] indicate that apoM may also affect the oxidative processes that increase the atherogenicity of LDL. Oxidized LDL particles, which have reduced affinity for LDL receptors, are instead removed by scavenger receptors in, for example, macrophages; this is a critical step for the generation of foam cells and thus of atherosclerotic lesions. ApoM$^+$-HDL was more efficient than apoM-HDL in preventing Cu^{2+}-induced oxidation of LDL in vitro [26], indicating that an antioxidative function of apoM may also contribute to its antiatherogenic effect. Recently it has been reported that apoM is a new adipokine which could be upregulated by calorie restriction and decreased with obesity [72]. ApoM was expressed in human subcutaneous and visceral adipose tissues and was released from adipose tissues into circulation, and plasma apoM concentrations were correlated to the apoM mRNA levels in these tissues. In adipose tissues apoM expression was inversely correlated to the adipocyte size, was lower in obese people than in lean individuals, and decreased in patients with metabolic syndrome and type 2 diabetes. Regardless of fat content, adipose tissues and apoM expression were positively correlated with systemic insulin sensitivity, independently of fat mass and plasma HDL cholesterol. In human multipotent adipose-derived stem cell adipocytes, apoM expression was enhanced by insulin-sensitizing peroxisome proliferator-activated receptor agonists and inhibited by TNFα, a cytokine causing insulin resistance. In obese individuals, apoM expression and secretion were increased by calorie restriction in adipose tissues.

7.9 Hepatic ApoM Expression and Liver Diseases

It has been demonstrated that the expressions of most apolipoproteins, including apoM, were downregulated in HepG2 cells infected with HBV [32]. And both apoM mRNA levels and apoM protein mass were significantly lower in human hepatocellular carcinoma (HCC) tissues than in there adjacent tissues [38]. Recently Zhang et al. reported that apoM could play a key role in normal autophagy activity in the liver and thereby further regulate the metabolism of lipids in the liver, particularly triglycerides [101]. In another study by using microarray analysis, apoM was found to be involved in the liver regeneration by regulating proliferation of liver sinusoidal endothelial cells (LSEC) [91]. LSEC has anti-fibrotic effect and plays an important role in liver regeneration after traumatic injury [23]. And S1P plays a significant role in the protection of cells from experimentally induced apoptosis [102] and stimulates hepatocytes proliferation through IL-6 and VEGF signaling [42]. ApoM knockout mice show a severe vascular adaptive remodeling of the hepatic sinusoidal vasculature after either 70% hepatectomy or bile duct ligation (BDL) [23]. The expression levels of α-smooth muscle actin and collagen were markedly increased in the liver of animals after BDL, while the expression levels of that in apoM transgenic mice (by 11-fold increased apoM expression) and control mice were significantly reduced. Additional experiments in an endothelial cell-specific S1P1 knockout mouse model confirmed these findings that S1P1 could be as a key S1P receptor mediating LSEC recovery and further liver regeneration. More recently it has been reported that apoM was related to the non-alcoholic fatty liver disease (NAFLD) [61]. NAFLD affects 25% of the population and can progress to cirrhosis with limited treatment options. As the liver secretes most of the blood plasma proteins, liver disease may affect the plasma proteome. Plasma proteome profiling of 48 patients with and without cirrhosis or NAFLD revealed six statistically significantly changing proteins including ALDOB, APOM, LGALS3BP, PIGR, VTN, and AFM, two of which are already linked to liver disease, whereas the importance of apoM in the process of NAFLD is still unknown [61].

7.10 ApoM and Renal Diseases

The high levels of expression of apoM in proximal tubular epithelium of the kidney suggest a physiological role of apoM in excretion or reabsorption of metabolites in the urine [28]. Megalin is a receptor located in tubular epithelial membranes that strongly binds to various substances in urine, including lipocalins, thereby mediating their reabsorption and preservation in the body [28]. Megalin deficient mice consequently excrete lipocalins (e.g., RBP, MUP-6, and vitamin D-binding protein) in urine [28]. It has been demonstrated that megalin has high affinity for apoM [28], suggesting that tubules-derived apoM may also be a ligand for megalin. It is therefore interesting that megalin-deficient mice (unlike normal mice) excrete apoM in urine [28]. Deletion of apoM gene in mouse could induce apoptosis in renal tissues, probably via the pathways of mitochondrial and endoplasmic reticulum stress [64], which causes glomerular cell damage and eventually glomerular sclerosis [66]. Plasma apoM levels have been reported to be lower in patients with chronic kidney diseases (CKD) at stages 3–5 than in CKD stages 1 + 2 patients and controls. Plasma apoM levels were further reduced in CKD patients with known cardiovascular diseases (CVD) compared to those without known CVD [10, 71]. Accordingly, when plasma apoM values were corrected for HDL-C, a significant difference persisted only between CKD stage 3 and stages 1 + 2 patients, whereas difference between CKD patients with and without known CVD disappeared. Recently it has been reported that apoM overexpression could reduce the severity of nephropathy in a mouse model of hyper IgA nephropathy [48]. Conversely, lack of apoM appears to further accelerate the disease. The change of S1P-signaling could be one of the underlying mechanisms. Thus, S1P1 or S1P3 antagonist, but not S1P2-targeting drugs, reversed the protective effect of apoM overexpression. As previously reported, the role of the S1P-albumin complex differs from that of the S1P-containing apoM particles. S1P-albumin promotes fibrogenesis, whereas S1P particle containing apoM could suppress these responses in vitro. Hence, apoM-S1P complexes could be used to treat IgA-induced nephropathy. In a study of patients with diabetic nephropathy, nephropathy patients without diabetes and healthy controls showed [100] that, surprisingly, patients with diabetic nephropathy had higher plasma apoM concentrations than those nephropathy patients without diabetes. In addition, this study did not identify any differences in plasma apoM levels among CKD stages 1–5. However, a study including 20 CKD patients showed that the HDL particles of CKD patients contained increased S1P but decreased apoM levels compared to controls [10]. The HDL of CKD patients with elevated S1P has cardioprotective effects in vitro. However, this beneficial effect of HDL in CKD has not been confirmed in animal models. Mild-to-moderate uremia was induced by subtotal nephrectomy in apoE-deficient mice that were either apoM-wild-type, apoM knockout, or apoM transgenic mice [9]. Uremia could increase plasma apoM by 25% but had no effect on S1P. ApoM knockout or apoM overexpression had no effect on uremic atherosclerosis. Together, these studies demonstrate the complexity of apoM/S1P in uremia and atherosclerosis.

7.11 Conclusions and Perspectives

Since its identification apoM has been extensively characterized with regard to gene and protein structure, while fundamental regulatory mechanisms have also been identified. Its highly selective expression, in hepatocytes and in renal tubular epithelium, indicates that apoM has its principal roles in hepatic lipoprotein metabolism and renal function. More recently the lipocalin structure of apoM has been demonstrated to be a carrier of S1P. HDL-apoM and S1P concentrations are inversely associated with atherosclerosis progression in rodents, and plasma S1P and apoM concentrations were probably inversely and independently associated with the mortality with type 2 diabetes mellitus. More detailed pathophysiological mechanisms behind apoM-S1P axis on the abnormal lipid metabolism, cardiovascular diseases, liver diseases, and kidney diseases need further investigations.

Acknowledgment This chapter was modified from the chapter 4 (Xu and Nilsson-Ehle. ApoM – A Novel Apolipoprotein with Antiatherogenic Properties, 2007: 89–109) published by our group in the book (High-Density Lipoproteins: From Basic Biology to Clinical Aspects, Editor: Christopher J. Fielding). All figures were from the papers published by our team in The Journal of Biological Chemistry ([85]; 274: 31286–31290) and Biochemical and Biophysical Research Communications ([105]; 421:152–156). The related contents have been licensed for reuse.

References

1. Aoki S, Yatomi Y, Ohta M, Osada M, Kazama F, Satoh K et al (2005) Sphingosine 1-phosphate-related metabolism in the blood vessel. J Biochem 138:47–55. https://doi.org/10.1093/jb/mvi100
2. Argraves KM, Argraves WS (2007) HDL serves as a S1P signaling platform mediating a multitude of cardiovascular effects. J Lipid Res 48:2325–2333. https://doi.org/10.1194/jlr.R700011-JLR200
3. Arkensteijn BW, Berbee JF, Rensen PC, Nielsen LB, Christoffersen C (2013) The apolipoprotein m-sphingosine-1-phosphate axis: biological relevance in lipoprotein metabolism, lipid disorders and atherosclerosis. Int J Mol Sci 14:4419–4431. https://doi.org/10.3390/ijms14034419
4. Badely M, Sepandi M, Samadi M, Parastouei K, Taghdir M (2019) The effect of whey protein on the components of metabolic syndrome in overweight and obese individuals; a systematic review and meta-analysis. Diabetes Metab Syndr 13:3121–3131. https://doi.org/10.1016/j.dsx.2019.11.001
5. Bardagjy AS, Steinberg FM (2019) Relationship between HDL functional characteristics and cardiovascular health and potential impact of dietary patterns: a narrative review. Nutrients 11. https://doi.org/10.3390/nu11061231

6. Bisgaard LS, Christoffersen C (2019) Apolipoprotein M/sphingosine-1-phosphate: novel effects on lipids, inflammation and kidney biology. Curr Opin Lipidol 30:212–217. https://doi.org/10.1097/MOL.0000000000000606
7. Blaho VA, Galvani S, Engelbrecht E, Liu C, Swendeman SL, Kono M et al (2015) HDL-bound sphingosine-1-phosphate restrains lymphopoiesis and neuroinflammation. Nature 523:342–346. https://doi.org/10.1038/nature14462
8. Borup A, Christensen PM, Nielsen LB, Christoffersen C (2015) Apolipoprotein M in lipid metabolism and cardiometabolic diseases. Curr Opin Lipidol 26:48–55. https://doi.org/10.1097/MOL.0000000000000142
9. Bosteen MH, Madsen Svarrer EM, Bisgaard LS, Martinussen T, Madsen M, Nielsen LB et al (2017) Effects of apolipoprotein M in uremic atherosclerosis. Atherosclerosis 265:93–101. https://doi.org/10.1016/j.atherosclerosis.2017.08.005
10. Brinck JW, Thomas A, Brulhart-Meynet MC, Lauer E, Frej C, Dahlback B et al (2018) High-density lipoprotein from end-stage renal disease patients exhibits superior cardioprotection and increase in sphingosine-1-phosphate. Eur J Clin Investig 48. https://doi.org/10.1111/eci.12866
11. Brites F, Martin M, Guillas I, Kontush A (2017) Antioxidative activity of high-density lipoprotein (HDL): mechanistic insights into potential clinical benefit. BBA Clin 8:66–77. https://doi.org/10.1016/j.bbacli.2017.07.002
12. Burg N, Swendeman S, Worgall S, Hla T, Salmon JE (2018) Sphingosine 1-phosphate receptor 1 signaling maintains endothelial cell barrier function and protects against immune complex-induced vascular injury. Arthritis Rheum 70:1879–1889. https://doi.org/10.1002/art.40558
13. Calayir E, Becker TM, Kratzer A, Ebner B, Panzenbock U, Stefujl J et al (2008) LXR-agonists regulate ApoM expression differentially in liver and intestine. Curr Pharm Biotechnol 9:516–521
14. Cantalupo A, Di Lorenzo A (2016) S1P Signaling and de novo biosynthesis in blood pressure homeostasis. J Pharmacol Exp Ther 358:359–370. https://doi.org/10.1124/jpet.116.233205
15. Charkoftaki G, Wang Y, McAndrews M, Bruford EA, Thompson DC, Vasiliou V et al (2019) Update on the human and mouse lipocalin (LCN) gene family, including evidence the mouse Mup cluster is result of an "volutionary bloom". Hum Genomics 13:11. https://doi.org/10.1186/s40246-019-0191-9
16. Christensen PM, Liu CH, Swendeman SL, Obinata H, Qvortrup K, Nielsen LB et al (2016) Impaired endothelial barrier function in apolipoprotein M-deficient mice is dependent on sphingosine-1-phosphate receptor 1. FASEB J 30:2351–2359. https://doi.org/10.1096/fj.201500064
17. Christoffersen C, Nielsen LB (2012) Apolipoprotein M – a new biomarker in sepsis. Crit Care 16:126. https://doi.org/10.1186/cc11320
18. Christoffersen C, Nielsen LB (2013) Apolipoprotein M: bridging HDL and endothelial function. Curr Opin Lipidol 24:295–300. https://doi.org/10.1097/MOL.0b013e328361f6ad
19. Christoffersen C, Nielsen LB, Axler O, Andersson A, Johnsen AH, Dahlback B (2006) Isolation and characterization of human apolipoprotein M-containing lipoproteins. J Lipid Res 47:1833–1843. https://doi.org/10.1194/jlr.M600055-JLR200
20. Christoffersen C, Obinata H, Kumaraswamy SB, Galvani S, Ahnstrom J, Sevvana M et al (2011) Endothelium-protective sphingosine-1-phosphate provided by HDL-associated apolipoprotein M. Proc Natl Acad Sci U S A 108:9613–9618. https://doi.org/10.1073/pnas.1103187108
21. Costa RH, Kalinichenko VV, Holterman AX, Wang X (2003) Transcription factors in liver development, differentiation, and regeneration. Hepatology 38:1331–1347. https://doi.org/10.1016/j.hep.2003.09.034
22. Darabi M, Guillas-Baudouin I, Le Goff W, Chapman MJ, Kontush A (2016) Therapeutic applications of reconstituted HDL: when structure meets function. Pharmacol Ther 157:28–42. https://doi.org/10.1016/j.pharmthera.2015.10.010
23. Ding BS, Liu CH, Sun Y, Chen Y, Swendeman SL, Jung B et al (2016) HDL activation of endothelial sphingosine-1-phosphate receptor-1 (S1P1) promotes regeneration and suppresses fibrosis in the liver. JCI Insight 1:e87058. https://doi.org/10.1172/jci.insight.87058
24. Duan J, Dahlback B, Villoutreix BO (2001) Proposed lipocalin fold for apolipoprotein M based on bioinformatics and site-directed mutagenesis. FEBS Lett 499:127–132. https://doi.org/10.1016/s0014-5793(01)02544-3
25. Duchateau PN, Pullinger CR, Orellana RE, Kunitake ST, Naya-Vigne J, O'Connor PM et al (1997) Apolipoprotein L, a new human high density lipoprotein apolipoprotein expressed by the pancreas. Identification, cloning, characterization, and plasma distribution of apolipoprotein L. J Biol Chem 272:25576–25582. https://doi.org/10.1074/jbc.272.41.25576
26. Elsoe S, Ahnstrom J, Christoffersen C, Hoofnagle AN, Plomgaard P, Heinecke JW et al (2012) Apolipoprotein M binds oxidized phospholipids and increases the antioxidant effect of HDL. Atherosclerosis 221:91–97. https://doi.org/10.1016/j.atherosclerosis.2011.11.031
27. Faber K, Axler O, Dahlback B, Nielsen LB (2004) Characterization of apoM in normal and genetically modified mice. J Lipid Res 45:1272–1278. https://doi.org/10.1194/jlr.M300451-JLR200
28. Faber K, Hvidberg V, Moestrup SK, Dahlback B, Nielsen LB (2006) Megalin is a receptor for apolipoprotein M, and kidney-specific megalin-deficiency confers urinary excretion of apolipoprotein M. Mol Endocrinol 20:212–218. https://doi.org/10.1210/me.2005-0209

29. Frej C, Mendez AJ, Ruiz M, Castillo M, Hughes TA, Dahlback B et al (2017) A shift in ApoM/S1P between HDL-particles in women with type 1 diabetes mellitus is associated with impaired anti-inflammatory effects of the ApoM/S1P complex. Arterioscler Thromb Vasc Biol 37:1194–1205. https://doi.org/10.1161/ATVBAHA.117.309275
30. Galvani S, Sanson M, Blaho VA, Swendeman SL, Obinata H, Conger H et al (2015) HDL-bound sphingosine 1-phosphate acts as a biased agonist for the endothelial cell receptor S1P1 to limit vascular inflammation. Sci Signal 8:ra79. https://doi.org/10.1126/scisignal.aaa2581
31. Grygiel-Gorniak B (2014) Peroxisome proliferator-activated receptors and their ligands: nutritional and clinical implications – a review. Nutr J 13:17. https://doi.org/10.1186/1475-2891-13-17
32. Gu JG, Zhu CL, Cheng DZ, Xie Y, Liu F, Zhou X (2011) Enhanced levels of apolipoprotein M during HBV infection feedback suppresses HBV replication. Lipids Health Dis 10:154. https://doi.org/10.1186/1476-511X-10-154
33. Han L, Shen WJ, Bittner S, Kraemer FB, Azhar S (2017) PPARs: regulators of metabolism and as therapeutic targets in cardiovascular disease. Part I: PPAR-alpha. Futur Cardiol 13:259–278. https://doi.org/10.2217/fca-2016-0059
34. He L, Wu P, Tan L, Le B, Du W, Shen T et al (2017) Characteristics of lipid metabolism including serum apolipoprotein M levels in patients with primary nephrotic syndrome. Lipids Health Dis 16:167. https://doi.org/10.1186/s12944-017-0556-9
35. Hong F, Pan S, Guo Y, Xu P, Zhai Y (2019) PPARs as nuclear receptors for nutrient and energy metabolism. Molecules 24. https://doi.org/10.3390/molecules24142545
36. Hui N, Barter PJ, Ong KL, Rye KA (2019) Altered HDL metabolism in metabolic disorders: insights into the therapeutic potential of HDL. Clin Sci 133:2221–2235. https://doi.org/10.1042/CS20190873
37. Hussain MA (1997) Transcription factors in the pathophysiology of diabetes mellitus. Eur J Endocrinol 136:469–470. https://doi.org/10.1530/eje.0.1360469
38. Jiang J, Wu C, Luo G, Zheng L, Chen L, Zhang X et al (2011) Expression of apolipoprotein M in human hepatocellular carcinoma tissues. Acta Histochem 113:53–57. https://doi.org/10.1016/j.acthis.2009.08.005
39. Jiang B, Zhang X, Di D, Luo G, Shi Y, Zhang J et al (2015) Hyperglycemia-induced downregulation of apolipoprotein M expression is not via the hexosamine pathway. Lipids Health Dis 14:110. https://doi.org/10.1186/s12944-015-0103-5
40. Karlsson H, Leanderson P, Tagesson C, Lindahl M (2005) Lipoproteomics I: mapping of proteins in low-density lipoprotein using two-dimensional gel electrophoresis and mass spectrometry. Proteomics 5:551–565. https://doi.org/10.1002/pmic.200300938
41. Karlsson H, Lindqvist H, Tagesson C, Lindahl M (2006) Characterization of apolipoprotein M isoforms in low-density lipoprotein. J Proteome Res 5:2685–2690. https://doi.org/10.1021/pr060180x
42. Kawasaki T, Murata S, Takahashi K, Nozaki R, Ohshiro Y, Ikeda N et al (2010) Activation of human liver sinusoidal endothelial cell by human platelets induces hepatocyte proliferation. J Hepatol 53:648–654. https://doi.org/10.1016/j.jhep.2010.04.021
43. Kumaraswamy SB, Linder A, Akesson P, Dahlback B (2012) Decreased plasma concentrations of apolipoprotein M in sepsis and systemic inflammatory response syndromes. Crit Care 16:R60. https://doi.org/10.1186/cc11305
44. Kurano M, Yatomi Y (2018) Sphingosine 1-phosphate and atherosclerosis. J Atheroscler Thromb 25:16–26. https://doi.org/10.5551/jat.RV17010
45. Kurano M, Hara M, Tsuneyama K, Sakoda H, Shimizu T, Tsukamoto K et al (2014) Induction of insulin secretion by apolipoprotein M, a carrier for sphingosine 1-phosphate. Biochim Biophys Acta 1841:1217–1226. https://doi.org/10.1016/j.bbalip.2014.05.002
46. Kurano M, Ikeda H, Iso ON, Hara M, Tsukamoto K, Yatomi Y (2018a) Regulation of the metabolism of apolipoprotein M and sphingosine 1-phosphate by hepatic PPARgamma activity. Biochem J 475:2009–2024. https://doi.org/10.1042/BCJ20180052
47. Kurano M, Tsuneyama K, Morimoto Y, Shimizu T, Jona M, Kassai H et al (2018b) Apolipoprotein M protects lipopolysaccharide-treated mice from death and organ injury. Thromb Haemost 118:1021–1035. https://doi.org/10.1055/s-0038-1641750
48. Kurano M, Tsuneyama K, Morimoto Y, Nishikawa M, Yatomi Y (2019) Apolipoprotein M suppresses the phenotypes of IgA nephropathy in hyper-IgA mice. FASEB J 33:5181–5195. https://doi.org/10.1096/fj.201801748R
49. Laurencikiene J, Ryden M (2012) Liver X receptors and fat cell metabolism. Int J Obes 36:1494–1502. https://doi.org/10.1038/ijo.2012.21
50. Lee M, Kim JI, Choi S, Jang Y, Sorn SR (2016) The effect of apoM polymorphism associated with HDL metabolism on obese Korean Adults. J Nutrigenet Nutrigenomics 9:306–317. https://doi.org/10.1159/000455948
51. Linden D, Lindberg K, Oscarsson J, Claesson C, Asp L, Li L et al (2002) Influence of peroxisome proliferator-activated receptor alpha agonists on the intracellular turnover and secretion of apolipoprotein (Apo) B-100 and ApoB-48. J Biol Chem 277:23044–23053. https://doi.org/10.1074/jbc.M110416200
52. Liu M, Seo J, Allegood J, Bi X, Zhu X, Boudyguina E et al (2014) Hepatic apolipoprotein M (apoM) overexpression stimulates formation of larger apoM/

sphingosine 1-phosphate-enriched plasma high density lipoprotein. J Biol Chem 289:2801–2814. https://doi.org/10.1074/jbc.M113.499913
53. Liu M, Allegood J, Zhu X, Seo J, Gebre AK, Boudyguina E et al (2015) Uncleaved ApoM signal peptide is required for formation of large ApoM/sphingosine 1-phosphate (S1P)-enriched HDL particles. J Biol Chem 290:7861–7870. https://doi.org/10.1074/jbc.M114.631101
54. Liu M, Frej C, Langefeld CD, Divers J, Bowden DW, Carr JJ et al (2019) Plasma apoM and S1P levels are inversely associated with mortality in African Americans with type 2 diabetes mellitus. J Lipid Res 60:1425–1431. https://doi.org/10.1194/jlr.P089409
55. Luo G, Shi Y, Zhang J, Mu Q, Qin L, Zheng L et al (2014) Palmitic acid suppresses apolipoprotein M gene expression via the pathway of PPARbeta/delta in HepG2 cells. Biochem Biophys Res Commun 445:203–207. https://doi.org/10.1016/j.bbrc.2014.01.170
56. Mancone C, Amicone L, Fimia GM, Bravo E, Piacentini M, Tripodi M et al (2007) Proteomic analysis of human very low-density lipoprotein by two-dimensional gel electrophoresis and MALDI-TOF/TOF. Proteomics 7:143–154. https://doi.org/10.1002/pmic.200600339
57. Mendel DB, Crabtree GR (1991) HNF-1, a member of a novel class of dimerizing homeodomain proteins. J Biol Chem 266:677–680
58. Mishima Y, Kurano M, Kobayashi T, Nishikawa M, Ohkawa R, Tozuka M et al (2018) Dihydrosphingosine 1-phosphate interacts with carrier proteins in a manner distinct from that of sphingosine 1-phosphate. Biosci Rep 38. https://doi.org/10.1042/BSR20181288
59. Mughal SA, Park R, Nowak N, Gloyn AL, Karpe F, Matile H et al (2013) Apolipoprotein M can discriminate HNF1A-MODY from type 1 diabetes. Diabet Med 30:246–250. https://doi.org/10.1111/dme.12066
60. Niu N, Zhu X, Liu Y, Du T, Wang X, Chen D et al (2007) Single nucleotide polymorphisms in the proximal promoter region of apolipoprotein M gene (apoM) confer the susceptibility to development of type 2 diabetes in Han Chinese. Diabetes Metab Res Rev 23:21–25. https://doi.org/10.1002/dmrr.641
61. Niu L, Geyer PE, Wewer Albrechtsen NJ, Gluud LL, Santos A, Doll S et al (2019) Plasma proteome profiling discovers novel proteins associated with non-alcoholic fatty liver disease. Mol Syst Biol 15: e8793. https://doi.org/10.15252/msb.20188793
62. Nojiri T, Kurano M, Tokuhara Y, Ohkubo S, Hara M, Ikeda H et al (2014) Modulation of sphingosine-1-phosphate and apolipoprotein M levels in the plasma, liver and kidneys in streptozotocin-induced diabetic mice. J Diabetes Investig 5:639–648. https://doi.org/10.1111/jdi.12232
63. Ouimet M, Barrett TJ, Fisher EA (2019) HDL and reverse cholesterol transport. Circ Res 124:1505–1518. https://doi.org/10.1161/CIRCRESAHA.119.312617
64. Pei W, Wu Y, Zhang X, Lv K, Zhang Y, Li Z et al (2018) Deletion of ApoM gene induces apoptosis in mouse kidney via mitochondrial and endoplasmic reticulum stress pathways. Biochem Biophys Res Commun 505:891–897. https://doi.org/10.1016/j.bbrc.2018.09.162
65. Peng J, Li XP (2018) Apolipoprotein A-IV: a potential therapeutic target for atherosclerosis. Prostaglandins Other Lipid Mediat 139:87–92. https://doi.org/10.1016/j.prostaglandins.2018.10.004
66. Qian Y, Feldman E, Pennathur S, Kretzler M, Brosius FC 3rd (2008) From fibrosis to sclerosis: mechanisms of glomerulosclerosis in diabetic nephropathy. Diabetes 57:1439–1445. https://doi.org/10.2337/db08-0061
67. Richter S, Shih DQ, Pearson ER, Wolfrum C, Fajans SS, Hattersley AT et al (2003) Regulation of apolipoprotein M gene expression by MODY3 gene hepatocyte nuclear factor-1alpha: haploinsufficiency is associated with reduced serum apolipoprotein M levels. Diabetes 52:2989–2995. https://doi.org/10.2337/diabetes.52.12.2989
68. Ruiz M, Frej C, Holmer A, Guo LJ, Tran S, Dahlback B (2017) High-density lipoprotein-associated apolipoprotein M limits endothelial inflammation by delivering sphingosine-1-phosphate to the sphingosine-1-phosphate receptor 1. Arterioscler Thromb Vasc Biol 37:118–129. https://doi.org/10.1161/ATVBAHA.116.308435
69. Sevvana M, Ahnstrom J, Egerer-Sieber C, Lange HA, Dahlback B, Muller YA (2009) Serendipitous fatty acid binding reveals the structural determinants for ligand recognition in apolipoprotein M. J Mol Biol 393:920–936. https://doi.org/10.1016/j.jmb.2009.08.071
70. Skupien J, Kepka G, Gorczynska-Kosiorz S, Gebska A, Klupa T, Wanic K et al (2007) Evaluation of apolipoprotein M serum concentration as a biomarker of HNF-1alpha MODY. Rev Diabet Stud 4:231–235. https://doi.org/10.1900/RDS.2007.4.231
71. Sorensen IM, Bertelsen M, Freese E, Lindhard K, Ullum H, Feldt-Rasmussen B et al (2018) Apolipoprotein M in patients with chronic kidney disease. Atherosclerosis 275:304–311. https://doi.org/10.1016/j.atherosclerosis.2018.06.815
72. Sramkova V, Berend S, Siklova M, Caspar-Bauguil S, Carayol J, Bonnel S et al (2019) Apolipoprotein M: a novel adipokine decreasing with obesity and upregulated by calorie restriction. Am J Clin Nutr 109:1499–1510. https://doi.org/10.1093/ajcn/nqy331
73. Stienstra R, Duval C, Muller M, Kersten S (2007) PPARs, obesity, and inflammation. PPAR Res 2007:95974. https://doi.org/10.1155/2007/95974
74. Sutter I, Park R, Othman A, Rohrer L, Hornemann T, Stoffel M et al (2014) Apolipoprotein M modulates erythrocyte efflux and tubular reabsorption of sphingosine-1-phosphate. J Lipid Res 55:1730–1737. https://doi.org/10.1194/jlr.M050021

75. Swendeman SL, Xiong Y, Cantalupo A, Yuan H, Burg N, Hisano Y et al (2017) An engineered S1P chaperone attenuates hypertension and ischemic injury. Sci Signal 10:eaal2722. https://doi.org/10.1126/scisignal.aal2722
76. Terao R, Honjo M, Aihara M (2017) Apolipoprotein M Inhibits angiogenic and inflammatory response by sphingosine 1-phosphate on retinal pigment epithelium cells. Int J Mol Sci 19. https://doi.org/10.3390/ijms19010112
77. Terao R, Honjo M, Totsuka K, Miwa Y, Kurihara T, Aihara M (2019) The role of sphingosine 1-phosphate receptors on retinal pigment epithelial cells barrier function and angiogenic effects. Prostaglandins Other Lipid Mediat 145:106365. https://doi.org/10.1016/j.prostaglandins.2019.106365
78. Tsai HC, Han MH (2016) Sphingosine-1-phosphate (S1P) and S1P signaling pathway: therapeutic targets in autoimmunity and inflammation. Drugs 76:1067–1079. https://doi.org/10.1007/s40265-016-0603-2
79. Tyden H, Lood C, Jonsen A, Gullstrand B, Kahn R, Linge P et al (2019) Low plasma concentrations of apolipoprotein M are associated with disease activity and endothelial dysfunction in systemic lupus erythematosus. Arthritis Res Ther 21:110. https://doi.org/10.1186/s13075-019-1890-2
80. Wang B, Tontonoz P (2018) Liver X receptors in lipid signalling and membrane homeostasis. Nat Rev Endocrinol 14:452–463. https://doi.org/10.1038/s41574-018-0037-x
81. Wang M, Luo GH, Liu H, Zhang YP, Wang B, Di DM et al (2019) Apolipoprotein M induces inhibition of inflammatory responses via the S1PR1 and DHCR24 pathways. Mol Med Rep 19:1272–1283. https://doi.org/10.3892/mmr.2018.9747
82. Wei J, Yu Y, Luo GH, Feng YH, Shi YP, Zhang J et al (2017) 17beta-estradiol regulates the expression of apolipoprotein M through estrogen receptor alpha-specific binding motif in its promoter. Lipids Health Dis 16:66. https://doi.org/10.1186/s12944-017-0458-x
83. Wilkins JT, Seckler HS (2019) HDL modification: recent developments and their relevance to atherosclerotic cardiovascular disease. Curr Opin Lipidol 30:24–29. https://doi.org/10.1097/MOL.0000000000000571
84. Wolfrum C, Poy MN, Stoffel M (2005) Apolipoprotein M is required for prebeta-HDL formation and cholesterol efflux to HDL and protects against atherosclerosis. Nat Med 11:418–422. https://doi.org/10.1038/nm1211
85. Xu N, Dahlback B (1999) A novel human apolipoprotein (apoM). J Biol Chem 274:31286–31290. https://doi.org/10.1074/jbc.274.44.31286
86. Xu N, Zhang XY, Dong X, Ekstrom U, Ye Q, Nilsson-Ehle P (2002) Effects of platelet-activating factor, tumor necrosis factor, and interleukin-1alpha on the expression of apolipoprotein M in HepG2 cells. Biochem Biophys Res Commun 292:944–950. https://doi.org/10.1006/bbrc.2002.6755
87. Xu N, Hurtig M, Zhang XY, Ye Q, Nilsson-Ehle P (2004a) Transforming growth factor-beta down-regulates apolipoprotein M in HepG2 cells. Biochim Biophys Acta 1683:33–37. https://doi.org/10.1016/j.bbalip.2004.04.001
88. Xu N, Nilsson-Ehle P, Ahren B (2004b) Correlation of apolipoprotein M with leptin and cholesterol in normal and obese subjects. J Nutr Biochem 15:579–582. https://doi.org/10.1016/j.jnutbio.2004.03.001
89. Xu N, Ahren B, Jiang J, Nilsson-Ehle P (2006a) Down-regulation of apolipoprotein M expression is mediated by phosphatidylinositol 3-kinase in HepG2 cells. Biochim Biophys Acta 1761:256–260. https://doi.org/10.1016/j.bbalip.2006.02.002
90. Xu X, Ye Q, Xu N, He X, Luo G, Zhang X et al (2006b) Effects of ischemia-reperfusion injury on apolipoprotein M expression in the liver. Transplant Proc 38:2769–2773. https://doi.org/10.1016/j.transproceed.2006.08.133
91. Xu CS, Chen XG, Chang CF, Wang GP, Wang WB, Zhang LX et al (2011) Analysis of time-course gene expression profiles of sinusoidal endothelial cells during liver regeneration in rats. Mol Cell Biochem 350:215–227. https://doi.org/10.1007/s11010-010-0701-5
92. Yang L, Li T, Zhao S, Zhang S (2019) Expression of apolipoprotein M and its association with adiponectin in an obese mouse model. Exp Ther Med 18:1685–1692. https://doi.org/10.3892/etm.2019.7755
93. Zeng Y, Liu X, Yan Z, Xie L (2017) Sphingosine 1-phosphate regulates proliferation, cell cycle and apoptosis of hepatocellular carcinoma cells via syndecan-1. Prog Biophys Mol Biol. https://doi.org/10.1016/j.pbiomolbio.2017.11.006
94. Zhang XY, Dong X, Zheng L, Luo GH, Liu YH, Ekstrom U et al (2003) Specific tissue expression and cellular localization of human apolipoprotein M as determined by in situ hybridization. Acta Histochem 105:67–72
95. Zhang XY, Jiao GQ, Hurtig M, Dong X, Zheng L, Luo GH et al (2004) Expression pattern of apolipoprotein M during mouse and human embryogenesis. Acta Histochem 106:123–128. https://doi.org/10.1016/j.acthis.2003.11.004
96. Zhang X, Jiang B, Luo G, Nilsson-Ehle P, Xu N (2007) Hyperglycemia down-regulates apolipoprotein M expression in vivo and in vitro. Biochim Biophys Acta 1771:879–882. https://doi.org/10.1016/j.bbalip.2007.04.020
97. Zhang X, Zhu Z, Luo G, Zheng L, Nilsson-Ehle P, Xu N (2008) Liver X receptor agonist downregulates hepatic apoM expression in vivo and in vitro. Biochem Biophys Res Commun 371:114–117. https://doi.org/10.1016/j.bbrc.2008.04.017
98. Zhang H, Pluhackova K, Jiang Z, Bockmann RA (2016a) Binding Characteristics of Sphingosine-1-Phosphate to ApoM hints to Assisted Release

Mechanism via the ApoM Calyx-Opening. Sci Rep 6:30655. https://doi.org/10.1038/srep30655
99. Zhang PH, Gao JL, Pu C, Feng G, Wang LZ, Huang LZ et al (2016b) A single-nucleotide polymorphism C-724/del in the proter region of the apolipoprotein M gene is associated with type 2 diabetes mellitus. Lipids Health Dis 15:142. https://doi.org/10.1186/s12944-016-0307-3
100. Zhang P, Gao J, Pu C, Feng G, Wang L, Huang L et al (2017) ApoM/HDL-C and apoM/apoA-I ratios are indicators of diabetic nephropathy in healthy controls and type 2 diabetes mellitus. Clin Chim Acta 466:31–37. https://doi.org/10.1016/j.cca.2017.01.006
101. Zhang X, Zhang P, Gao J, Huang Q (2018) Autophagy dysregulation caused by ApoM deficiency plays an important role in liver lipid metabolic disorder. Biochem Biophys Res Commun 495:2643–2648. https://doi.org/10.1016/j.bbrc.2017.12.148
102. Zheng DM, Kitamura T, Ikejima K, Enomoto N, Yamashina S, Suzuki S et al (2006) Sphingosine 1-phosphate protects rat liver sinusoidal endothelial cells from ethanol-induced apoptosis: role of intracellular calcium and nitric oxide. Hepatology 44:1278–1287. https://doi.org/10.1002/hep.21384
103. Zheng Z, Zeng Y, Zhu X, Tan Y, Li Y, Li Q et al (2019) ApoM-S1P modulates Ox-LDL-induced inflammation through the PI3K/Akt signaling pathway in HUVECs. Inflammation 42:606–617. https://doi.org/10.1007/s10753-018-0918-0
104. Zhu B, Luo GH, Feng YH, Yu MM, Zhang J, Wei J et al (2018) Apolipoprotein M protects against lipopolysaccharide-induced acute lung injury via sphingosine-1-phosphate signaling. Inflammation 41:643–653. https://doi.org/10.1007/s10753-017-0719-x
105. Di D, Wang Z, Liu Y, Luo G, Shi Y, Berggren-Söderlund M et al (2012) ABCA1 upregulating apolipoproein M expression mediates via the RXR/LXR pathway in HepG2 cells. Biochem Biophys Res Commun 421:152–156. https://doi.org/10.1016/j.bbrc.2012.04.022

Recent Advances in the Critical Role of the Sterol Efflux Transporters ABCG5/G8 in Health and Disease

8

Helen H. Wang, Min Liu, Piero Portincasa, and David Q.-H. Wang

Abstract

Cardiovascular disease is characterized by lipid accumulation, inflammatory response, cell death, and fibrosis in the arterial wall and is the leading cause of morbidity and mortality worldwide. Cholesterol gallstone disease is caused by complex genetic and environmental factors and is one of the most prevalent and costly digestive diseases in the USA and Europe. Although sitosterolemia is a rare inherited lipid storage disease, its genetic studies led to identification of the sterol efflux transporters ABCG5/G8 that are located on chromosome 2p21 in humans and chromosome 17 in mice. Human and animal studies have clearly demonstrated that ABCG5/G8 play a critical role in regulating hepatic secretion and intestinal absorption of cholesterol and plant sterols. Sitosterolemia is caused by a mutation in either the *ABCG5* or the *ABCG8* gene alone, but not in both simultaneously. Polymorphisms in the ABCG5/G8 genes are associated with abnormal plasma cholesterol metabolism and may play a key role in the genetic determination of plasma cholesterol concentrations. Moreover, *ABCG5/G8* is a new gallstone gene, *LITH9*. Gallstone-associated variants in *ABCG5/G8* are involved in the pathogenesis of cholesterol gallstones in European, Asian, and South American populations. In this chapter, we summarize the latest advances in the critical role of the sterol efflux transporters ABCG5/G8 in regulating hepatic secretion of biliary cholesterol, intestinal absorption of cholesterol and plant sterols, the classical reverse cholesterol transport, and the newly established transintestinal cholesterol excretion, as well as in the pathogenesis and pathophysiology of ABCG5/G8-related metabolic diseases such as sitosterolemia, cardiovascular disease, and cholesterol gallstone disease.

Keywords

Bile flow · Bile salts · Biliary lipid secretion · Gallstones · Cardiovascular disease · Cholesterol-lowering drugs · Coronary heart disease · Intestinal lipid absorption · *Lith* gene · Lithogenic bile · Reverse cholesterol transport · Statins · Stroke

H. H. Wang · D. Q.-H.Wang (✉)
Department of Medicine and Genetics, Division of Gastroenterology and Liver Diseases, Marion Bessin Liver Research Center, Einstein-Mount Sinai Diabetes Research Center, Albert Einstein College of Medicine, Bronx, NY, USA
e-mail: david.wang@einsteinmed.org

M. Liu
Department of Pathology and Laboratory Medicine, University of Cincinnati College of Medicine, Cincinnati, OH, USA

P. Portincasa
Department of Biomedical Sciences and Human Oncology, Clinica Medica "A. Murri", University of Bari Medical School, Bari, Italy

X.-C. Jiang (ed.), *Lipid Transfer in Lipoprotein Metabolism and Cardiovascular Disease*, Advances in Experimental Medicine and Biology 1276, https://doi.org/10.1007/978-981-15-6082-8_8

Abbreviations

ABC	ATP-binding cassette (transporter)
ACAT2	Acyl-CoA: cholesterol acyltransferase isoform 2
APO	Apolipoprotein
CSI	Cholesterol saturation index
CYP7A1	Cholesterol 7α-hydroxylase
CYP27A1	Sterol 27-hydroxylase
FABPpm	Plasma membrane-associated fatty acid-binding protein
FATP4	Fatty acid transport protein 4
FXR	Farnesoid X receptor
HDL	High-density lipoprotein
HMGCR	3-Hydroxy-3-methylglutaryl coenzyme A reductase
LDL	Low-density lipoprotein
LXR	Liver X receptor
MTTP	Microsomal triglyceride transfer protein
NPC1L1	Niemann-Pick C1 like 1 (protein)
PPAR-δ	Peroxisome proliferator-activated receptor-delta
QTL	Quantitative trait locus
SR-BI	Scavenger receptor class B type I
TICE	Transintestinal cholesterol excretion
VLDL	Very-low-density lipoprotein

8.1 Introduction

It is well-known that cholesterol is essential for all cells in the body because it is widely used as a key structural component for cell membranes and as a central substrate for the synthesis of other steroids, including bile salts, vitamin D, and sex hormones such as estradiol, progesterone, androsterone, and testosterone, as well as adrenocortical hormones such as cortisone and aldosterone [1]. It has been found that the liver and small intestine are two major organs for cholesterol biosynthesis. Furthermore, high cholesterol biosynthesis in the liver leads to more very-low-density lipoprotein (VLDL) secreted into plasma, which has a significant impact on plasma total and low-density lipoprotein (LDL) cholesterol concentrations. High dietary cholesterol also could contribute an increase in plasma cholesterol concentrations in most individuals. Elevated plasma total and LDL cholesterol levels are an important risk factor for the development of cardiovascular disease in humans [2].

Clinical studies and epidemiological investigations have clearly demonstrated that cardiovascular disease is a leading cause of death and disability not only in the USA but also in European and Asian countries. Therefore, the National Cholesterol Education Program Adult Treatment Panel III guidelines [3] along with the 2012 update and the American Heart Association and American College of Cardiology recommendations [4–7] have proposed a much lower target for plasma LDL cholesterol concentrations (i.e., <100 mg/dL) for individuals at high risk for adverse cardiovascular events. As a result, the total number of patients requiring more aggressive cholesterol-lowering treatment has significantly increased. Because the cholesterol carried in LDL particles is derived mainly from both de novo biosynthesis in the liver and intestinal absorption from the diet, a better understanding of the cellular and molecular mechanisms of elucidating the regulation of hepatic cholesterol biosynthesis and intestinal cholesterol absorption should lead to novel approaches to the treatment and the prevention of cardiovascular disease. Despite significant advances in the treatment of cardiovascular disease, a large number of residual risks in these patients are still being fully studied. Based on the genetic studies on patients with sitosterolemia [8–10], the ATP-binding cassette (ABC) sterol efflux transporters ABCG5 and ABCG8, encoded by the *ABCG5* and *ABCG8* genes, have been identified, which are located primarily on the canalicular membrane of hepatocytes and the apical membrane of enterocytes and play a key role in hepatic secretion and intestinal absorption of cholesterol and plant sterols [9, 11–13].

Cholesterol gallstone disease is caused by complex genetic and environmental factors. It is one of the most common and costly digestive diseases worldwide. In Western countries, 15–20% of the populations suffer from gallstones. At least 20 million Americans (~12%

of adults) have gallstones, leading to a considerable financial and social burden in the USA [14–19]. The prevalence of gallstones appears to be rising due to the epidemic of obesity that is associated with insulin resistance and the metabolic syndrome [16]. It is estimated that there are approximately 1 million new cases diagnosed each year [20–22]. Although most patients with gallstones are asymptomatic, one third of patients eventually develop clinical symptoms with or without complications [20]. The estimated 1,000,000 cholecystectomies are performed for gallstone disease every year. The annual medical cost of treating gallstones exceeded $6 billion in 2004 and even higher in 2019 [23]. The burden of gallstone disease is exacerbated by the fact that laparoscopic cholecystectomy remains the standard treatment for symptomatic gallstones worldwide [24]. In addition, unavoidable complications of gallstones result in 3000 deaths (~0.12% of all deaths) per year in the USA [14]. In general, persons with gallstone disease have increased overall, cardiovascular disease, and cancer mortality [18]. Most importantly, the prevalence of gallstones is increasing year by year because of the epidemic of obesity that is associated with insulin resistance, hyperlipidemia, and the metabolic syndrome.

To reduce the morbidity, mortality, and costs of health care associated with this disease, it is imperative to decipher the pathophysiology of cholesterol gallstone disease. This would facilitate the development of a novel, effective, and noninvasive therapy for patients with gallstone disease. Compelling evidence from the physical-chemical, pathophysiological, and genetic studies shows that cholesterol gallstone disease is determined by multiple *Lith* genes, which is a dominant trait. The principal pathogenic factor is persistent hepatic hypersecretion of cholesterol into bile, thereby contributing to the formation of cholesterol-supersaturated gallbladder bile. Clinical studies have found that cholesterol-supersaturated bile is an essential prerequisite for the precipitation of solid cholesterol monohydrate crystals and the formation of cholesterol gallstones [23]. Although it has been established that ABCG5/G8 play a key role in hepatic secretion and intestinal absorption of cholesterol and plant sterols [9, 11–13] and in the pathogenesis of sitosterolemia in patients [8–10], the *Abcg5/g8* has also been identified as the mouse gallstone gene, *Lith9*, on chromosome 17 by quantitative trait locus (QTL) linkage analysis [25–28]. Subsequently, the *ABCG5/G8* was found to be associated with cholesterol gallstone disease in patients, and two gallstone-associated variants in *ABCG5/G8* (*ABCG5*-R50C and *ABCG8*-D19H) were identified not only in Germans and Chileans but also in Chinese and Indians [29–34]. These findings indicate the importance of *ABCG5/G8* as *LITH9* in the pathogenesis of gallstones not only in mice but also in humans [14].

In this chapter, we summarize the latest advances in the critical role of the sterol efflux transporters ABCG5/G8 in regulating hepatic secretion of biliary cholesterol, intestinal absorption of cholesterol and plant sterols, and reverse cholesterol transport, as well as in the pathogenesis and pathophysiology of ABCG5/G8-related metabolic diseases such as sitosterolemia, cardiovascular disease, and cholesterol gallstone disease.

8.2 Chemistry of Cholesterol and Plant Sterols

By definition, a steroid is a biologically active organic compound with four rings arranged in a specific molecular configuration, including the sterols, hormones (such as anabolic steroids or corticosteroids), and glycosides. The steroid core structure is typically composed of 17 carbon atoms, bonded in 4 "fused" rings: 3 6-member cyclohexane rings, called the A, B, and C rings, and 1 5-member cyclopentane ring, named the D ring [1]. It is well-known that the basic chemical structure of steroids has a nucleus containing the four-ringed carbon skeleton of cyclopentenophenanthrene and the numbering of the carbon atoms in steroids [1]. Furthermore, sterols are various solid steroid alcohols that are widely distributed in human, animal, and plant lipids. It is often

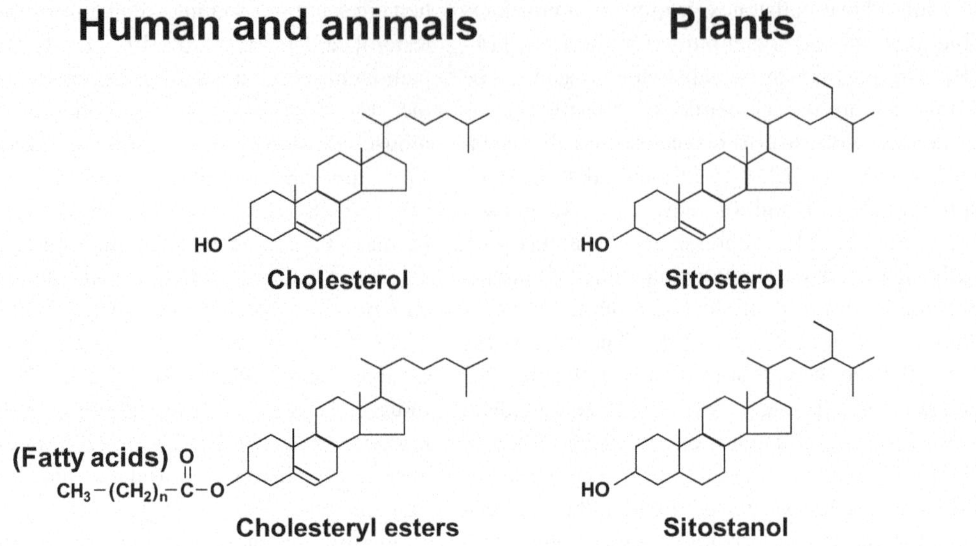

Fig. 8.1 All these substances have a nucleus containing the four-ringed carbon skeleton of cyclopentenophenanthrene and are known as steroids. The sterols are one of the steroids and they are widely distributed in humans, animals, and plants. It is often called cholesterol in humans and animals and phytosterols (also called plant sterols) in plants. Notably, the general structural formula for the sterols includes the designation of the four rings with a side chain at C-17 and two methyl groups at C-18 and C-19. Cholesterol is one of the most abundant steroids in bile. Its hydroxyl group on the third carbon can react with the COOH group of a fatty acid molecule to form a cholesteryl ester. Plant sterols (e.g., β-sitosterol and β-sitostanol) are naturally occurring. Their chemical structures are very similar to cholesterol but with structural modifications of the side chain. In addition, stanols are saturated sterols, having no double bonds in the sterol ring structure, e.g., β-sitostanol

called cholesterol in humans and animals, as well as phytosterols, or plant sterols, in plants.

As shown in Fig. 8.1, the basic chemical structure of the cholesterol molecule includes (i) the perhydrocyclopentenophenanthrene nucleus with its four fused rings, (ii) a single hydroxyl group at C-3, (iii) a double bond between C-5 and C-6, (iv) an eight-membered branched hydrocarbon chain attached to C-17 in the D ring, and (v) a methyl group (C-19) attached to C-10, and a second methyl group (C-18) attached to C-13. Furthermore, in the esterified form, a long-chain fatty acid, usually linoleic acid, is attached by ester linkage to the hydroxyl group at C-3 in the A ring. Similar to cholesterol in humans and animals, phytosterols, which encompass plant sterols and stanols, are phytosteroids, which occur in plants and vary only in carbon side chains and/or presence or absence of a double bond. Stanols are saturated sterols, having no double bonds in the sterol ring structure (Fig. 8.1).

8.3 Discovery of the Sterol Efflux Transporters ABCG5/G8

The ATP-binding cassette (ABC) transporters are a family of large proteins in cell membranes. Using the energy from the ATP hydrolysis, these ABC transporters can make an active transport of various compounds crossing the cell membranes against steep concentration gradients [35]. Hitherto, 48 ABC genes have been found in the human genome [36]. The major physiological functions of these ABC transporters are involved in an active transport of a wide variety of substrates across extracellular and intracellular membranes, which include lipids, amino acids, sugars, vitamins, metals, drugs (xenotoxins) and drug conjugates, and peptides for antigen presentation or other purposes [37]. Of the 48 human ABC proteins, a significant number are known to mediate the extrusion of lipids from membranes or the flipping of membrane lipids across the bilayer to generate and maintain membrane lipid asymmetry [38]. For example, the bile salt export pump, ABCB11, is responsible for hepatic secretion of biliary bile salts. Other members of the subfamily of ABC transporters such as ABCB4, ABCG1, ABCC2, and ABCA1 implicated in lipid transport play important roles in diverse biological processes involving hepatic phospholipid secretion, cell signaling, membrane lipid asymmetry, removal of potentially toxic compounds and metabolites, and apoptosis [39]. The importance of the ABC lipid transporters in cell physiology is revealed based on the finding that mutations in the genes encoding many of these proteins are responsible for severe inherited diseases. At least 14 ABC genes have been found to be associated with a defined human disease due to genetic defects [40]. Especially, several ABC transporters are involved in inborn errors relevant to metabolic disorders [41]. For example, Tangier disease is caused by mutations in the *ABCA1* gene, which is associated with defective efflux of cholesterol and phosphatidylcholine from the plasma membrane to the lipid acceptor protein, apolipoprotein A-I (apoA-I) [42]. In addition, relative phospholipid deficiency is caused mostly by missense mutations in the ABC subfamily B member 4 (*ABCB4*) gene, also known as the multidrug resistance protein 3 (*MDR3*) gene. The *ABCB4* gene encodes for an energy-dependent phospholipid efflux translocator at the canalicular membrane of the hepatocytes, which facilitates the transport of phospholipids from the inner to the outer canalicular membrane of hepatocytes for hepatic secretion into canalicular bile [43–46].

The half-transporters, ABCG5 and ABCGG8, are found to heterodimerize into a functional transport. The genes, *ABCG5* and *ABCGG8*, encoding these transporters are highly expressed in the liver and small intestine of both humans and mice [47–49]. The *ABCG5/G8* genes are located on chromosome 2p21 in humans and chromosome 17 in mice. The two proteins form heterodimers in the endoplasmic reticulum and then traffic to the canalicular membrane of hepatocytes and the apical membrane of enterocytes where they transport neutral sterols into bile and into the gut lumen, respectively [48]. Further cellular and molecular studies found that ABCG5/G8 play a critical role in regulating hepatic secretion and intestinal absorption of cholesterol and plant sterols. Mutations in either *ABCG5* or *ABCG8* cause sitosterolemia [8–10], which is an autosomal recessive disorder characterized by phytosterolemia, hypercholesterolemia, and premature coronary heart disease [50].

8.4 Physiological Functions of ABCG5/G8

Many studies have found that almost all the cells in the body need a continuous supply of cholesterol. As a result, a series of complex and sophisticated transport, biosynthetic, and regulatory mechanisms have developed in humans and animals [51, 52]. Under normal physiological conditions, cholesterol is obtained from the intestinal absorption of dietary and biliary cholesterol, as well as the newly synthesized de novo from acetyl CoA in the body. However, because human and animal tissues do not possess enzymes that can degrade the ring structure of this sterol, cholesterol cannot be metabolized to CO_2 and water in the body. Therefore, to prevent a potentially hazardous accumulation of cholesterol in the body, excess cholesterol must be metabolized to other compounds and/or excreted in the feces. This challenging task is usually accomplished by chemical modifications of certain substituent groups on the hydrocarbon tail or on the ring structure of the cholesterol molecule. Subsequently, excess cholesterol is excreted from the body essentially either as the unaltered molecule (i.e., in both unesterified and esterified forms) or after structural modifications to other sterol products such as bile salts and steroid hormones.

It has been recognized that the cholesterol molecule is a key lipid component of mostly all the cell membranes, as well as is the precursor of various steroid hormones such as the sex hormones (estrogen, progesterone, and testosterone) and corticosteroids (cortisone, corticosterone, cortisol, and aldosterone) [53–

56]. Moreover, during the biosynthesis of bile salts in the liver, cholesterol is mainly converted into bile salts. As a result, large amounts of biliary cholesterol and bile salts are simultaneously secreted to bile. This dramatically reduces plasma cholesterol concentrations and greatly enhances removal of excess amounts of cholesterol from the body.

Because cholesterol is virtually insoluble in an aqueous solution, e.g., water, the mechanisms for cholesterol solubilization in plasma and bile are complex. It is well-known that cholesterol is mainly carried by lipoproteins in plasma and by micelles and vesicles in bile. If excess cholesterol is accumulated in the artery wall, it leads to atherosclerosis and causes cardiovascular disease. If excess cholesterol cannot be dissolved in bile by bile salts and/or phospholipids, it precipitates as plate-like solid cholesterol monohydrate crystals, thus leading to the formation of cholesterol gallstones in the gallbladder and/or the bile duct.

Based on animal studies [57–59], several pathways have been identified for elucidating the net flow of cholesterol through the major tissue compartments of the human, which explains how the cholesterol pool in the body is kept essentially constant. New cholesterol is added to the pool mainly from two sources: the absorbed cholesterol from dietary and biliary origins across the epithelial cells of small intestinal tract and the newly synthesized cholesterol in a variety of different tissues in the body, predominantly in the liver and small intestine. The availability of dietary and biliary cholesterol to the body varies tremendously in different individuals, and the consumed amounts of dietary cholesterol also vary dramatically from day to day [57–68]. The total amount of cholesterol from the small intestine to the body also depends mainly on the absorption efficiency of intestinal cholesterol and the amount of cholesterol that is consumed daily. Additionally, bile cholesterol is reabsorbed by the small intestine, which provides about two thirds of the total daily amount of cholesterol originating from the intestine [2]. The rate of cholesterol biosynthesis in the liver varies extremely in different individuals. The absorbed cholesterol from the small intestine can regulate hepatic cholesterol synthesis, depending on the amount of daily food intake, through a negative regulatory mechanism.

Taken together, the regulatory mechanisms on cholesterol metabolism must be operative, which can accurately and sophisticatedly adjust the rate of cholesterol biosynthesis in the body and the rate of cholesterol excretion from the body to accommodate the varying amounts of cholesterol that are absorbed by the small intestine at different times. Basically, these regulatory mechanisms on cholesterol metabolism work well. Therefore, there is little net accumulation of excess cholesterol in the body, and yet sufficient cholesterol is always available to meet the metabolic needs of the various cells. However, delicate imbalances lead to an increase in plasma cholesterol concentration and/or hepatic cholesterol hypersecretion in humans [69–72]. In the cardiovascular system, this metabolic abnormality often causes an accumulation of excess cholesteryl esters within the wall of arteries, leading to clinically apparent atherosclerosis mainly in the heart and brain and causing cardiovascular disease [73–80]. In the biliary system, when an imbalance of cholesterol metabolism in bile occurs, gallbladder bile becomes supersaturated with cholesterol, thereby promoting the precipitation of plate-like solid cholesterol monohydrate crystals and, eventually, leading to clinically apparent cholesterol gallstone formation [81–91].

Because the sterol efflux transporters ABCG5/G8 play a key role in the regulation of cholesterol metabolism in bile and plasma and in the maintenance of cholesterol homeostasis in the body, we will discuss the regulatory mechanisms of ABCG5/G8 in (i) hepatic secretion of biliary cholesterol; (ii) intestinal absorption of cholesterol and plant sterols; (iii) reverse cholesterol transport; and (iv) transintestinal cholesterol excretion.

(a) Hepatic secretion of biliary cholesterol

Bile is composed mainly of water, organic solutes, and inorganic electrolytes. Cholesterol, phospholipids, and bile salts are three major lipid species in bile, which account for approximate 99% of total lipids by weight. Bilirubin is a minor solute and represents less than 1% of biliary lipids. Hepatic secretion of biliary cholesterol and its degradation product, bile salts, represents the major route for elimination of cholesterol from the liver and, eventually, from the body. After entering the intestinal lumen, bile salts play an important role in the emulsification of dietary lipids and the breakdown of large lipid globules into a suspension of droplets for intestinal absorption. In addition, bile salts promote the intestinal absorption of cholesterol, fatty acids, fat-soluble vitamins (A, D, E, and K), and certain drugs.

Hepatic secretion of biliary lipids is determined by four members of the family of ABC transporters on the canalicular membrane of hepatocytes: ABCB4 for phospholipids, ABCB11 for bile salts, ABCG5/G8 for cholesterol, and ABCC2 for bilirubin (Fig. 8.2). Most, if not all, bile salts enter the canalicular space as monomers, whereas biliary phospholipids and cholesterol could enter together as unilamellar vesicles. Bile salts play a critical role in promoting hepatic secretion of vesicles that are always found in hepatic bile by quasi-elastic light-scattering spectroscopy and electronic microscopy with rapid fixation techniques. These imaging studies have provided clear morphologic evidence of the vesicle formation and secretion on the outer surface of the canalicular membrane of hepatocytes during the bile formation.

Although biliary phospholipids are derived possibly from the cell membranes of hepatocytes, their compositions differ significantly. The cell membranes of hepatocytes contain a large amount of phosphatidylcholine (i.e., lecithin), phosphatidylethanolamine, phosphatidylinositol, phosphatidylserine, and sphingomyelin. The

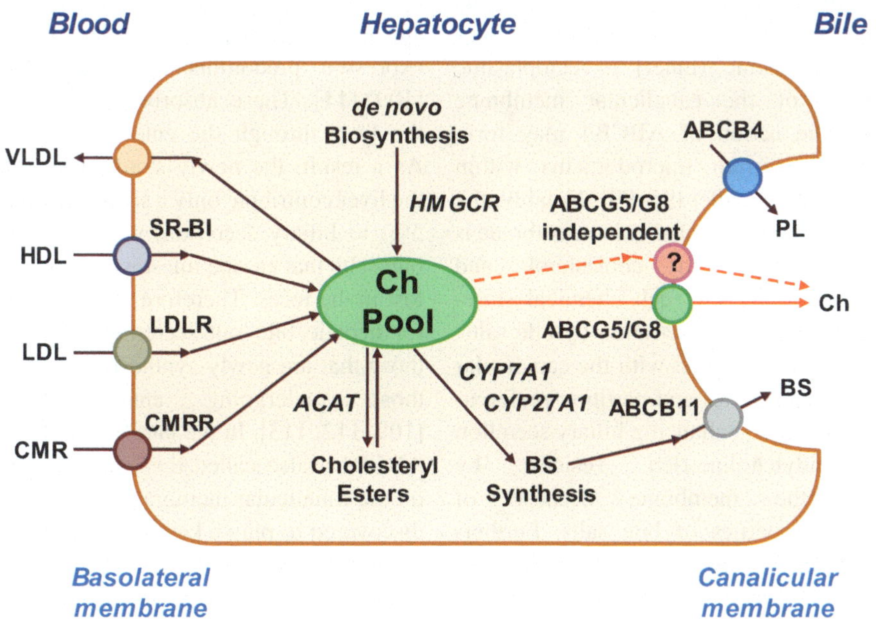

Fig. 8.2 This diagram of the hepatocyte shows the ABCG5/G8-dependent (red solid lines) and the ABCG5/G8-independent (red dashed lines) pathways for biliary cholesterol (Ch) secretion, as well as the ABCB4 and ABCB11 transporters for biliary phospholipid (PL) and bile salt (BS) secretion, respectively. Abbreviations: *ABC* ATP-binding cassette (transporter), *ACAT* acyl-coenzyme A:cholesterol acyltransferase, *CMR* chylomicron remnants, *CMRR* CMR receptor, *CYP7A1* cholesterol 7-α-hydroxylase, *CYP27A1* sterol 27-hydroxylase, *HDL* high-density lipoprotein, *HMGCR* 3-hydroxy-3-methylglutaryl-coenzyme A reductase, *LDL* low-density lipoprotein, *LDLR* LDL receptor, *SR-BI* scavenger receptor class B type I, i.e., HDL receptor, *VLDL* very-low-density lipoprotein

major source of the phosphatidylcholine molecules destined for hepatic secretion into bile is its synthesis in the liver. However, a fraction of biliary phosphatidylcholines may also originate from the surface phospholipid coat of HDL particles. In the early 1990s, it was first reported that hepatic phospholipid secretion is a protein-mediated process because deletion of the *Abcb4* gene completely inhibits hepatic phospholipid secretion in mice [43]. This important study provided clear evidence for the first time that a P-glycoprotein member of the multidrug resistance gene family, ABCB4, plays a key role in regulating hepatic secretion of biliary phospholipids [43]. Studies from cryoelectron microscopy with rapid fixation techniques found that the knockout of the *Abcb4* gene in mice dramatically reduces the formation and secretion of vesicles on the outer surface of the canalicular membrane of hepatocytes [92–94]. It is highly likely that ABCB4 could be responsible for the translocation or "flip" of phosphatidylcholines from the endoplasmic (inner) to ectoplasmic (outer) leaflet of the canalicular membrane bilayer, and the action of ABCB4 may form phosphatidylcholine-rich microdomains within the outer membrane leaflet [95–99]. Notably, the ectoplasmic leaflet of the canalicular membrane is composed mainly with cholesterol and sphingomyelin. However, such chemical structure is quite resistant to penetration by bile salts. Thus, bile salts may interact with the canalicular membrane of hepatocytes and partition preferentially into these areas, enhancing biliary secretion of phosphatidylcholine-rich vesicles by destabilizing the membrane because of detergent-like properties of bile salts. Furthermore, mutations in the *ABCB4* gene are the molecular defect underlying progressive familial intrahepatic cholestasis, type III in humans [99–104]. In addition, biliary phospholipids can dramatically solubilize excess cholesterol in bile through a vesicle mechanism. Low phospholipid-associated cholelithiasis is characterized mainly by the occurrence of gallbladder and intrahepatic microlithiasis in young adults associated with mutations in the *ABCB4* gene [105–107]. To study the pathogenesis of low phospholipid-associated cholelithiasis, gallstone phenotypes have been systematically investigated in the ABCB4 knockout mouse model. It is interesting to find that even on the chow diet containing trace amounts of (<0.02%) cholesterol, ABCB4 knockout mice can spontaneously develop gallstones that are composed mainly of needle-shaped anhydrous cholesterol crystals [98]. These anhydrous cholesterol crystals and gallstones are formed in phospholipid-deficient gallbladder bile with its relative biliary lipid composition that is located in the far left crystallization region of the phase diagram [108]. These studies support the concept that this gene is a monogenic risk factor for this "peculiar" form of cholesterol gallstones and a target for novel therapeutic strategies in humans.

After bile salts are secreted into bile and enter the intestine, approximately 95% of the secreted bile salts are absorbed through an active transport mechanism by a specific bile salt transporter, apical sodium-dependent bile salt transporter expressed predominantly in the distal ileum [109–111]. These absorbed bile salts return to the liver through the enterohepatic circulation. As a result, the newly synthesized bile salts in the liver contribute only a small fraction (less than 5%) to biliary secretion, which compensate for bile salts that escape intestinal absorption and are lost in the feces. Therefore, there are two sources for hepatic bile salt secretion, which consist of those that are newly synthesized in the liver and those undergoing enterohepatic cycling [109, 112, 113]. In the late 1990s, the transporter ABCB11, also called the bile salt export pump, on the canalicular membrane of hepatocytes, was discovered to play a key role in hepatic secretion of biliary bile salts [114–118]. Deletion of the *Abcb11* gene in mice completely impedes hepatic bile salt secretion. The cellular and molecular mechanisms by which bile salt secretion is coupled to cholesterol and phospholipid secretion are still under extensive investigations. Notably, hepatic secretion of bile salts could directly affect phospholipid vesicle secretion [119–122]. The relationship between bile salt secretion and cholesterol secretion has been found to be curvilinear. At a low hepatic bile salt secretion rate

(<10 μmol/h/kg), more biliary cholesterol is secreted per molecule of bile salts compared to that at a higher rate. Although biliary bile salt secretion is not often low in healthy individuals, it could be reduced during prolonged fasting, during the overnight period, and with substantial bile salt losses such as with a biliary fistula or ileal resection when the liver cannot sufficiently compensate with increased bile salt synthesis. In contrast, at a high bile salt secretion rate, for example, during and after eating, biliary saturation is less compared to that during the interprandial period. Although biliary organic anion secretion does not influence bile acid secretion, it inhibits hepatic secretion of biliary phospholipids and cholesterol because organic anions can bind bile salts within the bile canaliculi and curtail interactions with the canalicular membrane of hepatocytes.

Many animal and human studies have found that bile salts promote vesicle secretion by the hepatocytes, and these unilamellar vesicles are always found in freshly collected hepatic bile [123–128]. In the early 2000s, genetic studies in patients with sitosterolemia revealed that the efflux of biliary cholesterol from the canalicular membrane of hepatocytes to bile is a protein-mediated process [8, 9, 129–139], which is determined by the sterol efflux transporters ABCG5/G8. Mutations in either *ABCG5* or *ABCG8* significantly reduce biliary cholesterol secretion in patients with sitosterolemia. The key role of ABCG5/G8 in hepatic cholesterol secretion has been investigated in genetically modified mice [11, 12, 140–142]. Overexpression of the human *ABCG5/G8* gene in the liver increases the cholesterol content of gallbladder bile in transgenic mice. In contrast, hepatic secretion of biliary cholesterol is dramatically reduced in ABCG5/G8 double knockout mice, as well as in ABCG5 or ABCG8 single gene knockout mice. More interestingly, clinical studies found that sitosterolemia is caused by a mutation in either the *ABCG5* or the *ABCG8* gene alone, but not in both simultaneously, and hepatic cholesterol secretion is dramatically reduced, but not completely eliminated in these patients [50, 135, 143, 144]. To further study the cellular and molecular mechanisms underlying the key role of ABCG5/G8 in biliary sterol secretion, biliary cholesterol and sitostanol secretion is quantified for 6 h in ABCG8 knockout mice. Mass transport rate of [^{3}H]sitostanol from plasma HDL into bile is significantly faster than that of [^{14}C]cholesterol in wild-type mice; however, reduced amounts of [^{14}C]cholesterol and no [^{3}H]sitostanol are detected in bile of ABCG8 knockout mice [141]. These results indicate that knockout of the *Abcg8* gene alone significantly reduces but does not eliminate hepatic cholesterol secretion. In addition, biliary cholesterol secretion studies uncovered that hepatic cholesterol output is dramatically diminished, but cholesterol is still secreted into bile in mice with the targeted deletion of either *Abcg5* or *Abcg8* alone, or both [11–13, 141, 145]. In agreement with the human data, these mouse results imply that an ABCG5/G8-independent pathway could also be involved in the regulation of hepatic cholesterol secretion in both humans and mice. In addition, scavenger receptor class B type I (SR-BI), the HDL receptor, is expressed mainly in the sinusoidal and, perhaps, in the canalicular membrane of hepatocytes. In transgenic and knockout mice, biliary secretion of cholesterol varies in proportion to the hepatic expression of SR-BI, and the established contribution of SR-BI to sinusoidal uptake of HDL cholesterol is destined for secretion into bile [146–148]. These studies indicate that SR-BI could play a critical role in the reverse cholesterol transport in the body.

(b) Intestinal absorption of cholesterol and plant sterols

Cholesterol is the most abundant steroid in human and animal tissues and in the intestinal lumen. It is poorly soluble in an aqueous environment. In addition to a double bond at C-5 and C-6 nucleus and a hydroxyl group on the third carbon of the cholestene nucleus (Fig. 8.1), the β-configuration is connected with the angular methyl groups at C-10 and C-13, the hydrogen atom at C-8, and the side chain at C-17. The hydrogen atoms at C-9 and C-14 are in the α-configuration [149].

Phytosterols, also called plant sterols, refer to sterols that originate from plants. These are abundant in the intestine, but not in human and animal tissues. As shown in Fig. 8.1, plant sterols are naturally occurring, and their chemical structures are very similar to cholesterol, i.e., a Δ^5 double bond and a 3β-hydroxyl group, but with structural modifications of the side chain. Plant sterols have the same basic importance in plants as cholesterol in animals, playing a vital role in cell membrane function. Sitosterol and campesterol, which are 24-ethyl and the 24-methyl substituted variants of cholesterol, respectively, are the most abundant plant sterols [149]. They are consumed in the diet and may be absorbed in the intestine. However, they are often present only at very low plasma concentrations in human and animal tissues due to a poor (<5%) net absorption rate by the small intestine. Other sterols such as brassicasterol and isofucosterol may also originate from shellfish.

As shown in Fig. 8.3, within the intestinal lumen, the micellar solubilization of cholesterol and fatty acids facilitates movement through the diffusion barrier overlying the surface of the absorptive cells. In the presence of bile salts, mixed micelles deliver large amounts of the cholesterol molecules to the aqueous-membrane interface so that the uptake rate is greatly increased. Human and animal studies have found that the Niemann-Pick C1 like 1 (NPC1L1) protein, a sterol influx transporter, is expressed at the apical membrane of the enterocytes and can actively facilitate the uptake of cholesterol by promoting the passage of cholesterol across the brush border membrane of the enterocytes. Moreover, NPC1L1 plays a key role in the ezetimibe-sensitive cholesterol absorption pathway [150–154], which is highly likely to make the influx of cholesterol and likely plant sterols from the intestinal lumen into the cytoplasm of enterocytes. NPC1L1 could mediate cholesterol uptake via vesicular endocytosis, and ezetimibe may inhibit cholesterol absorption by suppressing the internalization of NPC1L1/cholesterol complex. In contrast, ABCG5/G8 are apical sterol export pumps promoting active efflux of cholesterol and plant sterols from the enterocytes back into the intestinal lumen for fecal excretion [8, 9, 12, 47, 48, 131, 141, 155–158]. The combined regulatory actions of NPC1L1 and ABCG5/G8 play a pivotal role in modulating the amount of cholesterol that reaches the lymph from the intestinal lumen. These findings imply that intestinal cholesterol absorption is a multistep process that is regulated by multiple genes at the enterocyte level and that the efficiency of cholesterol absorption is determined by the net effect between influx and efflux of intraluminal cholesterol molecules crossing the brush border membrane of the enterocyte [159]. In addition, 3-hydroxy-3-methylglutaryl coenzyme A reductase (HMGCR) is the rate-limiting enzyme for cholesterol biosynthesis in the body [160–165]. Cholesterol that is synthesized de novo from acetyl CoA in different organs (i.e., the liver and small intestine) is the second major source to the body [166–173]. The absorbed cholesterol molecules, as well as some that are newly synthesized from acetate by HMGCR within the enterocytes, are esterified to fatty acids by acyl-CoA:cholesterol acyltransferase isoform 2 (ACAT2) to form cholesteryl esters. Furthermore, there are three putative pathways for uptake of fatty acids and their transport across the apical membranes of enterocytes, either by simple passive diffusion mostly for short-chain fatty acids or by multiple transporters and proteins such as fatty acid transport protein 4 (FATP4), CD36 (also referred to as fatty acid translocase), and plasma membrane-associated fatty acid-binding protein (FABPpm; 43 kDa) largely for medium- and long-chain fatty acids. Finally, all of these lipids are used for the assembly of chylomicrons, which also requires the synthesis of apoB-48 and the activity of microsomal triglyceride transfer protein (MTTP). The core of chylomicrons secreted in lymph contains triglycerides and cholesteryl esters, and their surface is a monolayer containing phospholipids (mainly phosphatidylcholine), unesterified cholesterol, and apolipoproteins such as apoB-48, apoA-I, and apoA-IV [149].

Although daily intake of cholesterol and plant sterols from the diet is almost equal, the intestinal absorption efficiency is significantly lower in the latter compared to the former. For example, the

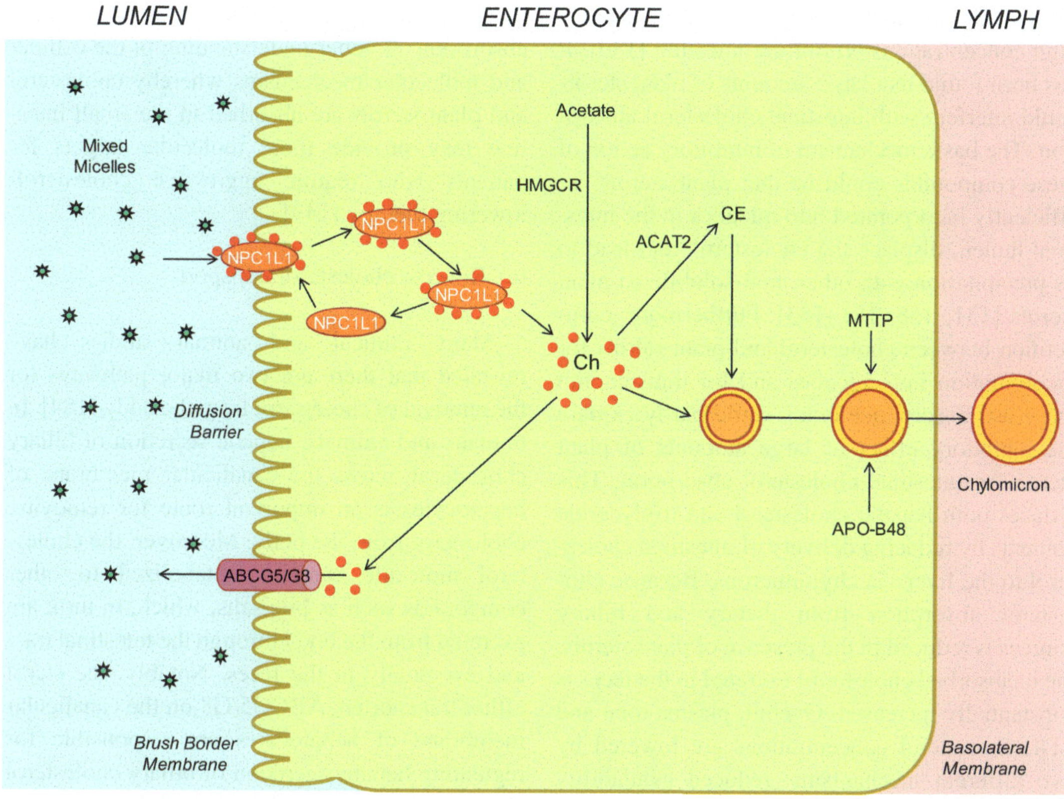

Fig. 8.3 Molecular and cellular mechanisms of intestinal cholesterol absorption. Within the intestinal lumen, the micellar solubilization of sterols facilitates movement through the diffusion barrier overlying the surface of the absorptive cells in the small intestine. In the presence of bile salts, mixed micelles deliver large amounts of the cholesterol (Ch) molecules to the aqueous-membrane interface so that the uptake rate is greatly increased. The Niemann-Pick C1 like 1 (NPC1L1) protein, a sterol influx transporter, is located at the apical membrane of the enterocyte and can actively facilitate the uptake of cholesterol by promoting the passage of cholesterol across the brush border membrane of the enterocyte. NPC1L1 appears to mediate cholesterol uptake via vesicular endocytosis, and ezetimibe may inhibit cholesterol absorption by suppressing the internalization of NPC1L1/cholesterol complex. In contrast, ABCG5/G8 promote active efflux of cholesterol from the enterocyte back into the intestinal lumen for fecal excretion. The combined regulatory effects of NPC1L1 and ABCG5/G8 play a critical role in modulating the amount of cholesterol that reaches the lymph from the intestinal lumen. The absorbed cholesterol molecules, as well as some that are newly synthesized from acetate by 3-hydroxy-3-methylglutaryl-CoA reductase (HMGCR) within the enterocytes, are esterified to fatty acids by acyl-CoA:cholesterol acyltransferase isoform 2 (ACAT2) to form cholesteryl esters (CE). All of these lipids are involved in the assembly of chylomicrons, which also requires the synthesis of apolipoprotein B-48 (apoB-48) and the activity of microsomal triglyceride transfer protein (MTTP). The core of chylomicrons secreted in lymph contains triglycerides and cholesteryl esters, and their surface is a monolayer containing phospholipids (mainly phosphatidylcholine), unesterified cholesterol, and apolipoproteins such as apoB-48, apoA-I, and apoA-IV

absorption efficiency of sitosterol and campesterol is 5–8% and 9–18%, respectively [174], compared with 30–60% of intestinal cholesterol absorption in humans [175–179]. It is highly likely that most of the plant sterols that do enter the enterocyte are rapidly pumped back into the intestinal lumen for excretion, as done by the actions of ABCG5/G8. In addition to poor net absorption, plant sterols are more efficiently secreted into bile compared to cholesterol. These combined mechanisms maintain plasma plant sterol concentrations at less than 1 mg/dL in humans. Because plant sterols are insoluble, they must be esterified and incorporated into triglycerides in margarines in order to achieve high concentrations within the intestine [180]. It has been found that large amounts of plant sterols could interfere with intestinal cholesterol absorption. The basic mechanism of inhibitory action of these compounds could be that plant sterols are efficiently incorporated into micelles in the intestinal lumen, displace the cholesterol, and lead to its precipitation with other, non-solubilized plant sterols [131, 158, 181–183]. Furthermore, competition between cholesterol and plant sterols for incorporation into micelles and for transfer into the brush border membrane could partly explain the inhibitory effect of large amounts of plant sterols on intestinal cholesterol absorption. This reduces both hepatic cholesterol and triglyceride contents by reducing delivery of intestinal cholesterol to the liver via chylomicrons. Because cholesterol absorption from dietary and biliary sources is reduced in the presence of plant sterols, the unabsorbed cholesterol excreted in the feces is substantially increased. Overall, plasma total and LDL cholesterol concentrations are lowered by two different mechanisms: reduced availability of cholesterol for incorporations into VLDL particles and increased expression of LDL receptor in the liver.

The identification of defective structures in the sterol efflux transporters ABCG5/G8 in patients with sitosterolemia indicates that cholesterol absorption is a selective process; also the activities of ABCG5/G8 provide an explanation for the selectivity against plant sterols so that plant sterols are absorbed poorly [159]. The NPC1L1 is also expressed at the apical membrane of enterocytes and plays a decisive role in the ezetimibe-sensitive cholesterol absorption pathway. As discussed above, intestinal cholesterol absorption is a multistep process that is regulated by multiple genes at the enterocyte level. The significant inter-individual differences in cholesterol absorption efficiency found in humans and the variations observed in inbred strains of mice strongly suggest that many additional genes may be involved in the regulation of intestinal cholesterol absorption. These differences also provide opportunities to apply advanced genetic techniques to identify the responsible genes that contribute to the regulation of intestinal lipid absorption. A better understanding of the cellular and molecular mechanisms whereby cholesterol and plant sterols are absorbed in the small intestine may provide more molecular targets for patients who require aggressive cholesterol-lowering therapy [149].

(c) Reverse cholesterol transport

Many clinical and animal studies have revealed that there are two major pathways for the removal of cholesterol from the body [184]. In humans and animals, hepatic secretion of biliary cholesterol across the canalicular membrane of hepatocytes is an important route for removing cholesterol from the body. Moreover, the cholesterol molecule can be metabolized to other compounds such as bile salts, which, in turn, are excreted from the body through the intestinal tract and eventually in the feces. Notably, the sterol efflux transporters ABCG5/G8 on the canalicular membrane of hepatocytes are responsible for regulating hepatic secretion of biliary cholesterol [11, 12, 140, 142, 185], and the bile salt export pump, ABCB11, plays a critical role in hepatic secretion of biliary bile salts [186]. These transporters in the liver has a vital impact on determining excretion of excess cholesterol from the body, either as unesterified cholesterol or as its metabolic products, bile salts.

In the mid-1960, the definition of the reverse cholesterol transport and the speculated role of HDL in promoting this process were first proposed [187]. Classically, the reverse cholesterol transport is a process involved in the removal of excess cholesterol that is accumulated in the peripheral tissues (e.g., macrophages in the aortae) by HDL, transporting it to the liver for excretion into the feces via the bile (Fig. 8.4). In the 1980s and 1990s, many results from animal studies strongly supported the concept that HDL could play a critical role in protecting against cardiovascular disease [188–191]. Subsequently, numerous clinical studies found that plasma HDL is the smallest lipoprotein particles and contains the highest proportion of apolipoproteins to lipids compared to LDL, VLDL, and chylomicrons.

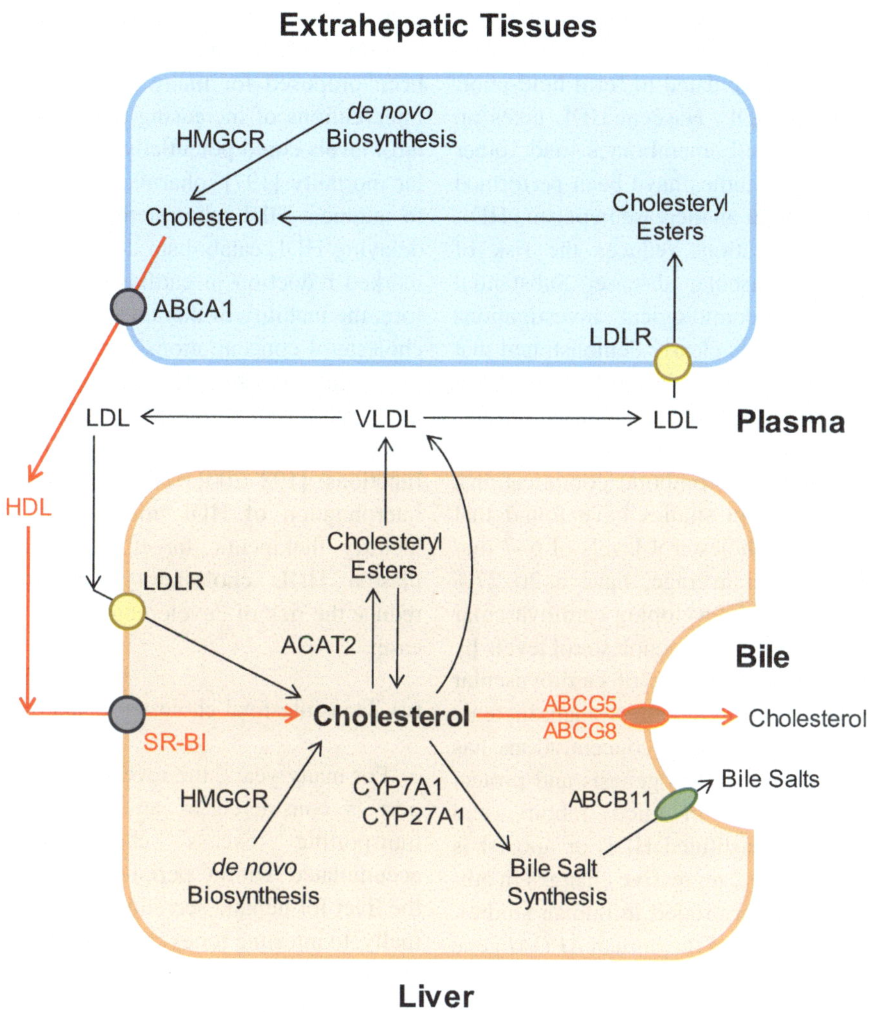

Fig. 8.4 The reverse cholesterol transport through the classical biliary route to the bile as secreted by ABCG5/G8 on the canalicular membrane of hepatocytes. The reverse cholesterol transport in the hepatocytes is shown in red lines with arrows indicating the direction of transport. Abbreviations: *ABC* ATP-binding cassette (transporter), *ACAT2* acyl-coenzyme A:cholesterol acyltransferase isoform 2, *CYP7A1* cholesterol 7-α-hydroxylase, *CYP27A1* sterol 27-hydroxylase, *HDL* high-density lipoprotein, *HMGCR* 3-hydroxy-3-methylglutaryl-CoA reductase, *LDL* low-density lipoprotein, *SR-BI* scavenger receptor class B type I, i.e., HDL receptor, *VLDL* very-low-density lipoprotein

Although the molecular and genetic mechanisms underlying its beneficial properties in humans are not fully understood, HDL is most widely recognized for its ability to promote cholesterol efflux from the macrophages and other cells in the extrahepatic tissues and transport cholesterol from the periphery to the liver for hepatic secretion and, subsequently, fecal excretion. Obviously, during the process of the reverse cholesterol transport, the deposition of cholesterol in the peripheral tissues, including the aortae, is greatly reduced [192–194]. Many animal studies have consistently found that HDL is protective on several processes that are involved in preventing atherosclerosis, at least in part by mediating the removal of cholesterol from lipid-laden macrophages through the reverse cholesterol transport [189, 195, 196].

The major HDL-associated apoA-I and apoA-II are secreted into plasma by the liver and intestine, where they are lipidated to form lipid-poor, discoidal, nascent HDL. Nascent HDL takes up cholesterol from cell membranes and other lipoproteins. Many studies have been performed to investigate whether an increase in plasma HDL cholesterol concentrations reduces the risk of developing cardiovascular disease. Substantial evidence from epidemiological investigations and clinical studies has clearly demonstrated that the level of plasma HDL cholesterol, especially at average to slightly above average concentrations, is inversely related to the incidence of cardiovascular disease and its thrombotic complications. Prospective population studies have found that humans with HDL cholesterol levels of 6–7 mg/dL, i.e., higher than average, have a 20–27% decrease in the risk of developing cardiovascular disease, and increasing HDL cholesterol levels by 1 mg/dL may reduce the risk of cardiovascular disease by 2% in men and 3% in women. Increasing plasma HDL cholesterol concentrations has been found to prevent atherogenesis and protect against atherosclerosis in mice, rabbits, and humans. When reconstituted HDL or apoA-I is provided exogenously, regressive changes in atherosclerotic plaques are found in human studies. Transgenic expression of the human *APOAI* gene increases HDL and suppresses atherosclerosis in APOE knockout mice, and genetic lowering of plasma HDL in mice reduces the appearance of macrophage-derived cholesterol in feces. Collectively, these results from human and animal studies have led to the idea that increasing plasma HDL may be a new strategy for the treatment and the prevention of cardiovascular disease.

Although most published studies attribute the atheroprotective properties of HDL to HDL_2, a lot of results also reveal that HDL_3 may be inversely related to the risk of developing cardiovascular disease. More recently, clinical studies of HDL metabolism have focused mainly on plasma total HDL cholesterol concentrations, but not on each HDL subclass. In addition, cardiovascular risk associated with HDL cholesterol levels is independent of plasma LDL cholesterol concentrations, as well as other lipid parameters (e.g., triglycerides and total cholesterol), and non-lipid risk factors. Although the concept has been proposed for many years that therapeutic interventions of increasing plasma HDL cholesterol levels could potentially reduce cardiovascular mortality [197], pharmacologic interventions to augment HDL cholesterol concentrations by delaying HDL catabolism do not translate into a marked reduction in cardiovascular risk. Therefore, the inability of therapies of increasing HDL cholesterol concentrations and new insights into the complexity of HDL composition and function have prompted researchers to further explore whether and how HDL exerts its cardioprotective functions [198–200]. Nevertheless, systematic interpretation of HDL metabolism could help identify therapeutic targets that may increase plasma HDL cholesterol concentrations and reduce the risk of developing cardiovascular disease.

(d) Transintestinal cholesterol excretion (TICE)

For many years, the reverse cholesterol transport is considered as an important route for transporting excess cholesterol that is accumulated within peripheral tissues back to the liver for hepatic secretion into bile and, eventually, to intestine for excretion in the feces. Some studies on patients with hepatobiliary and/or pancreatic disorders and several animal models with obstruction of the bile duct or cholestasis have found a novel non-biliary transport route likely for reverse cholesterol transport, independent of classical pathway of the reverse cholesterol transport through the liver. In the late 1950s, a secondary, non-biliary pathway was proposed, which was defined as the transintestinal cholesterol excretion (TICE) [201]. It is suggested that the TICE may contribute a new way to the reverse cholesterol transport. However, these studies were greatly criticized about the selection of patients and animal models because dramatic diminution or discontinuation of bile flow entering the small intestine could damage the normal physiological function of the epithelial cells of small intestine. Moreover, these alterations could lead to a remarkable reduction in intestinal lipid absorption because of a lack of bile salts. Such results with a striking increase in fecal neutral sterols were questioned because these studies were performed under conditions of severe hepatobiliary disease and inappropriate experimental approaches. Consequently, the TICE was not accepted even though this new concept challenged the classical view of the reverse cholesterol transport by showing that the small intestine is also highly likely to be involved in mass fecal neutral sterol excretion, independent of the biliary cholesterol excretion route. In the mid-2000s, using different mouse models with new experimental methods, some exciting data were reported that direct transintestinal excretion of plasma-derived cholesterol might contribute to the reverse cholesterol transport in mice [202, 203]. Based on the results from these mouse experiments, it is estimated that this non-biliary route may account for ~30% of total fecal neutral sterol excretion under basal conditions and could be regulated by several nuclear receptors such as liver X receptor (LXR), peroxisome proliferator-activated receptor-delta (PPAR-δ), and farnesoid X receptor (FXR) [204, 205]. Moreover, some results from animal studies suggest that this non-biliary route may be a novel therapeutic target to increase reverse cholesterol transport and, in this manner, confer protection against cardiovascular disease [205]. Although in vitro studies for examining the activity of this transintestinal route have been reported in explants from human small intestine mounted in Ussing chambers [206], the existence and importance of the TICE route in humans have not been established because of some difficult technical issues and methodology.

Interestingly, the contribution of TICE to total fecal neutral sterol excretion is recently studied in a small number of subjects [207]. Combining a cholesterol balance approach with stable isotopes that label cholesterol and bile salt molecules, the body cholesterol fluxes are analyzed in subjects with mild hypercholesterolemia. After 4 weeks of ezetimibe (10 mg/day) treatment for inhibiting the intestinal cholesterol influx transporter NPC1L1, the same studies are performed in the subjects eating a regular meal. Under basal conditions, the classical reverse cholesterol transport could contribute approximately 65% of daily fecal neutral sterol excretion, and it is likely that the TICE accounts for the remainder (~35%), as shown in Fig. 8.5. More interestingly, ezetimibe-treated subjects display a fourfold increase in total fecal neutral sterol excretion most likely through the TICE. To further confirm the results reported from human studies, chow-fed ABCG8 knockout and wild-type mice are treated with ezetimibe at 0 or 8 mg/kg/day for 2 weeks. As a result, most of the ezetimibe-modulated TICE flux is likely to be determined by the intestinal sterol efflux transporters ABCG5/G8. These studies suggest that TICE may exist in humans, and most of the ezetimibe-modulated TICE flux may be regulated by ABCG5/G8. For that reason, the TICE may be a new therapeutic target to enhance the removal of excess cholesterol from the body in patients at risk for cardiovascular disease. It is highly likely that the TICE may be an alternative route to the biliary route of the reverse cholesterol transport. However, it is imperative to explore the cellular and molecular mechanisms underlying the pivotal role of the TICE alone in the regulation of reverse cholesterol transport in humans [208]. More importantly, it is crucial to decipher whether the TICE could excrete more cholesterol from the body in patients with hypercholesterolemia, as well as how the TICE works together with the classical biliary route and whether it is fully independent from the latter. In addition, it is critical to elucidate whether there is a striking difference between the fasting state and the fed condition for the TICE to regulate plasma cholesterol, HDL, and LDL metabolism. More studies are also needed to investigate how the TICE is regulated in the normal physiological state, as well as under conditions of high plasma total and LDL cholesterol concentrations. With new experimental techniques, it is crucial for exploring whether the TICE is associated with the absorption efficiency of intestinal cholesterol because it is well-known that ABCG5/G8 is actively involved in regulating both the TICE and intestinal cholesterol absorption. Definitely, it is interesting to study whether abnormality in the molecular and genetic regulation of the TICE is associated with

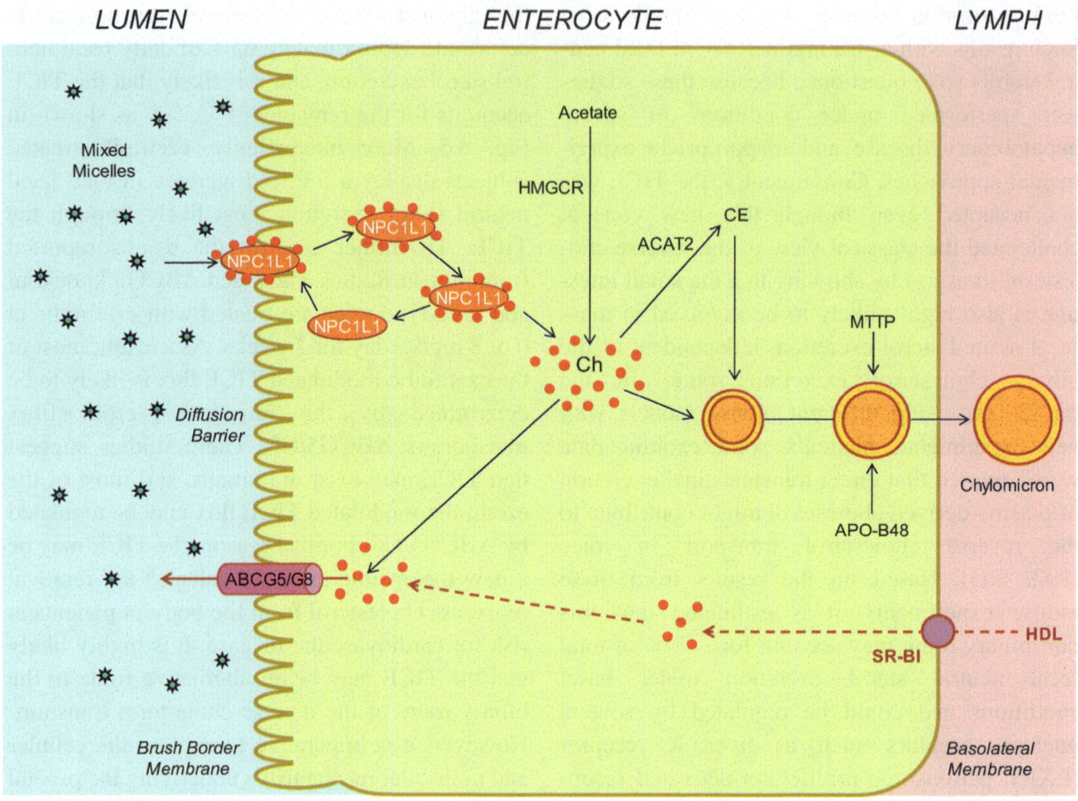

Fig. 8.5 Schematic diagram of the proposed transintestinal cholesterol excretion (TICE) pathway in the enterocytes, as showed in red dashed lines with arrows indicating the direction of transport. Abbreviations: *HDL* high-density lipoprotein, *SR-BI* scavenger receptor class B type I, i.e., HDL receptor. See Fig. 8.3 for other abbreviations

the prevalence of cardiovascular disease in humans [208]. Taken together, the TICE might provide a new target for the prevention and the treatment of cardiovascular disease.

8.5 Roles of ABCG5/G8 in Pathophysiology of Lipid-Related Metabolic Disorders

(a) Sitosterolemia

Sitosterolemia was first reported by Bhattacharyya and Connor in 1974 based on a clinical study on two sisters with tendon xanthomas and with elevated plant sterol concentrations in plasma [129]. Sitosterolemia is a rare inherited lipid storage disease characterized chemically by the accumulation of plant sterols and 5α-saturated stanols in plasma and tissues [134]. As analyzed by the sterol balance method, a large amount of dietary sitosterol is absorbed from the small intestine, thereby leading to the plant sterol accumulation in the body of patients with sitosterolemia. Further genetic studies find that sitosterolemia is a rare autosomal, recessively inherited lipid metabolic disorder [209]. However, the majority of heterozygous subjects are clinically and biochemically normal, and some heterozygotes display a slight, but not significant, increase in plasma sitosterol concentrations compared to normal healthy subjects [210]. Nevertheless, plasma sitosterol concentrations are 10- to 20-fold higher in homozygotes than in heterozygotes [211]. Therefore, the diagnosis of sitosterolemia is based mainly on a significant increase in the concentrations of plant sterols (sitosterol, campesterol, stigmasterol, and avenosterol) and 5α-stanols in plasma and tissues [212]. The clinical presentation in these patients includes tendon xanthomas, accelerated atherosclerosis particularly affecting males at a young age, hemolytic episodes, arthritis, and arthralgia [134]. The risk of premature atherosclerosis has been found in some young male patients who died because of acute myocardial infarctions associated with extensive coronary and aortic arteriosclerosis [139, 213].

Sitosterolemia is caused by a mutation in either the *ABCG5* or the *ABCG8* gene alone, but not in both simultaneously [8, 9, 136, 137, 214]. It is characterized mainly by increased intestinal absorption of cholesterol and sitosterol and diminished hepatic secretion of these sterols into bile [129, 209, 215]. In patients with sitosterolemia, the intestinal absorption of cholesterol is augmented by ~30%, from ~46% to ~60%; however, the intestinal absorption of sitosterol is dramatically increased by ~800%, from <5% to ~45% [50, 135, 143, 144]. Therefore, more cholesterol of intestinal origin, through the chylomicron pathway, reaches the liver for VLDL secretion into plasma, thereby increasing risk of developing cardiovascular disease in patients with sitosterolemia. Indeed, intestinal cholesterol absorption efficiency is also significantly increased in ABCG5/G8, ABCG5, and ABCG8 knockout mice [11–13, 141, 145].

Notably, several human studies on biliary lipid secretion have found that hepatic cholesterol secretion is markedly reduced and hepatic secretion of sitosterol and other plant sterols is almost totally inhibited [50, 135, 143, 144]. As a result, these patients often display hypercholesterolemia, tendon and tuberous xanthomas, premature development of atherosclerosis, and abnormal hematologic and liver function test results [134]. Further animal studies show that hepatic cholesterol output is dramatically reduced, but cholesterol is still secreted into bile in mice with the deletion of either *Abcg5* or *Abcg8* alone, or both [11–13, 141, 145]. These results clearly support the concept that the deletion of the *Abcg5/g8* double genes and *Abcg5* or *Abcg8* single gene significantly reduces, but does not eliminate, hepatic cholesterol secretion. In addition, consistent with the human results, these mouse data imply that an ABCG5/G8-independent pathway is also involved in hepatic cholesterol secretion, as discussed above.

The cholesterol molecules derived from HDL, but not LDL or VLDL, are an important source for hepatic secretion into bile because HDL promotes reverse cholesterol transport from peripheral tissues to the liver where the HDL-derived cholesterol is secreted preferentially into the bile [216]. After intravenous injection, HDL-derived [^{14}C]cholesterol, but not [^{3}H] sitostanol, is recovered in hepatic bile of ABCG5/G8 and ABCG8 knockout mice. This indicates that the ABCG5/G8-independent pathway is also able to regulate hepatic secretion of HDL-derived cholesterol, but not sitostanol. By contrast, ABCG5/G8 is involved in the regulation of hepatic secretion of both cholesterol and plant sterols. These results are consistent with the finding in sitosterolemic patients in whom only reduced amounts of cholesterol are found in bile and hepatic secretion of plant sterols is completely inhibited, leading to a significant increase in plasma plant sterol concentrations [135].

The treatment of sitosterolemia includes bile salt sequestrants such as cholestyramine, colestipol, and colesevelam in combination with the low-sterol diet [217–220]. Bile salt sequestrants bind bile salts in the intestine and increase the excretion of bile salts in the feces. This greatly diminishes the amount of bile salts returning to the liver and forces the liver to produce more bile salts to replace the bile salts lost in the feces. To synthetize more bile salts, the liver must convert more cholesterol into bile salts, thus leading to a reduction in plasma total and LDL cholesterol concentrations in sitosterolemic patients [221]. Moreover, ezetimibe, a potent intestinal cholesterol absorption inhibitor, has been used to treat patients with sitosterolemia [222–224] because ezetimibe can diminish plasma LDL cholesterol levels in patients with hypercholesterolemia by inhibiting the function of intestinal NPC1L1, the cholesterol influx transport protein [150, 153, 225–228].

(b) Cardiovascular disease

Atherosclerosis is characterized by lipid accumulation, inflammatory response, cell death, and fibrosis in the arterial wall, which is the pathological basis for cardiovascular disease, and the leading cause of morbidity and mortality in the USA and other industrialized nations [229]. Major risk factors for atherosclerosis include high plasma levels of LDL cholesterol and lipoprotein(a), as well as low plasma concentrations of HDL cholesterol [230]. Although genetic mechanisms underlying the pathogenesis of cardiovascular disease are largely unknown, accumulated evidence from human and animal studies has clearly demonstrated that cardiovascular disease may be determined by multiple genes disrupting cholesterol and lipoprotein metabolism [231–236]. Because mutations in either *ABCG5* or *ABCG8* cause phytosterolemia, hypercholesterolemia, and premature coronary heart disease in patients with sitosterolemia, this strongly suggests that defect or reduction in the ABCG5/G8 expression and function may be an important risk factor for the development of cardiovascular disease [141, 237–240]. Increased expression of *Abcg5/g8* attenuates Western-diet-induced hypercholesterolemia and atherosclerosis in LDL receptor knockout mice [241]. However, overexpression of *Abcg5/g8* in the liver, but not in the small intestine, does not reduce atherosclerosis development in LDL receptor or ApoE knockout mice fed the Western diet for 6 months [242]. This suggests that the increased hepatic secretion of biliary cholesterol could be absorbed back into the body, thus leading to unaltered atherosclerosis in these knockout mice. When these mice are fed ezetimibe, the potent intestinal cholesterol absorption inhibitor, total plasma cholesterol concentrations, and atherosclerosis are dramatically reduced in LDL receptor knockout mice overexpressing the human *ABCG5/G8* genes in the liver alone compared to LDL receptor knockout mice [243]. These mouse studies indicate that deletion of *Abcg5/g8* could play a determinant role in the development of hypercholesterolemia and atherosclerosis in mice fed the Western diet. In contrast, this suggests that ABCG5/G8 may be a novel target for the prevention and the treatment of cardiovascular disease. Furthermore, more studies are needed to explore whether dysfunction of ABCG5/G8 in the liver, or small intestine, or both sites is responsible for increased risk for the development of hypercholesterolemia and atherosclerosis in mice fed the Western diet.

In addition, it is interesting to investigate whether polymorphisms in the *ABCG5* and *ABCG8* genes are associated with plasma total and LDL cholesterol concentrations, increasing susceptibility to cardiovascular disease. Various polymorphisms (A632V, T400K, D19H, M429V, and C54Y) in the *ABCG8* and *ABCG5* (Q604E) genes have been found to be associated with several facets of cholesterol metabolism, including baseline cholesterol level, cholesterol kinetics, and individual responsiveness of plasma cholesterol to dietary and pharmaceutical interventions for hypercholesterolemia. For example, Tyr54Cys and Thr400Lys variations in the *ABCG8* gene may play a role in the genetic determination of plasma cholesterol levels and could influence the gender-specific response of plasma cholesterol levels after dietary changes [244]. More interestingly, low serum cholesterol concentrations and intestinal cholesterol absorption are found to be linked to the D19H polymorphism of the *ABCG8* gene, and characteristics of the insulin resistance syndrome in men are linked with the Q604E polymorphism of the *ABCG5* gene [245]. However, an association study between five common *ABCG5/G8* polymorphisms (p.Q604E, p.D19H, p.Y54C, p.T400K, and p.A632V) and plasma sterol levels was performed in 245 patients with hypercholesterolemia, and no significant associations were found [246]. Thus, most, but not all, studies reported that polymorphisms in the *ABCG5* and *ABCG8* genes may be associated with increased total and LDL-cholesterol concentrations [32]. Furthermore, a meta-analysis that comprised 3,364 subjects from 16 studies was carried out [246]. This study found that the *ABCG8* 632V variant is associated with a clinically irrelevant LDL-cholesterol reduction, whereas the 19H allele correlates with decreased cholesterol absorption and increased synthesis without affecting the lipid profile [246]. However, it is largely unknown whether small amounts of phytosterol exposure over a lifetime cause pathology in healthy humans with polymorphic variants in the *ABCG5* and *ABCG8* genes. Taken together, polymorphic variants in the *ABCG5* and *ABCG8* genes could increase or reduce the risk of these phenotypes, and loss of ABCG5/G8 function could cause more significant phenotypes, including premature atherosclerosis, platelet dysfunction, and thrombocytopenia, and perhaps, increased endocrine disruption and liver dysfunction [239]. Obviously, more studies are strongly needed to investigate how specific polymorphisms of the *ABCG5* and *ABCG8* genes confer to higher risk of these diseases.

Because elevated LDL cholesterol levels are a major causal factor for cardiovascular disease and have been a primary target of therapy for more than 30 years, the potent HMGCR inhibitors, statins, have been developed to lower plasma LDL cholesterol levels and reduce the risk of adverse cardiovascular events [247]. Moreover, reducing LDL cholesterol levels to below current guideline targets further inhibits atherogenesis and decreases adverse coronary events [4, 5, 248]. Many clinical studies have found that statins can reduce new adverse cardiovascular events and cardiovascular disease mortality by ~35%, but even aggressive statin therapy can not completely eliminate cardiovascular risk. Approximately 65% of the patients treated with statins still develop adverse cardiovascular events. Therefore, additional therapeutic interventions beyond statins are strongly needed to further reduce the risk of developing cardiovascular disease [249]. Overall, ABCG5/G8 may be an attractive target for the prevention and the treatment of hypercholesterolemia, and increasing their expression may reduce the risk of developing cardiovascular disease in humans.

(c) Cholesterol gallstone disease

Clinical investigations and animal studies have clearly established that hepatic hypersecretion of biliary cholesterol is the primary defect in the pathogenesis of cholesterol gallstone disease [14]. Hepatic cholesterol hypersecretion into bile may or may not be accompanied by normal, high, or low hepatic secretion rates of biliary bile salts and/or phospholipids [250]. Cholesterol-supersaturated bile is often defined as a state in which cholesterol cannot be dissolved in bile by biliary bile salts or phospholipids at equilibrium [70]. Therefore, the formation of supersaturated bile is often caused by (i) hepatic hypersecretion of biliary cholesterol; (ii) reduced hepatic bile salt and/or phospholipid secretion with normal biliary cholesterol secretion; or (iii) a combination of hepatic cholesterol hypersecretion with hyposecretion of these solubilizing lipids [251].

Genetic studies have been performed to investigate *Lith* genes in different strains of inbred mice fed the lithogenic diet for 8 weeks [26]. As shown in Fig. 8.6, *Lith9* is localized on mouse chromosome 17 and is co-localized with a genetic biomarker *D17Mit155* at approximately 55 centimorgans (cM). Genotyping and phenotyping studies have found that in the *Lith9* QTL region, *Abcg5/g8* is a strong candidate for this gallstone gene. Subsequently, *Abcg5/g8* is identified as a new gallstone gene, *Lith9*, by QTL studies in mice [25, 252, 253]. Based on mouse genetic analysis of the *Lith* genes, a genome-wide association study in a large cohort of patients with gallstones and a linkage study in affected sibling pairs have identified a common variant (D19H) of the sterol efflux transporters ABCG5/G8 as a key risk factor for cholesterol gallstone disease [29]. Indeed, *ABCG5/G8* is found to be associated with gallstones in patients, proving that it is human *LITH9*. Other *ABCG8* variants (T400K, D19H, A632V, M429V, and C54Y) and *ABCG5* variants (Q604E) have also been found to be associated with cholesterol gallstone disease in humans. Furthermore, many

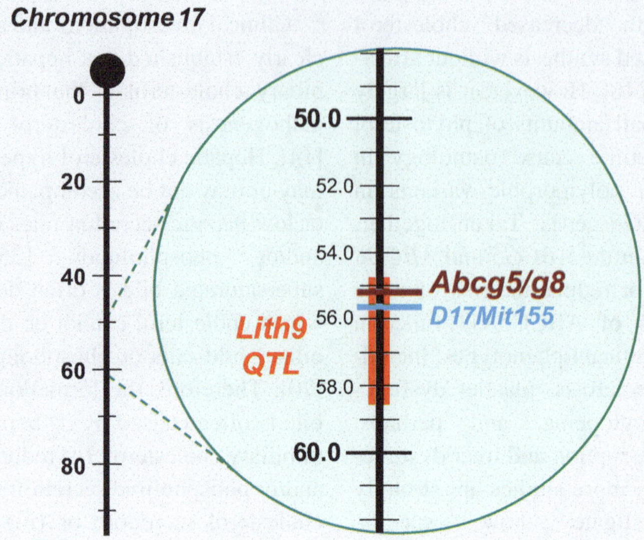

Fig. 8.6 As shown in the composite map, the quantitative trait locus (QTL) region of the *Lith9* gene is localized on chromosome 17 in mice. A vertical line represents chromosome 17, with the centromere at the top; genetic distances from the centromere (horizontal black lines) are indicated to the left of the chromosomes in centimorgans (cM). Chromosomes are drawn to scale, based on the estimated cM position of the most distally mapped locus taken from Mouse Genome Database. The gallstone gene, the *Lith9*, QTL region is represented by a vertical red bar, and the *Abcg5/g8* gene location is indicated by a horizontal borrow line. A genetic biomarker, D17Mit155, that is co-localized with *Lith9* is indicated by a horizontal blue line with the marker symbol to the right

research groups have reported that two gallstone-associated variants in *ABCG5/G8*, i.e., *ABCG5*-R50C and *ABCG8*-D19H, are involved in the pathogenesis of gallstones not only in Germans and Chileans but also in Chinese and Indians [29–34, 254, 255]. These studies strongly imply that *ABCG5*-R50C and *ABCG8*-D19H could play a central role in hepatic cholesterol hypersecretion, thereby leading to the formation of cholesterol-supersaturated bile in humans.

Because *Abcg5/g8* is *Lith9* in mice and two gallstone-associated variants in *ABCG5/G8* have been identified in humans, it is important to further investigate whether targeted disruption of the *Abcg5/g8* double genes or the *Abcg8* single gene protects against the formation of cholesterol gallstones in gallstone-susceptible C57BL/6J mice fed the lithogenic diet for 8 weeks [256]. It is surprising to find that despite a significant reduction in gallstone prevalence in ABCG5/G8 and ABCG8 knockout mice, classical parallelogram-shaped cholesterol monohydrate crystals and gallstones are still formed in these mice during the 8-week period of feeding the lithogenic diet. As discussed above, although sitosterolemia is caused by mutations in either the *ABCG5* or the *ABCG8* gene alone, but not in both simultaneously, hepatic cholesterol secretion is reduced, but not completely eliminated, in these patients [50, 135, 143, 144]. To explore the mechanism underlying the effect of ABCG5/G8 on hepatic cholesterol and plant sterol secretion, biliary cholesterol and sitostanol secretion is quantified for 6 h in ABCG8 knockout mice. Mass transport rate of [^{3}H]sitostanol from plasma HDL into bile is significantly faster than that of [^{14}C]cholesterol in wild-type mice; however, reduced amounts of [^{14}C]cholesterol and no [^{3}H] sitostanol are found in bile of ABCG8 knockout mice [141]. These results clearly exhibit that the deletion of the *Abcg8* gene alone significantly reduces, but does not eliminate, hepatic cholesterol secretion. In addition, biliary cholesterol studies show that hepatic cholesterol output is significantly reduced, but cholesterol is still secreted into bile in mice with the deletion of either *Abcg5* or *Abcg8* alone, or both [11–13, 141, 145].

Although ABCG5/G8 display a striking impact on hepatic cholesterol and plant sterol secretion, cholesterol is still secreted to bile in sitosterolemic patients with a defect in either ABCG5 or ABCG8 and in either ABCG5/G8 double or single gene knockout mice. This strongly suggests that in the defect of ABCG5/G8, an ABCG5/G8-independent pathway is essential for regulating hepatic secretion of biliary cholesterol, which is independent of the lithogenic mechanism of the ABCG5/G8 pathway. To decipher the effect of the ABCG5/G8-independent pathway on cholelithogenesis, the biliary and gallstone characteristics are investigated in wild-type as well as ABCG5/G8 and ABCG8 knockout mice fed the lithogenic diet or varying amounts of cholesterol, or injected intravenously with [^{3}H]sitostanol- and [^{14}C]cholesterol-labeled HDL. These studies show that ABCG5/G8 and ABCG8 knockout mice display the same biliary and gallstone phenotypes. Although both groups of knockout mice show a significant reduction in hepatic cholesterol output compared to wild-type mice, they still form gallstones. Especially, the ABCG5/G8-independent pathway plays an important role in the regulation of biliary cholesterol secretion, the transport of HDL-derived cholesterol from plasma to bile, and the formation of cholesterol gallstones, which works independently of the ABCG5/G8 pathway.

It is well-known that the LXR agonist T0901317 activates hepatic LXR, promoting biliary cholesterol secretion by stimulating hepatic *Abcg5/g8* expression in mice [145, 257, 258]. Additionally, LXR activation by T0901317 greatly promotes cholesterol crystallization and gallstone formation in mice fed the lithogenic diet [259]. However, this is not the case in ABCG5/G8 or ABCG8 knockout mice. This clearly implies that the hepatic LXR does not have an effect on the ABCG5/G8-independent pathway for regulating biliary cholesterol secretion, which is distinct from the ABCG5/G8 pathway that is effectively regulated by the hepatic LXR through a transcriptional mechanism. The LXR agonist dramatically increases biliary cholesterol secretion and gallstones in wild-type, but not ABCG5/G8 or ABCG8 knockout, mice. Taken together, these studies [256] provide clear evidence in support of the concepts that (i) the ABCG5/G8-independent pathway accounts for 30% to 40% of hepatic cholesterol output in the lithogenic state and plays a critical role in the regulation of biliary cholesterol secretion in response to high dietary cholesterol; (ii) in the absence of ABCG5/G8, it determines biliary cholesterol secretion and the formation of cholesterol gallstones; (iii) it modulates hepatic secretion of HDL-derived cholesterol, but not sitostanol; and (iv) its activity in the liver is not regulated by the LXR agonist through the LXR signaling cascade. These findings strongly support the existence of an ABCG5/G8-independent pathway for regulating hepatic cholesterol secretion. Moreover, these results imply that in the absence of ABCG5/G8, the ABCG5/G8-independent pathway is essential for the regulation of hepatic cholesterol secretion and also plays a vital role in determining the susceptibility to cholesterol gallstones, working independently of the ABCG5/G8 pathway in mice. However, further studies are strongly needed to observe if this pathway is also operational in humans. Nevertheless, both ABCG5/G8-dependent and ABCG5/G8-independent pathways could be potential therapeutic targets for cholesterol gallstone disease.

8.6 Conclusions and Future Directions

Accumulated evidence has clearly demonstrated that ABCG5/G8 play a key role not only in hepatic secretion of biliary cholesterol and plant sterols but also intestinal absorption of these two sterols. Moreover, ABCG5/G8 have an important impact on the classical reverse cholesterol transport and the TICE pathway. Obviously, mutations in either *ABCG5* or *ABCG8* are the major genetic mechanisms causing sitosterolemia. It is highly likely that gene therapy is a better option for curing this genetic disorder by repairing *ABCG5* or *ABCG8* gene mutations. Lowering plasma total and LDL cholesterol concentrations is also crucial to reduce the risk of cardiovascular disease in patients with sitosterolemia.

Many clinical studies have shown that statins can reduce the risk of developing cardiovascular disease; however, other lipid-lowering therapies are often used adjunctively when statin therapy is inadequate or as an alternative for patients who are intolerant of statins. More importantly, intensive lipid and pharmaceutical studies have led to significant development of new agents that could work on novel targets in the metabolic pathways of lipids and lipoproteins and that have the potential to serve as new alternative or adjunctive agents to the existing cholesterol-lowering drugs such as statins. Clinical trials in patients receiving these new classes of lipid-lowering agents, especially in individuals with monogenic disorders of lipid and/or lipoprotein metabolism, will certainly increase a great opportunity to identify the genotype that predicts response to lipid-lowering therapy and thus guides the choice of drug and dose for high-risk patients and, especially, for patients with the hardest-to-treat elevated plasma cholesterol concentrations due to intolerance to any statins and severe side effects of these drugs.

Although the pharmacogenomics of lipid-lowering drugs have greatly advanced and a few consistent trends on the therapy of cardiovascular disease have emerged, mainly relating to the genetic determinants of response to statins, many new cellular, molecular, genetic, and biochemical studies on lipid and lipoprotein metabolism are being extensively explored. Therefore, it is more interesting to investigate the cellular and molecular mechanisms of deciphering how ABCG5/G8 regulate cholesterol and lipoprotein metabolism in the plasma, liver, and intestine. In addition, the potential mechanisms underlying the removal of cholesterol from the body through the classical reverse cholesterol transport, i.e., the biliary route, and the TICE, i.e., the non-biliary routes, are desired to be revealed. Advances in the elucidation of lipid and lipoprotein metabolism, as well as the biliary and the non-biliary routes for removal of cholesterol and plant sterols from the body, will provide a great opportunity of finding new lipid-lowering strategies and proving that they are more effective in the prevention and therapeutic intervention of cardiovascular disease that affects millions worldwide.

The gallstone (*Lith*) gene map has been updated, which lists all known genetic loci that confer gallstone susceptibility, as well as candidate genes in inbred strains of mice. This would greatly help identify human *LITH* genes because genetic analysis of *Lith* genes in mouse models open the way for searching for the orthologous human *LITH* genes and for exploring their cholelithogenic effects in humans. Given that the ABCG5/G8-dependent and the ABCG5/G8-independent pathways are essential in the regulation of hepatic cholesterol secretion, both routes could be potential therapeutic targets for the prevention and the treatment of cholesterol gallstone disease. Deciphering the molecular and cellular mechanisms on the formation of cholesterol-supersaturated bile could be very helpful for exploring novel therapeutic approaches through modulating both the ABCG5/G8-dependent and the ABCG5/G8-independent pathways, thus greatly reducing the risk of developing gallstones.

More importantly, there should be a great development of the personalized medicine for the prevention and the treatment of cardiovascular disease and cholesterol gallstone disease because they are highly prevalent not only in the USA but also in European and Asian countries. The ideal application of lipid-lowering drugs and bile-cholesterol-desaturating drugs would be to identify patients at risk for either a suboptimal response with respect to efficacy or a marked adverse response to either a drug class or a specific drug. For that reason, individuals who would be predicted to have an unfavorable benefit-to-risk ratio can be identified and might be obtained from alternative methods more expeditiously and without the trial-and-error process that typically accompanies initiation and maintenance of such commonly used treatment. Obviously, it is imperative to understand the cellular and molecular mechanisms underlying the key role of ABCG5/G8 in regulating hepatic secretion of biliary cholesterol and plant sterols and intestinal absorption of these two sterols, as well as in modulating the classical reverse cholesterol transport and the TICE pathway, because it could provide novel insights into strategies for the prevention and the treatment of sitosterolemia, cardiovascular disease, and cholesterol gallstone disease.

Acknowledgments This work was supported in part by research grants DK101793, DK106249, DK114516, and AA025737 (to DQ-HW), as well as P30 DK041296 (to Marion Bessin Liver Research Center), all from the National Institutes of Health (US Public Health Service). This chapter was modified from the paper published by our group in Annals of Hepatology (Helen H. Wang, Gabriella Garruti, Min Liu, Piero Portincasa, David Q.-H. Wang. 2017; 16 (Suppl. 1): s28–s43). The related contents are re-used with the permission.

Conflict of Interest There is no conflict of interest to disclose for all authors.

References

1. Wang DQ, Neuschwander-Tetri BA, Portincasa P (2012) The biliary system. Morgan & Claypool Life Sciences, Princeton
2. Wang DQ (2007) Regulation of intestinal cholesterol absorption. Annu Rev Physiol 69:221–248
3. Panel TNCEPE (2001) Executive Summary of The Third Report of The National Cholesterol Education Program (NCEP) Expert Panel on Detection, Evaluation, and Treatment of High Blood Cholesterol in Adults (Adult Treatment Panel III). JAMA 285 (19):2486–2497
4. Grundy SM, Cleeman JI, Merz CN, Brewer HB Jr, Clark LT, Hunninghake DB, Pasternak RC, Smith SC Jr, Stone NJ (2004) Implications of recent clinical trials for the National Cholesterol Education Program Adult Treatment Panel III Guidelines. J Am Coll Cardiol 44(3):720–732
5. Smith SC Jr, Allen J, Blair SN, Bonow RO, Brass LM, Fonarow GC, Grundy SM, Hiratzka L, Jones D, Krumholz HM et al (2006) AHA/ACC guidelines for secondary prevention for patients with coronary and other atherosclerotic vascular disease: 2006 update: endorsed by the National Heart, Lung, and Blood Institute. Circulation 113(19):2363–2372
6. Greenland P, Alpert JS, Beller GA, Benjamin EJ, Budoff MJ, Fayad ZA, Foster E, Hlatky MA, Hodgson JM, Kushner FG et al (2010) 2010 ACCF/AHA guideline for assessment of cardiovascular risk in asymptomatic adults: a report of the American College of Cardiology Foundation/American Heart Association Task Force on Practice Guidelines. J Am Coll Cardiol 56(25):e50–103
7. Martin SS, Metkus TS, Horne A, Blaha MJ, Hasan R, Campbell CY, Yousuf O, Joshi P, Kaul S, Miller M et al (2012) Waiting for the National Cholesterol Education Program Adult Treatment Panel IV Guidelines, and in the meantime, some challenges and recommendations. Am J Cardiol 110 (2):307–313
8. Berge KE, Tian H, Graf GA, Yu L, Grishin NV, Schultz J, Kwiterovich P, Shan B, Barnes R, Hobbs HH (2000) Accumulation of dietary cholesterol in sitosterolemia caused by mutations in adjacent ABC transporters. Science 290(5497):1771–1775
9. Lee MH, Lu K, Hazard S, Yu H, Shulenin S, Hidaka H, Kojima H, Allikmets R, Sakuma N, Pegoraro R et al (2001) Identification of a gene, ABCG5, important in the regulation of dietary cholesterol absorption. Nat Genet 27(1):79–83
10. Lee MH, Lu K, Patel SB (2001) Genetic basis of sitosterolemia. Curr Opin Lipidol 12(2):141–149
11. Yu L, Hammer RE, Li-Hawkins J, Von Bergmann K, Lutjohann D, Cohen JC, Hobbs HH (2002) Disruption of Abcg5 and Abcg8 in mice reveals their crucial role in biliary cholesterol secretion. Proc Natl Acad Sci U S A 99(25):16237–16242
12. Yu L, Li-Hawkins J, Hammer RE, Berge KE, Horton JD, Cohen JC, Hobbs HH (2002) Overexpression of ABCG5 and ABCG8 promotes biliary cholesterol secretion and reduces fractional absorption of dietary cholesterol. J Clin Invest 110(5):671–680
13. Wang J, Mitsche MA, Lutjohann D, Cohen JC, Xie XS, Hobbs HH (2015) Relative roles of ABCG5/ABCG8 in liver and intestine. J Lipid Res 56 (2):319–330
14. Wang DQ, Portincasa P (2017) Gallstones: recent advances in epidemiology, pathogenesis, diagnosis and management. Nova Biomedical, New York
15. Lammert F, Gurusamy K, Ko CW, Miquel JF, Mendez-Sanchez N, Portincasa P, van Erpecum KJ, van Laarhoven CJ, Wang DQ (2016) Gallstones. Nat Rev Dis Primers 2:16024
16. Wang DQ, Cohen DE, Carey MC (2009) Biliary lipids and cholesterol gallstone disease. J Lipid Res 50(Suppl):S406–S411
17. Ruhl CE, Everhart JE (2011) Gallstone disease is associated with increased mortality in the United States. Gastroenterology 140(2):508–516
18. Peery AF, Crockett SD, Barritt AS, Dellon ES, Eluri S, Gangarosa LM, Jensen ET, Lund JL, Pasricha S, Runge T et al (2015) Burden of gastrointestinal, liver, and pancreatic diseases in the United States. Gastroenterology 149(7):1731–41 e3
19. Portincasa P, Wang DQ (2016) In: Podolsky DK, Camilleri M, Fitz JG, Kalloo AN, Shanahan F,

Wang TC (eds) Yamada's Atlas of gastroenterology. Wiley-Blackwell, Hoboken, pp 335–353
20. Portincasa P, Di Ciaula A, de Bari O, Garruti G, Palmieri VO, Wang DQ (2016) Management of gallstones and its related complications. Expert Rev Gastroenterol Hepatol 10(1):93–112
21. Di Ciaula A, Wang DQ, Portincasa P (2018) An update on the pathogenesis of cholesterol gallstone disease. Curr Opin Gastroenterol 34(2):71–80
22. Di Ciaula A, Wang DQ, Garruti G, Wang HH, Grattagliano I, de Bari O, Portincasa P (2014) Therapeutic reflections in cholesterol homeostasis and gallstone disease: a review. Curr Med Chem 21 (12):1435–1447
23. Wang DQ, Afdhal NH (2014) In: Feldman M, Friedman LS, Brandt L (eds) Sleisenger and Fordtran's gastrointestinal and liver disease. Elsevier Saunders, Philadelphia, pp 1100–1133
24. Portincasa P, Ciaula AD, Bonfrate L, Wang DQH (2012) Therapy of gallstone disease: what it was, what it is, what it will be. World J Gastrointest Pharmacol Ther 3(2):7–20
25. Wittenburg H, Lyons MA, Li R, Churchill GA, Carey MC, Paigen B (2003) FXR and ABCG5/ABCG8 as determinants of cholesterol gallstone formation from quantitative trait locus mapping in mice. Gastroenterology 125(3):868–881
26. Wang TY, Portincasa P, Liu M, Tso P, Wang DQ (2018) Mouse models of gallstone disease. Curr Opin Gastroenterol 34(2):59–70
27. Wang HH, Portincasa P, Afdhal NH, Wang DQ (2010) Lith genes and genetic analysis of cholesterol gallstone formation. Gastroenterol Clin N Am 39 (2):185–207. vii–viii
28. Wang DQ, Afdhal NH (2004) Genetic analysis of cholesterol gallstone formation: searching for Lith (gallstone) genes. Curr Gastroenterol Rep 6 (2):140–150
29. Grunhage F, Acalovschi M, Tirziu S, Walier M, Wienker TF, Ciocan A, Mosteanu O, Sauerbruch T, Lammert F (2007) Increased gallstone risk in humans conferred by common variant of hepatic ATP-binding cassette transporter for cholesterol. Hepatology 46(3):793–801
30. Wang Y, Jiang ZY, Fei J, Xin L, Cai Q, Jiang ZH, Zhu ZG, Han TQ, Zhang SD (2007) ATP binding cassette G8 T400K polymorphism may affect the risk of gallstone disease among Chinese males. Clinica Chimica Acta Int J Clin Chem 384(1–2):80–85
31. Kuo KK, Shin SJ, Chen ZC, Yang YH, Yang JF, Hsiao PJ (2008) Significant association of ABCG5 604Q and ABCG8 D19H polymorphisms with gallstone disease. Br J Surg 95(8):1005–1011
32. Rudkowska I, Jones PJ (2008) Polymorphisms in ABCG5/G8 transporters linked to hypercholesterolemia and gallstone disease. Nutr Rev 66(6):343–348
33. Katsika D, Magnusson P, Krawczyk M, Grunhage F, Lichtenstein P, Einarsson C, Lammert F, Marschall HU (2010) Gallstone disease in Swedish twins: risk is associated with ABCG8 D19H genotype. J Intern Med 268(3):279–285
34. von Kampen O, Buch S, Nothnagel M, Azocar L, Molina H, Brosch M, Erhart W, von Schonfels W, Egberts J, Seeger M et al (2013) Genetic and functional identification of the likely causative variant for cholesterol gallstone disease at the ABCG5/8 lithogenic locus. Hepatology 57(6):2407–2417
35. Borst P, Elferink RO (2002) Mammalian ABC transporters in health and disease. Annu Rev Biochem 71:537–592
36. Dean M, Allikmets R (1995) Evolution of ATP-binding cassette transporter genes. Curr Opin Genet Dev 5(6):779–785
37. Allikmets R, Gerrard B, Glavac D, Ravnik-Glavac M, Jenkins NA, Gilbert DJ, Copeland NG, Modi W, Dean M (1995) Characterization and mapping of three new mammalian ATP-binding transporter genes from an EST database. Mamm Genome 6 (2):114–117
38. Quazi F, Molday RS (2011) Lipid transport by mammalian ABC proteins. Essays Biochem 50 (1):265–290
39. Iida A, Saito S, Sekine A, Mishima C, Kitamura Y, Kondo K, Harigae S, Osawa S, Nakamura Y (2002) Catalog of 605 single-nucleotide polymorphisms (SNPs) among 13 genes encoding human ATP-binding cassette transporters: ABCA4, ABCA7, ABCA8, ABCD1, ABCD3, ABCD4, ABCE1, ABCF1, ABCG1, ABCG2, ABCG4, ABCG5, and ABCG8. J Hum Genet 47(6):285–310
40. van Meer G, Halter D, Sprong H, Somerharju P, Egmond MR (2006) ABC lipid transporters: extruders, flippases, or flopless activators? FEBS Lett 580(4):1171–1177
41. Stefkova J, Poledne R, Hubacek JA (2004) ATP-binding cassette (ABC) transporters in human metabolism and diseases. Physiol Res/Academia Scientiarum Bohemoslovaca 53(3):235–243
42. Marcil M, Brooks-Wilson A, Clee SM, Roomp K, Zhang LH, Yu L, Collins JA, van Dam M, Molhuizen HO, Loubster O et al (1999) Mutations in the ABC1 gene in familial HDL deficiency with defective cholesterol efflux. Lancet 354(9187):1341–1346
43. Smit JJ, Schinkel AH, Oude Elferink RP, Groen AK, Wagenaar E, van Deemter L, Mol CA, Ottenhoff R, van der Lugt NM, van Roon MA et al (1993) Homozygous disruption of the murine mdr2 P-glycoprotein gene leads to a complete absence of phospholipid from bile and to liver disease. Cell 75(3):451–462
44. Oude Elferink RP, Ottenhoff R, van Wijland M, Smit JJ, Schinkel AH, Groen AK (1995) Regulation of biliary lipid secretion by mdr2 P-glycoprotein in the mouse. J Clin Invest 95(1):31–38
45. Groen AK, Van Wijland MJ, Frederiks WM, Smit JJ, Schinkel AH, Oude Elferink RP (1995) Regulation of protein secretion into bile: studies in mice with a disrupted mdr2 p-glycoprotein gene. Gastroenterology 109(6):1997–2006

46. Oude Elferink RP, Ottenhoff R, van Wijland M, Frijters CM, van Nieuwkerk C, Groen AK (1996) Uncoupling of biliary phospholipid and cholesterol secretion in mice with reduced expression of mdr2 P-glycoprotein. J Lipid Res 37(5):1065–1075
47. Graf GA, Li WP, Gerard RD, Gelissen I, White A, Cohen JC, Hobbs HH (2002) Coexpression of ATP-binding cassette proteins ABCG5 and ABCG8 permits their transport to the apical surface. J Clin Invest 110(5):659–669
48. Graf GA, Yu L, Li WP, Gerard R, Tuma PL, Cohen JC, Hobbs HH (2003) ABCG5 and ABCG8 are obligate heterodimers for protein trafficking and biliary cholesterol excretion. J Biol Chem 278 (48):48275–48282
49. Zhang DW, Graf GA, Gerard RD, Cohen JC, Hobbs HH (2006) Functional asymmetry of nucleotide-binding domains in ABCG5 and ABCG8. J Biol Chem 281(7):4507–4516
50. Miettinen TA (1980) Phytosterolaemia, xanthomatosis and premature atherosclerotic arterial disease: a case with high plant sterol absorption, impaired sterol elimination and low cholesterol synthesis. Eur J Clin Investig 10(1):27–35
51. Turley SD, Dietschy JM (1988) In: Arias IM, Jakoby WB, Popper H, Schachter D, Shafritz DA (eds) The liver: biology and pathobiology. Raven Press, New York, pp 617–641
52. Cohen DE (2009) In: Arias IM, Alter HJ, Boyer JL, Cohen DE, Fausto N, Shafritz DA, Wolkoff AW (eds) The liver: biology and pathobiology. Wiley-Blackwell, Chichester, pp 271–285
53. Andersen JM, Dietschy JM (1977) Regulation of sterol synthesis in 16 tissues of rat. I. Effect of diurnal light cycling, fasting, stress, manipulation of enterohepatic circulation, and administration of chylomicrons and triton. J Biol Chem 252 (11):3646–3651
54. Andersen JM, Dietschy JM (1977) Regulation of sterol synthesis in 15 tissues of rat. II. Role of rat and human high and low density plasma lipoproteins and of rat chylomicron remnants. J Biol Chem 252 (11):3652–3659
55. Dietschy JM (1984) Regulation of cholesterol metabolism in man and in other species. Klin Wochenschr 62(8):338–345
56. Spady DK, Dietschy JM (1985) Rates of cholesterol synthesis and low-density lipoprotein uptake in the adrenal glands of the rat, hamster and rabbit in vivo. Biochim Biophys Acta 836(2):167–175
57. Dietschy JM, Turley SD (2002) Control of cholesterol turnover in the mouse. J Biol Chem 277 (6):3801–3804
58. Spady DK, Turley SD, Dietschy JM (1985) Rates of low density lipoprotein uptake and cholesterol synthesis are regulated independently in the liver. J Lipid Res 26(4):465–472
59. Spady DK, Turley SD, Dietschy JM (1985) Receptor-independent low density lipoprotein transport in the rat in vivo. Quantitation, characterization, and metabolic consequences. J Clin Invest 76 (3):1113–1122
60. Turley SD, Dietschy JM (2003) Sterol absorption by the small intestine. Curr Opin Lipidol 14(3):233–240
61. Wang DQ, Cohen DE (2008) In: Ballantyne CM (ed) Lipidology in the treatment and prevention of cardiovascular disease (Clinical lipidology: a companion to Braunwald's heart disease). Elsevier Saunders, Philadelphia, pp 26–44
62. Tso P, Fujimoto K (1991) The absorption and transport of lipids by the small intestine. Brain Res Bull 27 (3–4):477–482
63. Wang DQ, Lee SP (2008) Physical chemistry of intestinal absorption of biliary cholesterol in mice. Hepatology 48(1):177–185
64. Wang DQ, Paigen B, Carey MC (2001) Genetic factors at the enterocyte level account for variations in intestinal cholesterol absorption efficiency among inbred strains of mice. J Lipid Res 42(11):1820–1830
65. Bhattacharyya AK, Eggen DA (1987) Relationships between dietary cholesterol, cholesterol absorption, cholesterol synthesis, and plasma cholesterol in rhesus monkeys. Atherosclerosis 67(1):33–39
66. Trautwein EA, Forgbert K, Rieckhoff D, Erbersdobler HF (1999) Impact of beta-cyclodextrin and resistant starch on bile acid metabolism and fecal steroid excretion in regard to their hypolipidemic action in hamsters. Biochim Biophys Acta 1437(1):1–12
67. Turley SD, Daggy BP, Dietschy JM (1996) Effect of feeding psyllium and cholestyramine in combination on low density lipoprotein metabolism and fecal bile acid excretion in hamsters with dietary-induced hypercholesterolemia. J Cardiovasc Pharmacol 27 (1):71–79
68. Turley SD, Daggy BP, Dietschy JM (1991) Cholesterol-lowering action of psyllium mucilloid in the hamster: sites and possible mechanisms of action. Metab Clin Exp 40(10):1063–1073
69. Admirand WH, Small DM (1968) The physicochemical basis of cholesterol gallstone formation in man. J Clin Invest 47(5):1043–1052
70. Carey MC, Small DM (1978) The physical chemistry of cholesterol solubility in bile. Relationship to gallstone formation and dissolution in man. J Clin Invest 61(4):998–1026
71. Wang DQ, Carey MC (1996) Characterization of crystallization pathways during cholesterol precipitation from human gallbladder biles: identical pathways to corresponding model biles with three predominating sequences. J Lipid Res 37 (12):2539–2549
72. Wang DQ, Cohen DE, Lammert F, Carey MC (1999) No pathophysiologic relationship of soluble biliary proteins to cholesterol crystallization in human bile. J Lipid Res 40(3):415–425
73. Goldstein JL, Brown MS (1975) Lipoprotein receptors, cholesterol metabolism, and atherosclerosis. Arch Pathol 99(4):181–184

74. Brown MS, Goldstein JL (1984) How LDL receptors influence cholesterol and atherosclerosis. Sci Am 251 (5):58–66
75. Brown MS, Goldstein JL (1983) Lipoprotein metabolism in the macrophage: implications for cholesterol deposition in atherosclerosis. Annu Rev Biochem 52:223–261
76. Brown MS, Kovanen PT, Goldstein JL (1981) Regulation of plasma cholesterol by lipoprotein receptors. Science 212(4495):628–635
77. Small DM (1977) Cellular mechanisms for lipid deposition in atherosclerosis (first of two parts). N Engl J Med 297(16):873–877
78. Small DM (1988) George Lyman Duff memorial lecture. Progression and regression of atherosclerotic lesions. Insights from lipid physical biochemistry. Arteriosclerosis 8(2):103–129
79. Small DM, Shipley GG (1974) Physical-chemical basis of lipid deposition in atherosclerosis. Science 185(147):222–229
80. Goldstein JL, Brown MS (1987) Regulation of low-density lipoprotein receptors: implications for pathogenesis and therapy of hypercholesterolemia and atherosclerosis. Circulation 76(3):504–507
81. Portincasa P, Moschetta A, Di Ciaula A, Pontrelli D, Sasso RC, Wang HH, Wang DQ (2008) In: Borzellino G, Cordiano C (eds) Biliary lithiasis: basic science, current diagnosis and management. Springer, Milano, pp 19–49
82. Portincasa P, Moschetta A, Palasciano G (2006) Cholesterol gallstone disease. Lancet 368(9531):230–239
83. Portincasa P, Moschetta A, Palasciano G (2002) From lipid secretion to cholesterol crystallization in bile. Relevance in cholesterol gallstone disease. Ann Hepatol 1(3):121–128
84. Portincasa P, Moschetta A, van Erpecum KJ, Calamita G, Margari A, vanBerge-Henegouwen GP, Palasciano G (2003) Pathways of cholesterol crystallization in model bile and native bile. Dig Liver Dis 35(2):118–126
85. Afdhal NH, Smith BF (1990) Cholesterol crystal nucleation: a decade-long search for the missing link in gallstone pathogenesis. Hepatology 11 (4):699–702
86. Holan KR, Holzbach RT, Hermann RE, Cooperman AM, Claffey WJ (1979) Nucleation time: a key factor in the pathogenesis of cholesterol gallstone disease. Gastroenterology 77(4 Pt 1):611–617
87. Holzbach RT (1995) Cholesterol nucleation in bile. Ital J Gastroenterol 27(2):101–105
88. Holzbach RT (1990) Nucleation of cholesterol crystals in native bile. Hepatology 12(3 Pt 2):155S–159S. discussion 9S–61S
89. Holzbach RT (1986) Recent progress in understanding cholesterol crystal nucleation as a precursor to human gallstone formation. Hepatology 6(6):1403–1406
90. Holzbach RT (1984) Factors influencing cholesterol nucleation in bile. Hepatology 4(5 Suppl):173S–176S
91. Holzbach RT, Busch N (1991) Nucleation and growth of cholesterol crystals. Kinetic determinants in supersaturated native bile. Gastroenterol Clin N Am 20(1):67–84
92. Crawford JM, Mockel GM, Crawford AR, Hagen SJ, Hatch VC, Barnes S, Godleski JJ, Carey MC (1995) Imaging biliary lipid secretion in the rat: ultrastructural evidence for vesiculation of the hepatocyte canalicular membrane. J Lipid Res 36(10):2147–2163
93. Crawford AR, Smith AJ, Hatch VC, Oude Elferink RP, Borst P, Crawford JM (1997) Hepatic secretion of phospholipid vesicles in the mouse critically depends on mdr2 or MDR3 P-glycoprotein expression. Visualization by electron microscopy. J Clin Invest 100(10):2562–2567
94. Crawford JM (1996) Role of vesicle-mediated transport pathways in hepatocellular bile secretion. Semin Liver Dis 16(2):169–189
95. Oude Elferink RP, Beuers U (2011) Targeting the ABCB4 gene to control cholesterol homeostasis. Expert Opin Ther Targets 15(10):1173–1182
96. Langheim S, Yu L, von Bergmann K, Lutjohann D, Xu F, Hobbs HH, Cohen JC (2005) ABCG5 and ABCG8 require MDR2 for secretion of cholesterol into bile. J Lipid Res 46(8):1732–1738
97. Dikkers A, Tietge UJ (2010) Biliary cholesterol secretion: more than a simple ABC. World J Gastroenterol 16(47):5936–5945
98. Lammert F, Wang DQ, Hillebrandt S, Geier A, Fickert P, Trauner M, Matern S, Paigen B, Carey MC (2004) Spontaneous cholecysto- and hepatolithiasis in Mdr2-/- mice: a model for low phospholipid-associated cholelithiasis. Hepatology 39(1):117–128
99. Oude Elferink RP, Paulusma CC (2007) Function and pathophysiological importance of ABCB4 (MDR3 P-glycoprotein). Pflugers Arch – Eur J Physiol 453 (5):601–610
100. Davit-Spraul A, Gonzales E, Baussan C, Jacquemin E (2010) The spectrum of liver diseases related to ABCB4 gene mutations: pathophysiology and clinical aspects. Semin Liver Dis 30(2):134–146
101. Gonzales E, Davit-Spraul A, Baussan C, Buffet C, Maurice M, Jacquemin E (2009) Liver diseases related to MDR3 (ABCB4) gene deficiency. Front Biosci 14:4242–4256
102. Stapelbroek JM, van Erpecum KJ, Klomp LW, Houwen RH (2010) Liver disease associated with canalicular transport defects: current and future therapies. J Hepatol 52(2):258–271
103. Paulusma CC, Elferink RP, Jansen PL (2010) Progressive familial intrahepatic cholestasis type 1. Semin Liver Dis 30(2):117–124
104. Jansen PL, Sturm E (2003) Genetic cholestasis, causes and consequences for hepatobiliary transport. Liver Int 23(5):315–322
105. Rosmorduc O, Hermelin B, Boelle PY, Parc R, Taboury J, Poupon R (2003) ABCB4 gene

mutation-associated cholelithiasis in adults. Gastroenterology 125(2):452–459
106. Rosmorduc O, Hermelin B, Poupon R (2001) MDR3 gene defect in adults with symptomatic intrahepatic and gallbladder cholesterol cholelithiasis. Gastroenterology 120(6):1459–1467
107. Rosmorduc O, Poupon R (2007) Low phospholipid associated cholelithiasis: association with mutation in the MDR3/ABCB4 gene. Orphanet J Rare Dis 2:29
108. Wang DQ, Carey MC (1996) Complete mapping of crystallization pathways during cholesterol precipitation from model bile: influence of physical-chemical variables of pathophysiologic relevance and identification of a stable liquid crystalline state in cold, dilute and hydrophilic bile salt-containing systems. J Lipid Res 37(3):606–630
109. Carey MC, Cahalane MJ (1988) In: Arias IM, Jakoby WB, Popper H, Schachter D, Shafritz DA (eds) The liver: biology and pathobiology. Raven Press, New York, pp 573–616
110. Hofmann AF, Hagey LR (2008) Bile acids: chemistry, pathochemistry, biology, pathobiology, and therapeutics. Cell Mol Life Sci 65(16):2461–2483
111. Chiang JY (2009) Bile acids: regulation of synthesis. J Lipid Res 50(10):1955–1966
112. Hofmann AF (2009) The enterohepatic circulation of bile acids in mammals: form and functions. Front Biosci 14:2584–2598
113. Hofmann AF (2009) In: Arias IM, Alter HJ, Boyer JL, Cohen DE, Fausto N, Shafritz DA, Wolkoff AW (eds) The liver: biology and pathobiology. Wiley-Blackwell, Chichester, pp 290–304
114. Suchy FJ, Ananthanarayanan M (2006) Bile salt excretory pump: biology and pathobiology. J Pediatr Gastroenterol Nutr 43(Suppl 1):S10–S16
115. Trauner M, Boyer JL (2003) Bile salt transporters: molecular characterization, function, and regulation. Physiol Rev 83(2):633–671
116. Gerloff T, Stieger B, Hagenbuch B, Madon J, Landmann L, Roth J, Hofmann AF, Meier PJ (1998) The sister of P-glycoprotein represents the canalicular bile salt export pump of mammalian liver. J Biol Chem 273(16):10046–10050
117. Wang R, Salem M, Yousef IM, Tuchweber B, Lam P, Childs SJ, Helgason CD, Ackerley C, Phillips MJ, Ling V (2001) Targeted inactivation of sister of P-glycoprotein gene (spgp) in mice results in nonprogressive but persistent intrahepatic cholestasis. Proc Natl Acad Sci U S A 98(4):2011–2016
118. Wang R, Lam P, Liu L, Forrest D, Yousef IM, Mignault D, Phillips MJ, Ling V (2003) Severe cholestasis induced by cholic acid feeding in knockout mice of sister of P-glycoprotein. Hepatology 38 (6):1489–1499
119. Cohen DE, Leighton LS, Carey MC (1992) Bile salt hydrophobicity controls vesicle secretion rates and transformations in native bile. Am J Physiol 263 (3 Pt 1):G386–G395
120. Cohen DE, Leonard MR, Carey MC (1994) In vitro evidence that phospholipid secretion into bile may be coordinated intracellularly by the combined actions of bile salts and the specific phosphatidylcholine transfer protein of liver. Biochemistry 33 (33):9975–9980
121. Fuchs M, Carey MC, Cohen DE (1997) Evidence for an ATP-independent long-chain phosphatidylcholine translocator in hepatocyte membranes. Am J Physiol 273(6 Pt 1):G1312–G1319
122. Carey MC, Cohen DE (1995) Update on physical state of bile. Ital J Gastroenterol 27(2):92–100
123. Somjen GJ, Marikovsky Y, Lelkes P, Gilat T (1986) Cholesterol-phospholipid vesicles in human bile: an ultrastructural study. Biochim Biophys Acta 879 (1):14–21
124. Mazer NA, Carey MC, Kwasnick RF, Benedek GB (1979) Quasielastic light scattering studies of aqueous biliary lipid systems. Size, shape, and thermodynamics of bile salt micelles. Biochemistry 18 (14):3064–3075
125. Mazer NA, Benedek GB, Carey MC (1980) Quasielastic light-scattering studies of aqueous biliary lipid systems. Mixed micelle formation in bile salt-lecithin solutions. Biochemistry 19(4):601–615
126. Mazer NA, Schurtenberg P, Carey MC, Preisig R, Weigand K, Kanzig W (1984) Quasi-elastic light scattering studies of native hepatic bile from the dog: comparison with aggregative behavior of model biliary lipid systems. Biochemistry 23 (9):1994–2005
127. Somjen GJ, Gilat T (1985) Contribution of vesicular and micellar carriers to cholesterol transport in human bile. J Lipid Res 26(6):699–704
128. Somjen GJ, Gilat T (1983) A non-micellar mode of cholesterol transport in human bile. FEBS Lett 156 (2):265–268
129. Bhattacharyya AK, Connor WE (1974) Beta-sitosterolemia and xanthomatosis. A newly described lipid storage disease in two sisters. J Clin Invest 53 (4):1033–1043
130. Berge KE (2003) Sitosterolemia: a gateway to new knowledge about cholesterol metabolism. Ann Med 35(7):502–511
131. Berge KE, von Bergmann K, Lutjohann D, Guerra R, Grundy SM, Hobbs HH, Cohen JC (2002) Heritability of plasma noncholesterol sterols and relationship to DNA sequence polymorphism in ABCG5 and ABCG8. J Lipid Res 43(3):486–494
132. Salen G, Patel S, Batta AK (2002) Sitosterolemia. Cardiovas Drug Rev 20(4):255–270
133. Salen G, Shefer S, Nguyen L, Ness GC, Tint GS, Batta AK (1997) Sitosterolemia. Subcell Biochem 28:453–476
134. Salen G, Shefer S, Nguyen L, Ness GC, Tint GS, Shore V (1992) Sitosterolemia. J Lipid Res 33 (7):945–955
135. Salen G, Shore V, Tint GS, Forte T, Shefer S, Horak I, Horak E, Dayal B, Nguyen L, Batta AK

et al (1989) Increased sitosterol absorption, decreased removal, and expanded body pools compensate for reduced cholesterol synthesis in sitosterolemia with xanthomatosis. J Lipid Res 30(9):1319–1330
136. Lu K, Lee MH, Yu H, Zhou Y, Sandell SA, Salen G, Patel SB (2002) Molecular cloning, genomic organization, genetic variations, and characterization of murine sterolin genes Abcg5 and Abcg8. J Lipid Res 43(4):565–578
137. Lu K, Lee MH, Hazard S, Brooks-Wilson A, Hidaka H, Kojima H, Ose L, Stalenhoef AF, Mietinnen T, Bjorkhem I et al (2001) Two genes that map to the STSL locus cause sitosterolemia: genomic structure and spectrum of mutations involving sterolin-1 and sterolin-2, encoded by ABCG5 and ABCG8, respectively. Am J Hum Genet 69 (2):278–290
138. Nguyen LB, Shefer S, Salen G, Ness GC, Tint GS, Zaki FG, Rani I (1990) A molecular defect in hepatic cholesterol biosynthesis in sitosterolemia with xanthomatosis. J Clin Invest 86(3):923–931
139. Salen G, Horak I, Rothkopf M, Cohen JL, Speck J, Tint GS, Shore V, Dayal B, Chen T, Shefer S (1985) Lethal atherosclerosis associated with abnormal plasma and tissue sterol composition in sitosterolemia with xanthomatosis. J Lipid Res 26 (9):1126–1133
140. Yu L, Gupta S, Xu F, Liverman AD, Moschetta A, Mangelsdorf DJ, Repa JJ, Hobbs HH, Cohen JC (2005) Expression of ABCG5 and ABCG8 is required for regulation of biliary cholesterol secretion. J Biol Chem 280(10):8742–8747
141. Wang HH, Patel SB, Carey MC, Wang DQ (2007) Quantifying anomalous intestinal sterol uptake, lymphatic transport, and biliary secretion in Abcg8(-/-) mice. Hepatology 45(4):998–1006
142. Klett EL, Lu K, Kosters A, Vink E, Lee MH, Altenburg M, Shefer S, Batta AK, Yu H, Chen J et al (2004) A mouse model of sitosterolemia: absence of Abcg8/sterolin-2 results in failure to secrete biliary cholesterol. BMC Med 2:5
143. Lutjohann D, Bjorkhem I, Beil UF, von Bergmann K (1995) Sterol absorption and sterol balance in phytosterolemia evaluated by deuterium-labeled sterols: effect of sitostanol treatment. J Lipid Res 36 (8):1763–1773
144. Gould RG, Jones RJ, LeRoy GV, Wissler RW, Taylor CB (1969) Absorbability of beta-sitosterol in humans. Metab Clin Exp 18(8):652–662
145. Plosch T, Bloks VW, Terasawa Y, Berdy S, Siegler K, Van Der Sluijs F, Kema IP, Groen AK, Shan B, Kuipers F et al (2004) Sitosterolemia in ABC-transporter G5-deficient mice is aggravated on activation of the liver-X receptor. Gastroenterology 126(1):290–300
146. Acton S, Rigotti A, Landschulz KT, Xu S, Hobbs HH, Krieger M (1996) Identification of scavenger receptor SR-BI as a high density lipoprotein receptor. Science 271(5248):518–520
147. Wang DQ, Carey MC (2002) Susceptibility to murine cholesterol gallstone formation is not affected by partial disruption of the HDL receptor SR-BI. Biochim Biophys Acta 1583(2):141–150
148. Kozarsky KF, Donahee MH, Rigotti A, Iqbal SN, Edelman ER, Krieger M (1997) Overexpression of the HDL receptor SR-BI alters plasma HDL and bile cholesterol levels. Nature 387(6631):414–417
149. Wang DQ, Cohen DE (2015) In: Ballantyne CM (ed) Clinical lipidology: a companion to Braunwald's heart disease. Elsevier Saunders, Philadephia, pp 25–42
150. Altmann SW, Davis HR Jr, Zhu LJ, Yao X, Hoos LM, Tetzloff G, Iyer SP, Maguire M, Golovko A, Zeng M et al (2004) Niemann-Pick C1 Like 1 protein is critical for intestinal cholesterol absorption. Science 303(5661):1201–1204
151. Davis HR Jr, Altmann SW (2009) Niemann-Pick C1 Like 1 (NPC1L1) an intestinal sterol transporter. Biochim Biophys Acta 1791(7):679–683
152. Davis HR Jr, Basso F, Hoos LM, Tetzloff G, Lally SM, Altmann SW (2008) Cholesterol homeostasis by the intestine: lessons from Niemann-Pick C1 Like 1 [NPC1L1]. Atheroscler Suppl 9(2):77–81
153. Davis HR Jr, Zhu LJ, Hoos LM, Tetzloff G, Maguire M, Liu J, Yao X, Iyer SP, Lam MH, Lund EG et al (2004) Niemann-Pick C1 Like 1 (NPC1L1) is the intestinal phytosterol and cholesterol transporter and a key modulator of whole-body cholesterol homeostasis. J Biol Chem 279(32):33586–33592
154. Garcia-Calvo M, Lisnock J, Bull HG, Hawes BE, Burnett DA, Braun MP, Crona JH, Davis HR Jr, Dean DC, Detmers PA et al (2005) The target of ezetimibe is Niemann-Pick C1-Like 1 (NPC1L1). Proc Natl Acad Sci U S A 102(23):8132–8137
155. Graf GA, Cohen JC, Hobbs HH (2004) Missense mutations in ABCG5 and ABCG8 disrupt heterodimerization and trafficking. J Biol Chem 279 (23):24881–24888
156. Hubacek JA, Berge KE, Cohen JC, Hobbs HH (2001) Mutations in ATP-cassette binding proteins G5 (ABCG5) and G8 (ABCG8) causing sitosterolemia. Hum Mutat 18(4):359–360
157. Duan LP, Wang HH, Ohashi A, Wang DQ (2006) Role of intestinal sterol transporters Abcg5, Abcg8, and Npc1l1 in cholesterol absorption in mice: gender and age effects. Am J Physiol Gastrointest Liver Physiol 290(2):G269–G276
158. Duan LP, Wang HH, Wang DQ (2004) Cholesterol absorption is mainly regulated by the jejunal and ileal ATP-binding cassette sterol efflux transporters Abcg5 and Abcg8 in mice. J Lipid Res 45 (7):1312–1323
159. Lammert F, Wang DQ (2005) New insights into the genetic regulation of intestinal cholesterol absorption. Gastroenterology 129(2):718–734
160. Brown MS, Brannan PG, Bohmfalk HA, Brunschede GY, Dana SE, Helgeson J, Goldstein JL (1975) Use of mutant fibroblasts in the analysis of the regulation

of cholesterol metabolism in human cells. J Cell Physiol 85(2 Pt 2 Suppl 1):425–436
161. Brown MS, Dana SE, Dietschy JM, Siperstein MD (1973) 3-Hydroxy-3-methylglutaryl coenzyme A reductase. Solubilization and purification of a cold-sensitive microsomal enzyme. J Biol Chem 248 (13):4731–4738
162. Brown MS, Dana SE, Goldstein JL (1975) Receptor-dependent hydrolysis of cholesteryl esters contained in plasma low density lipoprotein. Proc Natl Acad Sci U S A 72(8):2925–2929
163. Brown MS, Dana SE, Goldstein JL (1975) Cholesterol ester formation in cultured human fibroblasts. Stimulation by oxygenated sterols. J Biol Chem 250 (10):4025–4027
164. Brown MS, Dana SE, Goldstein JL (1974) Regulation of 3-hydroxy-3-methylglutaryl coenzyme A reductase activity in cultured human fibroblasts. Comparison of cells from a normal subject and from a patient with homozygous familial hypercholesterolemia. J Biol Chem 249(3):789–796
165. Brown MS, Dana SE, Goldstein JL (1973) Regulation of 3-hydroxy-3-methylglutaryl coenzyme A reductase activity in human fibroblasts by lipoproteins. Proc Natl Acad Sci U S A 70 (7):2162–2166
166. Dietschy JM, Spady DK (1984) Measurement of rates of cholesterol synthesis using tritiated water. J Lipid Res 25(13):1469–1476
167. Dietschy JM, Spady DK (1984) Regulation of low density lipoprotein uptake and degradation in different animals species. Agents Actions Suppl 16:177–190
168. Dietschy JM, Spady DK, Stange EF (1983) Quantitative importance of different organs for cholesterol synthesis and low-density-lipoprotein degradation. Biochem Soc Trans 11(6):639–641
169. Dietschy JM, Turley SD, Spady DK (1993) Role of liver in the maintenance of cholesterol and low density lipoprotein homeostasis in different animal species, including humans. J Lipid Res 34 (10):1637–1659
170. Turley SD, Spady DK, Dietschy JM (1997) Identification of a metabolic difference accounting for the hyper- and hyporesponder phenotypes of cynomolgus monkey. J Lipid Res 38(8):1598–1611
171. Turley SD, Spady DK, Dietschy JM (1997) Regulation of fecal bile acid excretion in male golden Syrian hamsters fed a cereal-based diet with and without added cholesterol. Hepatology 25 (4):797–803
172. Turley SD, Spady DK, Dietschy JM (1995) Role of liver in the synthesis of cholesterol and the clearance of low density lipoproteins in the cynomolgus monkey. J Lipid Res 36(1):67–79
173. Spady DK, Dietschy JM (1983) Sterol synthesis in vivo in 18 tissues of the squirrel monkey, guinea pig, rabbit, hamster, and rat. J Lipid Res 24 (3):303–315
174. Heinemann T, Axtmann G, von Bergmann K (1993) Comparison of intestinal absorption of cholesterol with different plant sterols in man. Eur J Clin Investig 23(12):827–831
175. McNamara DJ, Kolb R, Parker TS, Batwin H, Samuel P, Brown CD, Ahrens EH Jr (1987) Heterogeneity of cholesterol homeostasis in man. Response to changes in dietary fat quality and cholesterol quantity. J Clin Invest 79(6):1729–1739
176. Kesaniemi YA, Miettinen TA (1987) Cholesterol absorption efficiency regulates plasma cholesterol level in the Finnish population. Eur J Clin Investig 17(5):391–395
177. Sehayek E, Nath C, Heinemann T, McGee M, Seidman CE, Samuel P, Breslow JL (1998) U-shape relationship between change in dietary cholesterol absorption and plasma lipoprotein responsiveness and evidence for extreme interindividual variation in dietary cholesterol absorption in humans. J Lipid Res 39(12):2415–2422
178. Miettinen TA, Tilvis RS, Kesaniemi YA (1990) Serum plant sterols and cholesterol precursors reflect cholesterol absorption and synthesis in volunteers of a randomly selected male population. Am J Epidemiol 131(1):20–31
179. Bosner MS, Lange LG, Stenson WF, Ostlund RE Jr (1999) Percent cholesterol absorption in normal women and men quantified with dual stable isotopic tracers and negative ion mass spectrometry. J Lipid Res 40(2):302–308
180. Hallikainen MA, Sarkkinen ES, Gylling H, Erkkila AT, Uusitupa MI (2000) Comparison of the effects of plant sterol ester and plant stanol ester-enriched margarines in lowering serum cholesterol concentrations in hypercholesterolaemic subjects on a low-fat diet. Eur J Clin Nutr 54(9):715–725
181. Gremaud G, Dalan E, Piguet C, Baumgartner M, Ballabeni P, Decarli B, Leser ME, Berger A, Fay LB (2002) Effects of non-esterified stanols in a liquid emulsion on cholesterol absorption and synthesis in hypercholesterolemic men. Eur J Nutr 41(2):54–60
182. Maki KC, Davidson MH, Umporowicz DM, Schaefer EJ, Dicklin MR, Ingram KA, Chen S, McNamara JR, Gebhart BW, Ribaya-Mercado JD et al (2001) Lipid responses to plant-sterol-enriched reduced-fat spreads incorporated into a National Cholesterol Education Program Step I diet. Am J Clin Nutr 74(1):33–43
183. Miettinen TA, Puska P, Gylling H, Vanhanen H, Vartiainen E (1995) Reduction of serum cholesterol with sitostanol-ester margarine in a mildly hypercholesterolemic population. N Engl J Med 333 (20):1308–1312
184. Wang DQ, Neuschwander-Tetri BA, Portincasa P (2017) The biliary system. Morgan & Claypool Life Sciences, Princeton
185. Kosters A, Frijters RJ, Schaap FG, Vink E, Plosch T, Ottenhoff R, Jirsa M, De Cuyper IM, Kuipers F, Groen AK (2003) Relation between hepatic

expression of ATP-binding cassette transporters G5 and G8 and biliary cholesterol secretion in mice. J Hepatol 38(6):710–716
186. Wang HH, Lammert F, Schmitz A, Wang DQ (2010) Transgenic overexpression of Abcb11 enhances biliary bile salt outputs, but does not affect cholesterol cholelithogenesis in mice. Eur J Clin Investig 40 (6):541–551
187. Glomset JA, Janssen ET, Kennedy R, Dobbins J (1966) Role of plasma lecithin: cholesterol acyltransferase in the metabolism of high density lipoproteins. J Lipid Res 7(5):638–648
188. Rader DJ (2014) New therapies for coronary artery disease: genetics provides a blueprint. Sci Transl Med 6(239):239ps4
189. Tuteja S, Rader DJ (2014) High-density lipoproteins in the prevention of cardiovascular disease: changing the paradigm. Clin Pharmacol Ther 96(1):48–56
190. Rader DJ, Hovingh GK (2014) HDL and cardiovascular disease. Lancet 384(9943):618–625
191. Rader DJ (2014) Spotlight on HDL biology: new insights in metabolism, function, and translation. Cardiovasc Res 103(3):337–340
192. Rosenson RS, Brewer HB Jr, Davidson WS, Fayad ZA, Fuster V, Goldstein J, Hellerstein M, Jiang XC, Phillips MC, Rader DJ et al (2012) Cholesterol efflux and atheroprotection: advancing the concept of reverse cholesterol transport. Circulation 125 (15):1905–1919
193. Lewis GF, Rader DJ (2005) New insights into the regulation of HDL metabolism and reverse cholesterol transport. Circ Res 96(12):1221–1232
194. Rader DJ, Alexander ET, Weibel GL, Billheimer J, Rothblat GH (2009) The role of reverse cholesterol transport in animals and humans and relationship to atherosclerosis. J Lipid Res 50(Suppl):S189–S194
195. Khera AV, Rader DJ (2010) Future therapeutic directions in reverse cholesterol transport. Curr Atheroscler Rep 12(1):73–81
196. Degoma EM, Rader DJ (2011) Novel HDL-directed pharmacotherapeutic strategies. Nat Rev Cardiol 8 (5):266–277
197. Rader DJ (2006) Molecular regulation of HDL metabolism and function: implications for novel therapies. J Clin Invest 116(12):3090–3100
198. Rosenson RS, Brewer HB Jr, Ansell B, Barter P, Chapman MJ, Heinecke JW, Kontush A, Tall AR, Webb NR (2013) Translation of high-density lipoprotein function into clinical practice: current prospects and future challenges. Circulation 128 (11):1256–1267
199. Toth PP, Barter PJ, Rosenson RS, Boden WE, Chapman MJ, Cuchel M, D'Agostino RB Sr, Davidson MH, Davidson WS, Heinecke JW et al (2013) High-density lipoproteins: a consensus statement from the National Lipid Association. J Clin Lipidol 7(5):484–525
200. Birner-Gruenberger R, Schittmayer M, Holzer M, Marsche G (2014) Understanding high-density lipoprotein function in disease: recent advances in proteomics unravel the complexity of its composition and biology. Prog Lipid Res 56C:36–46
201. Stanley MM, Pineda EP, Cheng SH (1959) Serum cholesterol esters and intestinal cholesterol secretion and absorption in obstructive jaundice due to cancer. N Engl J Med 261:368–373
202. Kruit JK, Plosch T, Havinga R, Boverhof R, Groot PH, Groen AK, Kuipers F (2005) Increased fecal neutral sterol loss upon liver X receptor activation is independent of biliary sterol secretion in mice. Gastroenterology 128(1):147–156
203. Temel RE, Sawyer JK, Yu L, Lord C, Degirolamo C, McDaniel A, Marshall S, Wang N, Shah R, Rudel LL et al (2010) Biliary sterol secretion is not required for macrophage reverse cholesterol transport. Cell Metab 12(1):96–102
204. de Boer JF, Schonewille M, Boesjes M, Wolters H, Bloks VW, Bos T, van Dijk TH, Jurdzinski A, Boverhof R, Wolters JC et al (2017) Intestinal farnesoid X receptor controls transintestinal cholesterol excretion in mice. Gastroenterology 152 (5):1126–1138. e6
205. de Boer JF, Schonewille M, Dikkers A, Koehorst M, Havinga R, Kuipers F, Tietge UJ, Groen AK (2017) Transintestinal and biliary cholesterol secretion both contribute to macrophage reverse cholesterol transport in rats-brief report. Arterioscler Thromb Vasc Biol 37(4):643–646
206. Le May C, Berger JM, Lespine A, Pillot B, Prieur X, Letessier E, Hussain MM, Collet X, Cariou B, Costet P (2013) Transintestinal cholesterol excretion is an active metabolic process modulated by PCSK9 and statin involving ABCB1. Arterioscler Thromb Vasc Biol 33(7):1484–1493
207. Jakulj L, van Dijk TH, de Boer JF, Kootte RS, Schonewille M, Paalvast Y, Boer T, Bloks VW, Boverhof R, Nieuwdorp M et al (2016) Transintestinal cholesterol transport is active in mice and humans and controls ezetimibe-induced fecal neutral sterol excretion. Cell Metab 24(6):783–794
208. Wang DQ, Portincasa P, Tso P (2017) Transintestinal cholesterol excretion (TICE): a secondary, non-biliary pathway contributing to reverse cholesterol transport. Hepatology
209. Shulman RS, Bhattacharyya AK, Connor WE, and Fredrickson DS. Beta-sitosterolemia and xanthomatosis. N Engl J Med 1976;294(9):482-3.
210. Hidaka H, Nakamura T, Aoki T, Kojima H, Nakajima Y, Kosugi K, Hatanaka I, Harada M, Kobayashi M, Tamura A et al (1990) Increased plasma plant sterol levels in heterozygotes with sitosterolemia and xanthomatosis. J Lipid Res 31 (5):881–888
211. Cobb MM, Salen G, Tint GS (1997) Comparative effect of dietary sitosterol on plasma sterols and cholesterol and bile acid synthesis in a sitosterolemic homozygote and heterozygote subject. J Am Coll Nutr 16(6):605–613

212. Connor WE, Lin DS, Pappu AS, Frohlich J, Gerhard G (2005) Dietary sitostanol and campestanol: accumulation in the blood of humans with sitosterolemia and xanthomatosis and in rat tissues. Lipids 40 (9):919–923
213. Salen G, Kwiterovich PO Jr, Shefer S, Tint GS, Horak I, Shore V, Dayal B, Horak E (1985) Increased plasma cholestanol and 5 alpha-saturated plant sterol derivatives in subjects with sitosterolemia and xanthomatosis. J Lipid Res 26(2):203–209
214. Patel SB, Salen G, Hidaka H, Kwiterovich PO, Stalenhoef AF, Miettinen TA, Grundy SM, Lee MH, Rubenstein JS, Polymeropoulos MH et al (1998) Mapping a gene involved in regulating dietary cholesterol absorption. The sitosterolemia locus is found at chromosome 2p21. J Clin Invest 102 (5):1041–1044
215. Salen G, Tint GS, Shefer S, Shore V, Nguyen L (1992) Increased sitosterol absorption is offset by rapid elimination to prevent accumulation in heterozygotes with sitosterolemia. Arterioscler Thromb 12(5):563–568
216. Robins SJ, Fasulo JM (1997) High density lipoproteins, but not other lipoproteins, provide a vehicle for sterol transport to bile. J Clin Invest 99 (3):380–384
217. Belamarich PF, Deckelbaum RJ, Starc TJ, Dobrin BE, Tint GS, Salen G (1990) Response to diet and cholestyramine in a patient with sitosterolemia. Pediatrics 86(6):977–981
218. Nguyen LB, Cobb M, Shefer S, Salen G, Ness GC, Tint GS (1991) Regulation of cholesterol biosynthesis in sitosterolemia: effects of lovastatin, cholestyramine, and dietary sterol restriction. J Lipid Res 32 (12):1941–1948
219. Cobb MM, Salen G, Tint GS, Greenspan J, Nguyen LB (1996) Sitosterolemia: opposing effects of cholestyramine and lovastatin on plasma sterol levels in a homozygous girl and her heterozygous father. Metab Clin Exp 45(6):673–679
220. Parsons HG, Jamal R, Baylis B, Dias VC, Roncari D (1995) A marked and sustained reduction in LDL sterols by diet and cholestyramine in beta-sitosterolemia. Clin Invest Med 18(5):389–400
221. Hidaka H, Kojima H, Kawabata T, Nakamura T, Konaka K, Kashiwagi A, Kikkawa R, Shigeta Y (1995) Effects of an HMG-CoA reductase inhibitor, pravastatin, and bile sequestering resin, cholestyramine, on plasma plant sterol levels in hypercholesterolemic subjects. J Atheroscler Thromb 2(1):60–65
222. Salen G, Starc T, Sisk CM, Patel SB (2006) Intestinal cholesterol absorption inhibitor ezetimibe added to cholestyramine for sitosterolemia and xanthomatosis. Gastroenterology 130(6):1853–1857
223. Salen G, von Bergmann K, Lutjohann D, Kwiterovich P, Kane J, Patel SB, Musliner T, Stein P, Musser B (2004) Ezetimibe effectively reduces plasma plant sterols in patients with sitosterolemia. Circulation 109(8):966–971
224. Tsubakio-Yamamoto K, Nishida M, Nakagawa-Toyama Y, Masuda D, Ohama T, Yamashita S (2010) Current therapy for patients with sitosterolemia – effect of ezetimibe on plant sterol metabolism. J Atheroscler Thromb 17(9):891–900
225. Iyer SP, Yao X, Crona JH, Hoos LM, Tetzloff G, Davis HR Jr, Graziano MP, Altmann SW (2005) Characterization of the putative native and recombinant rat sterol transporter Niemann-Pick C1 Like 1 (NPC1L1) protein. Biochim Biophys Acta 1722 (3):282–292
226. Sudhop T, Lutjohann D, Kodal A, Igel M, Tribble DL, Shah S, Perevozskaya I, von Bergmann K (2002) Inhibition of intestinal cholesterol absorption by ezetimibe in humans. Circulation 106 (15):1943–1948
227. Sudhop T, von Bergmann K (2002) Cholesterol absorption inhibitors for the treatment of hypercholesterolaemia. Drugs 62(16):2333–2347
228. Pandor A, Ara RM, Tumur I, Wilkinson AJ, Paisley S, Duenas A, Durrington PN, Chilcott J (2009) Ezetimibe monotherapy for cholesterol lowering in 2,722 people: systematic review and meta-analysis of randomized controlled trials. J Intern Med 265(5):568–580
229. Lusis AJ (2000) Atherosclerosis. Nature 407 (6801):233–241
230. Emerging Risk Factors Collaboration, Di Angelantonio E, Sarwar N, Perry P, Kaptoge S, Ray KK, Thompson A, Wood AM, Lewington S, Sattar N et al (2009) Major lipids, apolipoproteins, and risk of vascular disease. JAMA 302(18):1993–2000
231. Allen JK (2000) Genetics and cardiovascular disease. Nurs Clin North Am 35(3):653–662
232. Chico TJ, Milo M, Crossman DC (2010) The genetics of cardiovascular disease: new insights from emerging approaches. J Pathol 220(2):186–197
233. Cymet T (2010) Genetics of cardiovascular disease. Compr Ther 36:18–19
234. Hajar R (2020) Genetics in cardiovascular disease. Heart Views 21(1):55–56
235. Kathiresan S, Srivastava D (2012) Genetics of human cardiovascular disease. Cell 148(6):1242–1257
236. Katsuya T, Harrap SB, Ogihara T (2011) Genetics of hypertension and cardiovascular disease. Int J Hypertens 2010:951254
237. Sabeva NS, Liu J, Graf GA (2009) The ABCG5 ABCG8 sterol transporter and phytosterols: implications for cardiometabolic disease. Curr Opin Endocrinol Diabetes Obes 16(2):172–177
238. Klett EL, Patel S (2003) Genetic defenses against noncholesterol sterols. Curr Opin Lipidol 14 (4):341–345
239. Patel SB, Graf GA, Temel RE (2018) ABCG5 and ABCG8: more than a defense against xenosterols. J Lipid Res 59(7):1103–1113
240. Patel SB (2014) Recent advances in understanding the STSL locus and ABCG5/ABCG8 biology. Curr Opin Lipidol 25(3):169–175

241. Wilund KR, Yu L, Xu F, Hobbs HH, Cohen JC (2004) High-level expression of ABCG5 and ABCG8 attenuates diet-induced hypercholesterolemia and atherosclerosis in Ldlr-/- mice. J Lipid Res 45(8):1429–1436
242. Wu JE, Basso F, Shamburek RD, Amar MJ, Vaisman B, Szakacs G, Joyce C, Tansey T, Freeman L, Paigen BJ et al (2004) Hepatic ABCG5 and ABCG8 overexpression increases hepatobiliary sterol transport but does not alter aortic atherosclerosis in transgenic mice. J Biol Chem 279 (22):22913–22925
243. Basso F, Freeman LA, Ko C, Joyce C, Amar MJ, Shamburek RD, Tansey T, Thomas F, Wu J, Paigen B et al (2007) Hepatic ABCG5/G8 overexpression reduces apoB-lipoproteins and atherosclerosis when cholesterol absorption is inhibited. J Lipid Res 48 (1):114–126
244. Hubacek JA, Berge KE, Stefkova J, Pitha J, Skodova Z, Lanska V, Poledne R (2004) Polymorphisms in ABCG5 and ABCG8 transporters and plasma cholesterol levels. Physiol Res/Academia Scientiarum Bohemoslovaca 53(4):395–401
245. Gylling H, Hallikainen M, Pihlajamaki J, Agren J, Laakso M, Rajaratnam RA, Rauramaa R, Miettinen TA (2004) Polymorphisms in the ABCG5 and ABCG8 genes associate with cholesterol absorption and insulin sensitivity. J Lipid Res 45(9):1660–1665
246. Jakulj L, Vissers MN, Tanck MW, Hutten BA, Stellaard F, Kastelein JJ, Dallinga-Thie GM (2010) ABCG5/G8 polymorphisms and markers of cholesterol metabolism: systematic review and meta-analysis. J Lipid Res 51(10):3016–3023
247. Cholesterol Treatment Trialists' (CTT) Collaboration, Baigent C, Blackwell L, Emberson J, Holland LE, Reith C, Bhala N, Peto R, Barnes EH, Keech A et al (2010) Efficacy and safety of more intensive lowering of LDL cholesterol: a meta-analysis of data from 170,000 participants in 26 randomised trials. Lancet 376(9753):1670–1681
248. Grundy SM, Cleeman JI, Merz CN, Brewer HB Jr, Clark LT, Hunninghake DB, Pasternak RC, Smith SC Jr, Stone NJ (2004) Implications of recent clinical trials for the National Cholesterol Education Program Adult Treatment Panel III guidelines. Circulation 110 (2):227–239
249. Barter PJ, Brewer HB Jr, Chapman MJ, Hennekens CH, Rader DJ, Tall AR (2003) Cholesteryl ester transfer protein: a novel target for raising HDL and inhibiting atherosclerosis. Arterioscler Thromb Vasc Biol 23(2):160–167
250. Wang HH, Portincasa P, Wang DQ (2008) Molecular pathophysiology and physical chemistry of cholesterol gallstones. Front Biosci 13:401–423
251. Wang HH, Li T, Portincasa P, Ford DA, Neuschwander-Tetri BA, Tso P, Wang DQ (2017) New insights into the role of Lith genes in the formation of cholesterol-supersaturated bile. Liver Res 1 (1):42–53
252. Wittenburg H, Lyons MA, Li R, Kurtz U, Mossner J, Churchill GA, Carey MC, Paigen B (2005) Association of a lithogenic Abcg5/Abcg8 allele on Chromosome 17 (Lith9) with cholesterol gallstone formation in PERA/EiJ mice. Mamm Genome 16(7):495–504
253. Wittenburg H, Lyons MA, Paigen B, Carey MC (2003) Mapping cholesterol gallstone susceptibility (Lith) genes in inbred mice. Dig Liver Dis 35(Suppl 3):S2–S7
254. Jiang ZY, Parini P, Eggertsen G, Davis MA, Hu H, Suo GJ, Zhang SD, Rudel LL, Han TQ, Einarsson C (2008) Increased expression of LXR alpha, ABCG5, ABCG8, and SR-BI in the liver from normolipidemic, nonobese Chinese gallstone patients. J Lipid Res 49(2):464–472
255. von Schonfels W, Buch S, Wolk M, Aselmann H, Egberts JH, Schreiber S, Krawczak M, Becker T, Hampe J, Schafmayer C (2013) Recurrence of gallstones after cholecystectomy is associated with ABCG5/8 genotype. J Gastroenterol 48(3):391–396
256. Wang HH, Li X, Patel SB, Wang DQ (2016) Evidence that the adenosine triphosphate-binding cassette G5/G8-independent pathway plays a determinant role in cholesterol gallstone formation in mice. Hepatology 64(3):853–864
257. Repa JJ, Berge KE, Pomajzl C, Richardson JA, Hobbs H, Mangelsdorf DJ (2002) Regulation of ATP-binding cassette sterol transporters ABCG5 and ABCG8 by the liver X receptors alpha and beta. J Biol Chem 277(21):18793–18800
258. Yu L, York J, von Bergmann K, Lutjohann D, Cohen JC, Hobbs HH (2003) Stimulation of cholesterol excretion by the liver X receptor agonist requires ATP-binding cassette transporters G5 and G8. J Biol Chem 278(18):15565–15570
259. Uppal H, Zhai Y, Gangopadhyay A, Khadem S, Ren S, Moser JA, Xie W (2008) Activation of liver X receptor sensitizes mice to gallbladder cholesterol crystallization. Hepatology 47(4):1331–1342

Proprotein Convertase Subtilisin/Kexin-Type 9 and Lipid Metabolism

9

Shoudong Guo, Xiao-dan Xia, Hong-mei Gu, and Da-wei Zhang

Abstract

Plasma levels of cholesterol, especially low-density lipoprotein cholesterol (LDL-C), are positively correlated with the risk of cardiovascular disease. Buildup of LDL in the intima promotes the formation of foam cells and consequently initiates atherosclerosis, one of the main underlying causes of cardiovascular disease. Hepatic LDL receptor (LDLR) is mainly responsible for the clearance of plasma LDL. Mutations in LDLR cause familiar hypercholesterolemia and increase the risk of premature coronary heart disease. Proprotein convertase subtilisin/kexin-type 9 (PCSK9) promotes LDLR degradation and thereby plays a critical role in the regulation of plasma cholesterol metabolism. PCSK9 can bind to LDLR and reroute the receptor to lysosomes for degradation, increasing both circulating LDL-C levels and the risk of cardiovascular disease. PCSK9 is mainly regulated by sterol response element binding protein 2 (SREBP2) at the transcriptional level. Furthermore, many proteins have been identified as interacting with PCSK9, regulating plasma cholesterol levels. Pharmacotherapeutic inhibition of PCSK9 dramatically reduces plasma levels of LDL cholesterol and significantly reduces cardiovascular events. In this article, we summarize the latest advances in PCSK9, mainly focusing on the structure, function, and regulation of the protein, the underlying molecular mechanisms, and its pharmacotherapeutic applications.

Keywords

Hypercholesterolemia · Low-density lipoprotein receptor · Statin · Atherosclerosis · Proprotein convertase subtilisin/kexin-type 9

Shoudong Guo and Xiao-dan Xia contributed equally with all other contributors.

S. Guo
Departments of Pediatrics, Group on the Molecular and Cell Biology of Lipids, University of Alberta, Edmonton, AB, Canada

Institute of Lipid Metabolism and Atherosclerosis, Innovative Drug Research Centre, School of Pharmacy, Weifang Medical University, Weifang, China

X.-d. Xia
Departments of Pediatrics, Group on the Molecular and Cell Biology of Lipids, University of Alberta, Edmonton, AB, Canada

Department of Orthopedics, the sixth affiliated hospital (Qingyuan People's Hospital), Guangzhou Medical University, Qingyuan, China

H. m. Gu D. w. Zhang (✉)
Departments of Pediatrics, Group on the Molecular and Cell Biology of Lipids, University of Alberta, Edmonton, AB, Canada
e-mail: dzhang@ualberta.ca

X.-C. Jiang (ed.), *Lipid Transfer in Lipoprotein Metabolism and Cardiovascular Disease*, Advances in Experimental Medicine and Biology 1276, https://doi.org/10.1007/978-981-15-6082-8_9

Abbreviations

ADH	autosomal dominant hypercholesterolemia
Apo	apolipoprotein
ARH	autosomal recessive hypercholesterolemia
BACE1	β-site amyloid precursor protein-cleaving enzyme 1
bHLH	basic helix-loop-helix
CAP1	cyclase-associated protein 1
CAT	catalytic domain
COPII	the coat protein complex II
CSF	cerebrospinal fluid
CM	C-terminal module
CTD	C-terminal domain
CVD	cardiovascular disease
EGF-A	the epidermal growth factor precursor homology domain A
Epac2	exchange protein activated by cAMP-2
ER	endoplasmic reticulum
ERGIC	ER-Golgi intermediate compartment
FH	familial hypercholesterolemia
GPC3	glypican-3
HDL	high-density lipoprotein
HINFP	histone nuclear factor P
HMGCR	3-hydroxyl-3-methylglutaryl-CoA reductase
HNF1	hepatocyte nuclear factor 1
HSPG	heparan sulfate proteoglycan
INSIG	insulin-induced gene protein
LDL-C	low-density lipoprotein cholesterol
LDLR	LDL receptor
Lp(a)	lipoprotein (a)
miRNA	microRNA
PC	proprotein convertase
PCSK9	proprotein convertase subtilisin kexin-like 9
PLTP	phospholipid transfer protein
Rap1	ras-related protein-1
RISC	RNA-induced silencing complex
S1P	site-1 proteinase
S2P	site-2 proteinase
SCAP	SREBP cleavage activating protein
siRNA	small interfering RNA
SREBP-2	sterol regulatory element binding protein 2
Surf4	Surfeit 4
TLP	Toll-like receptor
UTR	untranslated region
VLDLR	very low-density lipoprotein receptor

9.1 Introduction

Cardiovascular disease (CVD) is the primary cause of morbidity and mortality worldwide. Atherosclerosis, characterized by the progressive accumulation of lipid and fibrous deposits in the vessel wall, is the most important underlying cause [1]. Plasma levels of cholesterol, especially low-density lipoprotein cholesterol (LDL-C), are positively correlated with the risk of atherosclerosis [2]. Cholesterol homeostasis in humans is regulated by well-balanced mechanisms of intestinal uptake, endogenous synthesis and metabolism, transport in lipoprotein particles, and biliary excretion. In humans, LDLs are the major cholesterol transport vesicle in the blood, carrying approximately 65–70% of plasma total cholesterol [2]. Elevated plasma LDL-C levels, such as those in patients with autosomal dominant hypercholesterolemia (ADH), lead to a progressive buildup of lipids in the inner walls of the arteries, promoting the formation of foam cells and consequently initiating atherosclerosis [3, 4].

LDL is produced as a metabolic by-product of very low-density lipoprotein (VLDL), a triglyceride-rich lipoprotein produced exclusively by the liver [5]. The LDL receptor (LDLR) in the liver is the protein primarily responsible for removal of LDL from circulation [2, 6]. Mutations in the LDLR cause familial hypercholesterolemia (FH), an inherited disorder associated with elevated circulating levels of LDL-C, which causes tendon and skin xanthomas, arcus cornea, and/or cardiovascular deposits and leads to increased risk in coronary heart disease and mortality [2, 6]. FH is the most common ADH and accounts for approximately 67% of case reports. The second most common ADH is caused by mutations in apolipoprotein B100 (apoB100) (~14% of case reports), the ligand for LDLR. ApoB100 is synthesized and lipidated in the liver and then secreted in plasma as VLDL. It is the main structural protein on VLDL and LDL. Recently, a third form of ADH was identified, which is caused by selected missense mutations in proprotein convertase subtilisin/kexin type 9 (PCSK9) and accounts for 2.3% of ADH [7, 8]. Gain-of-function mutations of PCSK9 cause higher plasma LDL-C levels and lead to accelerated atherosclerosis and premature coronary heart disease [7–10]. Thus, genetic defects in these three genes contribute to approximately 83.3% of ADH. The causes for the remaining 26.7% have yet to be determined. Interestingly, unlike defects in LDLR and apoB, certain mutations in PCSK9 lead to loss of function, resulting in reduced plasma levels of LDL-C and enhanced protection from coronary heart disease [11–13].

9.2 PCSK9 Structure

PCSK9, first known as neural apoptosis regulated convertase 1 (NARC-1), is a member of the subtilisin-like serine protease family that includes seven basic amino acid-specific proprotein convertases (PC): PC1, PC2, furin, PC4, PC5/6, PACE4, and PC7; it also includes two members, site-1 protease and PCSK9, that cleave at the carboxyl terminus of non-basic residues [14]. The human PCSK9 gene is located in chromosome 1p32.3 and covers 39.91 kb with 13 exons. PCSK9 is highly conserved among different species including human, mouse, rat, hamster, monkey, chimpanzee, *S. cerevisiae*, chicken, zebrafish, and frog. It is a 692-amino acid secretory glycoprotein that consists of a signal sequence (amino acids 1–30), followed by a prodomain (amino acids 31–152), a catalytic domain (CAT, amino acids 153–425), and a cysteine- and histidine-rich C-terminal domain (CTD). The CTD domain contains an exposed hinge region (residues 422–439) and three repeat modules: module 1 (CM1: amino acids 457–528), module 2 (CM2: amino acids 534–601), and module 3 (CM3: amino acids 608–692) (Fig. 9.1) [15–17]. PCSK9 is synthesized as a zymogen (~75 kDa) and undergoes autocatalytic cleavage in the endoplasmic reticulum (ER) at the carboxy terminus of FAQ152↓SIPK site to form the mature form (~62 kDa) (Fig. 9.1). After autocleavage, the prodomain is tightly associated with the rest of the protein.

The crystal structures of PCSK9 reveal that the overall domain structure of PCSK9 is similar to other subtilisin-like serine proteases in a wide range of pH conditions (from pH 5–10) [15, 18–20]. The prodomain of PCSK9 consists of one four- to five-stranded antiparallel β-sheet flanked by two α helices. The catalytic domain contains a classical serine protease catalytic triad of Asp186, His226, and Ser386 and shows a similar structure as other subtilisin-like family members such as yeast Kexin and mouse furin [21, 22]. It is composed of a seven-stranded parallel β-sheet core with α helices on each side. However, unlike other convertases that contain the negatively charged substrate-binding groove [21], the substrate-binding pocket in PCSK9 is mostly neutral. The β-sheet of the C-terminal prodomain associates with the catalytic site tightly through hydrophobic and electrostatic interactions, which blocks further substrate accessibility and thereby shields further catalytic activity [19]. The C-terminal domain of PCSK9 is connected to the catalytic domain through a flexible liner region as well as through hydrogen bonds and hydrophobicity interaction. The C-terminal domain is made up of three subdomains, each containing six antiparallel β-domains without helices in a similar cylindrical shape secured through three structurally conserved disulfide bonds. The C-terminal domain is unique among the subtilisin-like serine protease family and displays structural homology to resistin that is related to type II diabetes [23]. Furthermore, the C-terminal domain is enriched in cysteine and histidine residues and contains multiple potential protein-protein interaction motifs [15].

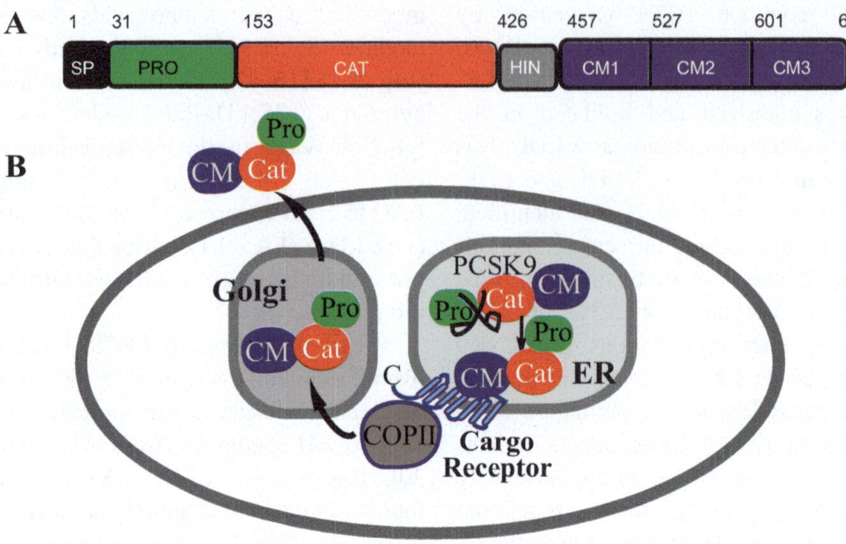

Fig. 9.1 *PCSK9 structure and secretion.* (**a**) Schematic of PCSK9. SP, signal peptide; PRO, prodomain; CAT, catalytic domain; HIN, hinge region; CM1, CM2, and CM3: three modules of C-terminal domain. Numbers on the top of the wild-type PCSK9 indicate the amino acid sequence of each domain. (**b**) PCSK9 secretion. PCSK9 undergoes autocleavage in the ER between the prodomain and the catalytic domain. The cleaved prodomain is tightly associated with the catalytic domain and is secreted together from cells

9.3 PCSK9 Secretion

PCSK9 is mainly expressed in the liver, intestine, and, to a lesser extent, in the kidneys, skin, and brain [24]. PCSK9 is secreted in plasma. However, the protein is undetectable in plasma of mice lacking hepatic PCSK9 [25], suggesting that circulating PCSK9 is mainly secreted from the liver. The autocatalytic cleavage of PCSK9 is required for its secretion. The cleaved N-terminal prodomain is tightly associated with the catalytic domain, and they are secreted together from cells [14, 26]. Structurally, it has been reported that the autocleavage of PCSK9 in the ER triggers a conformational change of the N-terminal α helix of the catalytic domain to permit PCSK9 secretion [15]. Further, the C-terminal PCSK9 has been implicated in its secretion. Loss-of-function mutations such as E498K and S462P located in the C-terminus of PCSK9 damage its secretion [27–29]. Biochemical studies reveal that removal of the whole C-terminus of PCSK9 (amino acids 456 to 692), the CM2 (amino acid 534 to 601), or the CM2 and CM3 (amino acids 534 to 692) does not significantly affect PCSK9 secretion. However, deletion of either CM1 (amino acids 457–528) or CM3 (amino acids 608–692) markedly impairs PCSK9 secretion [16, 17]. In addition, we found that the hinge region that connects the C-terminal domain to the catalytic domain played an important role in PCSK9 secretion. Deletion of the hinge region, in whole or in part, dramatically reduced PCSK9 secretion [30].

Secretion of certain abundant proteins such as N-acylglycotripeptides and amylase chymotrypsinogen is regulated by a default bulk flow pathway [31, 32]. Even so, there is emerging evidence suggesting that cargo receptors can facilitate ER-to-Golgi transport of secretory proteins in mammalian cells [33–40]. Cargo receptors are transmembrane proteins containing an ER lumen-exposed domain that binds cargo proteins within the lumen and a cytoplasmic domain that interacts with coat protein II complex (COPII) components, thereby sorting cargos into COPII vesicles. Several potential cargo receptors function as cargo-sorters for transport between the ER and Golgi: the ER-Golgi intermediate compartment (ERGIC)-53 family, the p24 family, and the

Erv family [33, 34]. It has been documented that the COPII adaptor proteins SEC24A and SEC24B facilitate the ER-to-Golgi transport of PCSK9 [41]. However, PCSK9 is located in the lumen of the ER and is unable to interact directly with SEC24, which is located in the cytosol. Thus, a cargo receptor is required. Recently, Emmer et al. reported that Surfeit locus protein 4 (SURF4) was implicated in PCSK9 secretion [42]. They found that SURF4 co-immunoprecipitated with PCSK9 and knockout of the cargo receptor significantly reduced secretion of PCSK9 overexpressed in HEK293 cells. SURF4 is a polytopic transmembrane protein containing seven putative transmembrane domains with a lumen-exposed N-terminus and a cytosolic C-terminus [43]. The protein is ubiquitously expressed and mainly localized in the ER [43]. However, we found that knockdown of SURF4 expression in cultured human-hepatoma-derived cell lines, HepG2 and Huh7, increased endogenous PCSK9 expression and secretion, indicating a negligible role for Surf4 in PCSK9 secretion in cultured hepatocytes [44]. This discrepancy might be caused by different cell lines used in the two studies. We investigated the secretion of PCSK9 endogenously expressed in HepG2 and Huh7 cells, while Emmer et al. studied the effect of Surf4 on the secretion of PCSK9 overexpressed in HEK293 cells that do not express endogenous PCSK9. In addition, conflicting data on the role of sortilin in PCSK9 secretion has been reported. Gustafsen et al. observed that plasma levels of PCSK9 were reduced in sortilin$^{-/-}$ mice but increased in sortilin-overexpressing mice. Circulating PCSK9 levels were also positively correlated with plasma levels of sortilin. Thus, the authors argued that sortilin interacted with PCSK9 in the trans-Golgi network and then facilitated its secretion [45]. Conversely, studies from Butkinaree et al. showed that knockdown of sortilin in cultured human hepatocytes or knockout of sortilin in mice had no detectable effect on PCSK9 secretion [46]. Nevertheless, these conflicting findings reveal the complexity of the molecular mechanisms of PCSK9 secretion.

9.4 PCSK9 Function

PCSK9 plays a central role in maintaining cholesterol homeostasis. Gain-of-function mutations lead to higher plasma LDL-C levels and accelerate premature coronary heart disease [7–9, 47, 48]. On the other hand, loss-of-function mutations result in low concentrations of LDL-C and protection from coronary heart disease [11–13, 49–53]. Overexpression of recombinant PCSK9 in mouse liver causes a significant reduction in hepatic LDLR protein levels without any effect on its mRNA levels, producing severe hypercholesterolemia [26, 54, 55]. On the other hand, knockdown or knockout of *PCSK9* expression in mice leads to increased levels of LDLR protein in the liver and accelerated LDL clearance [56, 57]. The natural gain-of-function mutation, D374Y, has a significantly increased binding affinity for LDLR and promotes LDLR degradation much more efficiently than the wild-type protein [15, 58], leading to a severe form of hypercholesterolemia [7]. Consistently, the FH mutation LDLR-H306Y binds PCSK9 with a higher affinity and exhibits enhanced sensitivity to PCSK9 as compared to the wild-type receptor [59]. Taken together, these findings demonstrate that the role of PCSK9 in homeostatic control of plasma LDL-C levels depends upon PCSK9-promoted degradation of LDLR, preventing clearance of LDL-C by the cells [26, 54–58, 60–64].

Studies in cultured cells and parabiotic mice demonstrate that PCSK9 promotes degradation of LDLR in an adaptor protein autosomal recessive hypercholesterolemia (ARH)-dependent manner in hepatocytes and lymphocytes [58, 60, 61, 65]. However, ARH is not required for PCSK9-promoted LDLR degradation in fibroblasts [65, 66]. McNutt et al. [59] showed that PCSK9 caused LDLR degradation primarily through interaction with the receptor on the cell surface. However, overexpression of PCSK9 in cultured cells and mouse liver also induces LDLR degradation intracellularly [55, 67]. For instance, the gain-of-function mutation R499H enhances PCSK9-promoted LDLR degradation intracellularly [68]. Similarly, mutations D129G and A168E impair PCSK9 secretion but enhance the ability of PCSK9 to induce LDLR degradation intracellularly, thereby causing hypercholesterolemia [69]. Poirier et al. [70] observed that, upon dose and incubation period, PCSK9 could act both intracellularly and extracellularly to promote LDLR degradation in cultured cells and mouse primary hepatocytes.

PCSK9's action on the LDLR is also cell-type specific. Increased plasma levels of PCSK9 in mice through infusion of purified PCSK9 or transgenic overexpression in the kidneys preferentially promoted LDLR degradation in the liver but not in the adrenal glands [71–73]. Consistently, the adrenal function of a human subject with no detectable plasma PCSK9 is normal [74]. Gustafsen et al. [75] recently reported that the prodomain of PCSK9 bound to the trisulfated heparan sulfate disaccharide repeats in heparan sulfate proteoglycans (HSPG) of the liver. Heparin mimetics such as sulfated oligosaccharides dextran sulfate and pentosan sulfate can suppress PCSK9-mediated LDLR degradation in HepG2 cells. The authors proposed that HSPG functioned as a coreceptor for PCSK9, capturing plasma PCSK9 and then presenting it to hepatic LDLR for the following degradation process. In cultured cells, the expression of PCSK9 in some cell types, such as human hepatoma cells (HepG2 and HuH7), dramatically reduces LDLR levels [55, 58, 60, 61]. On the other hand, PCSK9 appears to have no effect on LDLR expression in Chinese hamster ovarian cells (CHO-K1), monkey kidney cells (COS-7), and rat liver cells (McArdle RH7777) [55, 58, 60, 76]. The molecular mechanism of the cell type specific action of PCSK9 on LDLR is unknown. The dissociation of PCSK9 from LDLR after endocytosis may be responsible for the inability of PCSK9 to promote LDLR degradation in human skin fibroblasts SV-589 [77].

PCSK9-promoted degradation of LDLR requires binding of PCSK9 to LDLR and internalization of the receptor but does not require the proteolytic activity of PCSK9 [58, 60, 78]. Normally, the extracellular domain of the cell surface LDLR (neutral pH) adopts an extended linear open conformation that favors interactions between the receptor and LDL [79]. Upon ligand binding to the ligand binding repeats of LDLR, the receptors are internalized via clathrin-coated pits and delivered to endosomes [80, 81]. In the low pH environment of the endosome, LDLR undergoes a conformational change to form a close conformation that promotes the release of the bound LDL that is delivered to lysosomes for degradation and signals recycling of LDLR to the cell surface [79].

PCSK9 interacts with the EGF-A of LDLR at the cell surface, which is different from the LDL binding site on the receptor. Thus, the binding sites of PCSK9 and LDL on the receptor are not in proximity, and the binding of one ligand is unlikely to block the accessibility of another one to LDLR. We found that replacement of Leu at position 318 in the EGF-A of LDLR with Asp as it is in VLDLR markedly reduced PCSK9 binding to the receptor, indicating the important role of this residue in PCSK9 binding. Further, we observed that mutations G293H, D299V, L318D, and L318H in EGF-A reduced PCSK9 binding to LDLR at a neutral pH, while mutations R329P and E332G reduced PCSK9 binding at both neutral pH and acidic pH 6.0. Thus, EGF-A of the LDLR is critical for PCSK9 binding at the cell surface (neutral pH) and at the acidic endosomal environment (pH 6.0), but different determinants contribute to efficient PCSK9 binding in different pH environments [82].

Several lines of evidence demonstrate that PCSK9/LDLR complex enters cells via clathrin-coated pits. Knockdown of clathrin heavy chain markedly reduces PCSK9-promoted LDLR degradation in human hepatoma-derived cell lines, Huh7 and HepG2 cell [65, 83, 84]. Conversely, Jang et al. reported that knockdown of clathrin heavy chain did not affect PCSK9-promoted LDLR degradation in HepG2 cells; instead, the authors found that the PCSK9/LDLR complex entered cells via caveolae-dependent endocytosis [85]. The reasons for this discrepancy are unclear. It is of note that different approaches were used in the two studies. Romagnuolo et al. [83] overexpressed PCSK9 in HepG2 cells and then knocked down the expression of clathrin heavy chain, while Jang et al. silenced the expression of clathrin heavy chain and then supplied cells with various doses of recombinant flag-tagged PCSK9 purified from HEK293 cells overexpressing PCSK9 [85].

While LDL binds to the receptor that is much weaker at the acidic endosome compared to that at the neutral cell surface, PCSK9 binds the receptor with a much higher affinity at the endosomal pH value than at the neutral pH. Consequently, the receptor is transported from the endosome to the lysosome for degradation, rather than being recycled (Fig. 9.2) [60]. The binding of PCSK9 to LDLR interferes with the acid-dependent conformational change of the receptor, but disrupting the pH-dependent conformational change in the LDLR is not sufficient to trigger LDLR degradation [61]. We also demonstrated that YWTD repeats, and a minimum of three ligand-binding repeats in the LDLR that were not required for PCSK9 binding at neutral pH were essential for efficient LDLR degradation induced by PCSK9 [61, 82]. Furthermore, we reported that the C-terminal domain of PCSK9 was essential for PCSK9-promoted degradation of LDLR, but was not required for binding to LDLR at the neutral pH value [61]. The X-ray crystallographic structure of PCSK9-LDLR complex shows that YWTD repeats of LDLR interact with the prodomain of PCSK9 [86]. Several biochemical studies indicate that the negatively charged ligand binding repeats (LR) of LDLR may interact with the positively charged C-terminal domain of PCSK9 in the acidic endosomal environment to enhance PCSK9 binding [87–89]. Consistently, we found that mutation of Asp at position 172 in the linker between the LR4 and LR5 of LDLR to Asn and replacement of Asp at position 203 in the LR5 to Asn significantly reduced PCSK9 binding

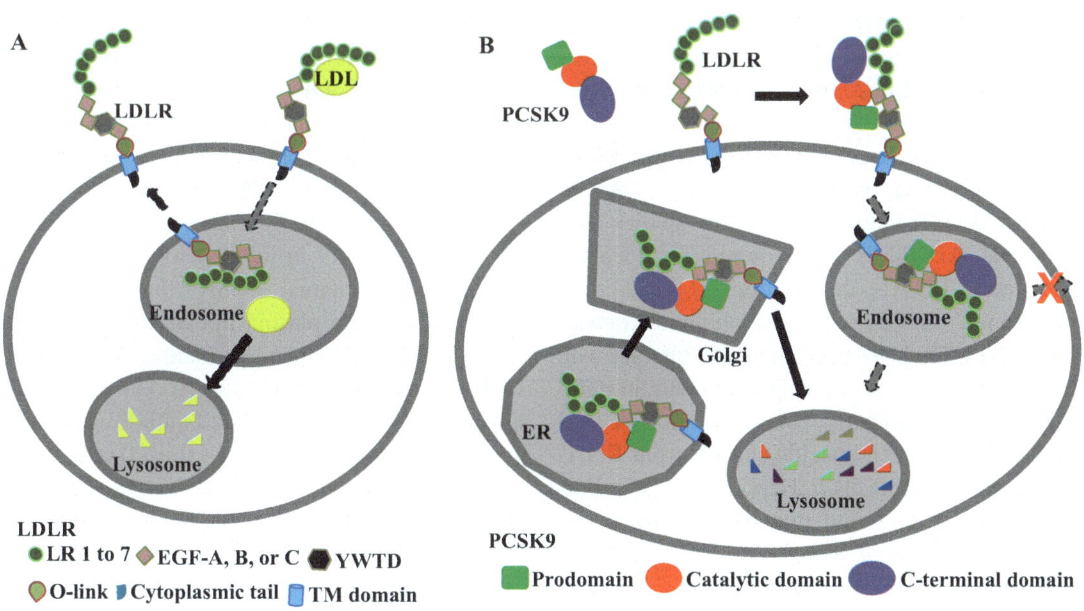

Fig. 9.2 *Recycling pathway of the LDLR.* (**a**) LDLR-mediated LDL uptake. Upon binding LDL, LDLR is internalized via clathrin-coated pits and delivered to endosomes, where the bound LDL is released from the receptor and delivered to lysosomes for degradation. LDLR recycles to the cell surface. (**b**) *PCSK9-mediated LDLR degradation.* Extracellularly, PCSK9 interacts with EGF-A of LDLR at the cell surface and enters cells via endocytosis. The LDLR/PCSK9 complex is then delivered to the endosome, where PCSK9 binds the receptor with a much higher affinity in the acidic environment. Consequently, the receptor-PCSK9 complex traffics from the endosome to the lysosome for degradation, rather than being recycled. For the intracellular pathway, PCSK9 binds to LDLR in the ER or during the secretory pathway and then reroutes the receptor to the lysosome for degradation

[90]. This further confirms the important role of the negatively charged amino acid residues within the LR in PCSK9 binding to LDLR.

Neither PCSK9 nor LDLR contains a lysosomal targeting signal. Removal of the C-terminal cytoplasmic tail of LDLR does not damage PCSK9-promoted LDLR degradation [82, 91, 92]. Thus, it is believed that co-factor(s) might be required for this process. Recently, DeVay et al. [93] reported that both amyloid precursor protein (APP) and amyloid precursor-like protein 2 (APLP2) co-immunoprecipitated with the full length but not the C-terminal deletion mutation PCSK9 at pH6.0 (endosomal environment), but not at pH7.4 (cytosol environment). The authors further showed that knockdown of APLP2 but not APP suppressed PCSK9-promoted LDLR degradation in HepG2 cells. This finding indicates that APLP2 binds to the C-terminus of PCSK9 and thereby targets the PCSK9/LDLR complex to lysosomes for degradation [93]. However, two independent groups reported that PCSK9 efficiently promoted LDLR degradation in *Aplp2*$^{-/-}$ mice [46, 94]. Butkinaree et al. [46] further showed that knockdown of APLP2 in both HepG2 and Huh7 cells had no significant effect on the ability of PCSK9 to enhance LDLR degradation. These studies suggest that APLP2 is not required for PCSK9-promoted LDLR degradation. In addition, glypican-3 (GPC3) and phospholipid transfer protein (PLTP) have been shown to interact with PCSK9 by co-immunoprecipitation and an unbiased mass spectrometry. Silence of either GPC3 or PLTP using their specific short hairpin RNAs increased LDLR levels in HepG2 cells [95]. Further, Jang et al. found that the Src homology 3 binding domain of adenylyl cyclase-associated protein 1 (CAP1) interacted with the C-terminal domain of PCSK9. The knockdown of expression of CAP1 increased LDLR levels in HepG2 cells, and haploid deficiency of Cap1 in mice led to increased hepatic LDLR levels and reduced plasma LDL-C. More interestingly, the author reported that CAP1 mediated endocytosis of the PCSK9/LDLR complex in a caveolae-dependent manner since addition of PCSK9 could not induce LDLR degradation in caveolin-deficient cells [85]. This finding contradicts several previous findings that clathrin is required for PCSK9-induced endocytosis and subsequent lysosomal degradation of the receptor [65, 84]. Nevertheless, presently, the mechanism by which binding of PCSK9 to LDLR reroutes the receptor to the lysosome for degradation is not well understood and is believed to be complex.

In addition to its regulatory role in plasma LDL-C levels via the LDLR pathway, PCSK9 regulates apoB secretion. The lack of PCSK9 in the liver of *Ldlr*$^{-/-}$/*Apobec1*$^{-/-}$ mice significantly reduces apoB100 secretion [96]. Gain-of-function PCSK9 mutation D374Y markedly increases the secretion of apoB-containing lipoprotein in transgenic mice expressing physiological levels of PCSK9 [97]. In human PCSK9 transgenic mice, the expression of microsomal triglyceride transfer protein *(MTP)* and lipogenic genes is significantly increased. Consequently, secretion of apoB48 and production of chylomicrons are increased in both LDLR-dependent and -independent manners [73, 98]. On the other hand, lymphatic apoB secretion is markedly reduced in *Pcsk9*$^{-/-}$ mice [99]. Consistently, plasma levels of PCSK9 and apoB-48 containing lipoproteins are positively correlated in men with insulin resistance [100]. Together, these findings suggest an important role of PCSK9 in the development of postprandial dyslipidemia. PCSK9 has also been shown to regulate plasma lipoprotein(a) (Lp(a)) levels in a LDLR-dependent manner. Lp(a) is an LDL-like particle that contains Apo(a) covalently linked to apoB by a disulfide bond. Several epidemiological studies show that (Lp(a)) is an independent risk factor for cardiovascular disease [101–104]. LDLR can mediate but is not required for the clearance of Lp(a) [105, 106]. PCSK9 is associated with Lp(a) in human plasma [107], and inhibition of PCSK9 reduces plasma Lp(a) levels and cardiovascular events [83, 108, 109].

It has been reported that PCSK9 binds to and stimulates degradation of several LDLR family members such as VLDLR and ApoER2 [110], as well as CD36 [111], but to a far lesser extent when compared to its binding to LDLR [60, 82]. PCSK9 can act on CD36 in HepG2 and 3T3-L1 cells, but not on HL-1 or THP-1 cells. In vivo, the levels of CD36 in the small intestine and the heart which highly express CD36 are comparable between wild-type and *Pcsk9*$^{-/-}$ mice, while the expression of CD36 in the liver and mouse adipose tissue is significantly increased [111]. The lack of PCSK9 in mice also markedly increases the expression of VLDLR in perigonadal depots and promotes accumulation of visceral fat [112]. On the other hand, Liu et al. reported that PCSK9 cannot promote degradation of VLDLR and apoER2 in the adult mouse brain [113]. Additionally, PCSK9 has been shown to promote LDLR-related protein 1 (LRP-1) degradation in mouse B16F1 melanoma cells [92] but not in mouse hepatocytes [58]. The expression of LDLR in CHO cells can suppress PCSK9's effect on LRP-1 [92]. Thus, it is possible that the relative high expression of hepatic LDLR competitively suppresses the effect of PCSK9 on LRP1 in the liver. Controversial data on the role of PCSK9 in neuron function and pathogenesis of Alzheimer's disease has also been reported in the literature. Knockdown of PCSK9 expression increases levels of ApoER2 and protects against apoptosis in cerebellar granule neurons [114], and silence of PCSK9 alleviates middle cerebral artery occlusion-induced cerebral histological injury and neuronal apoptosis in mice fed a high fat diet, probably through the regulation of apoER2 expression [115]. These findings indicate a pro-apoptosis role of PCSK9 in neurons. Consistently, inhibition of PCSK9 reduces Aβ aggregation and neuroinflammation, alleviating dendritic spine loss in a cardiac ischemic/reperfusion injury rat model [116]. On the other hand, Jonas et al. showed that PCSK9 promoted degradation of unacetylated β-site amyloid precursor protein-cleaving enzyme 1 (BACE1). The authors observed an increase in the levels of Aβ and BACE1 in the brain of *Pcsk9*$^{-/-}$ mice [117]. Thus, further studies are needed to elucidate these potential functions of PCSK9.

9.5 Regulation of PCSK9

Transcription of *PCSK9* is mainly controlled by the sterol regulatory element binding protein 2 (SREBP2) that regulates expression of genes involved in cholesterol metabolism such as *LDLR* and 3-hydroxy-3-methylglutaryl-CoA reductase (*HMGCR*), the rate-limiting enzyme in the cholesterol biosynthesis pathway [118, 119]. SREBP2 is a master regulator of cellular cholesterol homeostasis. It contains an NH_2-terminal transcriptionally active domain that belongs to the basic helix-loop-helix (bHLH) class, two transmembrane alpha-helixes, and a COOH-terminal regulatory domain [120]. The transcriptional activity of SREBP2 is strictly regulated by cellular cholesterol levels [121]. The COOH-terminal regulatory domain of newly synthesized SREBP2 immediately forms a complex with SREBP cleavage-activating protein (SCAP) in the ER. When ER cholesterol content is lower than 5% of total ER lipids, SCAP is separated from an ER-resident protein, insulin-induced gene protein (INSIG), and escorts SREBP2 to the Golgi, where SREBP2 is cleaved by S1P and site-2 protease (S2P) sequentially to liberate the transcriptionally active form. The active form then travels to the nucleus, where it recognizes and binds to sterol regulatory element (SRE) located in the promoter region of its target genes, promoting their transcription. Conversely, higher ER cholesterol content (more than the 5% threshold values) promotes formation of SCAP/INSIG complex, blocking transport of SREBP2 to the Golgi and the subsequent processing of SREBP2. Consequently, transcription of SREBP2 target genes is suppressed [121]. SREBP2 binds to an SRE motif in the promoter of *PCSK9,* stimulating its transcription. mRNA levels of *Pcsk9* are increased six- to ninefold in mice transgenic expressing the active form of SREBP2 but reduced in *Scap*$^{-/-}$ mice [118]. Additionally, Li et al. [122] identified a histone nuclear factor P (HINFP) recognition motif present within 20 bp upstream of the SRE motif and found that HINFP functioned as a co-activator for the transcriptional activity of SREBP2 through promoting the histone H4 acetylation of *PCSK9* promoter. Resistin, a small cysteine-rich protein secreted from macrophages and adipose tissue, increases mRNA levels of PCSK9 via SREBP2

[123]. Fibroblast growth factor 21 can suppress expression and activity of SREBP2 in mouse liver and reduce expression of PCSK9 [124].

Hepatocyte nuclear factor 1 (HNF1) also regulates expression of *PCSK9* at the transcriptional level. The promoter region of *PCSK9* contains a highly conserved HNF1 binding site at the upstream of SRE. Berberine, a plant-derived cholesterol-lowering compound, inhibits PCSK9 expression mainly through interfering with the HNF1's action [125, 126]. Additionally, stimulation of mTORC1 by insulin reduces activity of HNF1α and consequently suppresses PCSK9 expression. An opposite phenotype is observed when mTORC1 is inhibited by rapamycin or knockdown of hepatic insulin receptor [127]. Further, E2F2, a transcription factor that regulates the G1/S transition during the cell cycle, binds to the *PCSK9* promoter region. Feeding and high cellular cholesterol levels can stimulate E2F2 and consequently increase *PCSK9* expression [128]. Tao et al. [129] reported that forkhead transcription factor FoxO3 can recruit deacetylase Sirt6 to the proximal promoter region of *PCSK9*, which deacetylates histone H3 and consequently suppresses *PCSK9* expression. These transcriptional factors can regulate the expression of *PCSK9* separately and/or cooperatively since mutation of the HNF1 site reduces the action of both HNF1 and SREBP2 on *PCSK9* transcription [125]

At the post-transcriptional level, the 3'--untranslated region (UTR) of PCSK9 contains putative microRNA (miRNA) binding sites for miR-191, miR-222, and miR-224. Expression of these miRNAs significantly reduces mRNA and protein levels of *PCSK9* in HepG2 cells [130]. Posttranslationally, PCSK9 is N-glycosylated at Asn533 in the C-terminal domain and sulfated in the prodomain and catalytic domain. However, inhibition of the glycosylation and sulfation has no effect on PCSK9 autocleavage, secretion, and activity [24, 26, 131]. PCSK9 is also partially phosphorylated at Ser47 and Ser688 in a cell-type-dependent manner [132]. This posttranslational modification, however, is not necessary for PCSK9 function since PCSK9 can be efficiently processed and secreted from HEK293 cells and CHOK1 cells, in which PCSK9 is either poorly or not phosphorylated [132]. On the other hand, PCSK9 is phosphorylated at Ser as positions 47, 666, 668, and 688 by Farm20C in hepatocytes, and this phosphorylation significantly increases PCSK9 secretion and its ability to stimulate LDLR degradation [133]. These findings indicate a cell-type specific effect of PCSK9 phosphorylation relating to its function. In addition, PCSK9 is cleaved by furin at RFHR218↓QA to generate a truncated form that can be secreted to an extracellular milieu such as a culture medium and serum but loses the ability to promote LDLR degradation [131, 134]. The lost-of-function mutation A443T shows an increased susceptibility to furin cleavage [131].

PCSK9-promoted LDLR degradation is regulated by different cofactors. Circulating PCSK9 binds to LDL, but not to HDL or VLDL, through its N-terminal region (amino acid residues 31 to 52). Kosenko et al. observed that approximately 40% of plasma PCSK9 stays in its LDL-bound form [135]. Plasma levels of LDL are much higher than those of PCSK9. The reason why more than half of plasma PCSK9 remains as the LDL-free form is unclear. Further, the physiological significance of this association is unknown, but the binding of LDL inhibits PCSK9's ability to bind and degrade LDLR [135, 136]. LDL also can suppress PCSK9-mediated LDLR degradation through a direct association with cell surface heparin-like molecules, interfering with HSPG-facilitated binding of PCSK9 to LDLR [75, 137]. In addition, GRP94 can bind to the C-terminus of PCSK9 and block its binding to LDLR in the ER, protecting the early degradation of LDLR. The lack of GRP94 in mouse liver leads to a significant reduction in hepatic LDLR levels and an increase in plasma LDL-C levels [138]. The C-terminal domain of PCSK9 also directly interacts with annexin A2, which subsequently inhibits the extracellular PCSK9-promoted LDLR degradation. The high expression of annexin 2 in fibroblasts and COS-7 cells may account for PCSK9-resistance in these cells

[139]. On the other hand, the progestin and adipoQ receptor 3 associate with the prodomain of PCSK9 and the YWTD domain of LDLR probably in the early endosome, enhancing their interaction and consequently promoting PCSK9-mediated LDLR degradation [97]. In addition, matrix metalloproteinase-2 can associate with and cleave PCSK9, inhibiting PCSK9-promoted LDLR degradation [140].

The half-life of circulating PCSK9 is very short. Approximately 90% of PCSK9 is cleared from the blood within 15 min in the wild-type mice with a half-life of five min [71, 73]. Conversely, the half-life of PCSK9 in *Ldlr*$^{-/-}$ mice is 15 min [71]. *Ldlr*$^{-/-}$ mice also show a tenfold increase in plasma levels of PCSK9, whereas LDLR transgenic mice clear PCSK9 much faster compared to the wild-type mice [73]. We have shown that the PCSK9/LDLR complex was delivered to the lysosome for degradation after endocytosis [61]. Thus, PCSK9 may be quickly removed from circulation and then delivered for lysosomal degradation via the hepatic LDLR pathway. The LDLR-independent mechanism of PCSK9 clearance is currently unclear. Spotlitu et al. reported that hepatic glucagon receptor signaling activated the exchange protein activated by cAMP-2 (Epac2) and the ras-related protein-1 (Rap1) pathway, and then enhanced the lysosomal degradation of PCSK9 in a LDLR-independent pathway [141]. It is also possible that the other LDLR family members, such as VLDLR, may mediate PCSK9 clearance when LDLR is absent.

9.6 Pharmacotherapeutic Inhibition of PCSK9 and Perspectives

Plasma levels of LDL-C are positively correlated with the risk of atherosclerosis [2]. Statins reduce cardiovascular events by 20% to 40%. Evidence is also mounting that people with severe dyslipidemia or who are at high cardiovascular risk fail to achieve LDL-C targets even with high-intensity statin treatment [142]. Further, 15% of statin-treated people show statin intolerance [142]. Thus, there is an urgent need for an alternative strategy to reduce plasma LDL-C.

Gain-of-function PCSK9 mutations such as S127R, F216L, and D374Y are associated with an increase in plasma levels of mean LDL-C and the incidence of coronary heart disease [7, 143]. Conversely, subjects carrying loss-of-function PCSK9 mutations Y142X or C679X display a 40% reduction in plasma levels of mean LDL-C and an 88% reduction in the risk of coronary heart disease. Loss-of-function mutation R46L reduces plasma levels of LDL-C and the incidence of coronary heart disease by 21% and 47%, respectively, as shown in the Atherosclerosis Risk in Communities study and the Dallas Heart Study [11, 49]. Data from the Copenhagen General Population Study and the Copenhagen City Heart Study also shows that loss-of-function PCSK9 mutations R46L, R237W, I474V, and E670G are associated with a significant reduction in mean LDL-C (18%) and cardiovascular mortality [144]. A 15-year follow-up study of 4232 subjects (2039 men and 2193 women, all 60 years old at recruitment) demonstrates that serum levels of PCSK9 are positively associated with the future risk of cardiovascular disease [145]. Knockout of PCSK9 increases, while overexpression of PCSK9 reduces the development of atherosclerosis in *apoE*$^{-/-}$ mice [146]. Further, statins increase expression of *LDLR* and *PCSK9*. Elevated circulating PCSK9 levels then promote LDLR degradation, attenuating the lipid-lowering effect of statins. Plasma PCSK9 levels are increased in patients treated with atorvastatin, and *Pcsk9*$^{-/-}$ mice display hypersensitivity to statin treatment [56]. Together, these findings strongly indicate the potential of PCSK9 inhibition as a lipid-lowering strategy.

Currently, two monoclonal anti-PCSK9 antibody therapies, Repatha (evolocumab) and Praluent (alirocumab), are approved in the USA, Canada, Europe, and China for patients who have hereditary high cholesterol such as heterozygous and homozygous FH patients and high-risk patients intolerant to statins or experiencing poor LDL-C-lowering response even with high-intensity statin therapy. Both antibodies are against the catalytic domain of PCSK9 and block binding of plasma PCSK9 to LDLR, increasing hepatic clearance of LDL and reducing plasma levels of LDL-C. Subcutaneous administration of 150 mg alirocumab biweekly lowers plasma levels of LDL-C approximately 60% in patients and reduces the rate of main cardiovascular events from 3.3% to 1.7% [147]. Alirocumab at a dose of 75 mg once every two weeks also reduces the incidence of recurrent ischemic cardiovascular events in patients who have a previous acute coronary syndrome and are treated with maximally tolerated statin dosages [148]. Similarly, the FOURIER trial shows that evolocumab at a dose of 140 mg biweekly or 420 mg monthly leads to a 60% reduction in plasma levels of LDL-C and significantly reduces the risk of the primary end point (9.8% vs. 11.3%) and the main secondary end point (5.9% vs. 7.4%) as compared to the placebo group [149]. The two inhibitors do not show significant major side effects.

Inclisiran (ALN-PCSsc) is a chemically modified small interfering RNA (siRNA) inhibitor that targets PCSK9 mRNA and suppresses translation of PCSK9. It reduces hepatic PCSK9 production and plasma PCSK9 levels. A subcutaneous injection of 500 mg Inclisiran once every 6 months in patients with atherosclerotic cardiovascular disease on high-intensity statin therapy reduces plasma levels of LDL-C by approximately 50%. No major side effect has been reported in both Phase I and II trials [150, 151].

The monoclonal antibodies against PCSK9 show an impressive lipid-lowering effect in heterozygous and homozygous FH patients, high-risk patients intolerant to statins, and patients with poor LDL-C-lowering response even with maximally tolerated statin dosages [147, 152, 153]. Furthermore, when adding to the statin therapy, PCSK9 inhibitors can markedly reduce cardiovascular events, such as myocardial infarction and ischemic stroke, with no significant adverse side effects [147, 152]. However, this therapy requires injections of large amounts of antibodies to achieve clinical efficacy, with extremely high production costs. Given that the treatment of patients with hypercholesterolemia is lifelong, and it is predicted that PCSK9 inhibitors would cost approximately $592 billion but reduce cardiovascular care costs by only $29 billion on US health care spending over 5 years if used for all eligible patients at current pricing [154]; this treatment will place a high burden on the healthcare system. Inclisiran might reduce costs since it requires only two injections per year. However, siRNAs are small RNA duplexes that have 20–30 nucleotides. They interact with the RNA-induced silencing complex (RISC) in the cytosol. After cleavage of the sense strand by the endonuclease Argonaute 2 in the RISC, the antisense strand remains binding to RISC and guides the complex to the target mRNA for Argonaute 2-mediated cleavage. siRNAs silence target genes more specifically as compared to miRNA, since the antisense strand of siRNA duplexes theoretically only binds to mRNA that completely matches to it. However, it has been reported that siRNA can cause off-target translational inhibition [155]. In addition, duplex siRNA can trigger innate immune response in Toll-like receptor (TLR)-dependent and -independent mechanisms [156]. Considering the lifelong use of PCSK9 inhibitors, it is important to monitor the long-term safety of Inclisiran. Therefore, the need for more effective, more specific, and more cost-efficient therapies to lower LDL-C is urgent.

Crystallographic studies of PCSK9-EGF-AB complex show that the interaction face between the catalytic domain of PCSK9 and the EGF-A of LDLR is relatively flat and big, making it impossible to design a specific inhibitor to block the interaction between PCSK9 and LDLR [20]. Therefore, mechanistic studies of PCSK9 regulation, its secretion, and its ability to promote LDLR degradation are necessary. Questions need to be elucidated including, but not limited to the following:

1. PCSK9 is a serine proteinase. Currently, the only physiological substrate of PCSK9 is itself. Can PCSK9 cleave other proteins?
2. PCSK9 is expressed extrahepatically and most likely retained inside cells in the kidneys and the intestine. What are the physiological functions of PCSK9 in these tissues?

3. How does circulating PCSK9 preferentially stimulate degradation of hepatic LDLR?
4. Why is only hepatic PCSK9 efficiently secreted into circulation? What is the molecular machinery system that assists PCSK9 secretion?
5. It is believed that PCSK9 needs assistance from other proteins to efficiently redirect LDLR to the lysosome for degradation. What are these co-factors?
6. PCSK9 is expressed in the brain and present in cerebrospinal fluid (CSF). CSF PCSK9 levels are increased in patients with AD [157]. What are the physiological and pathophysiological roles of brain PCSK9?

Answering these questions will not only deepen and widen our understanding of the physiological and pathophysiological role of PCSK9 but also provide a foundation for the future development of PCSK9-specific small inhibitors that can lower plasma LDL cholesterol efficiently, specifically, and cost-effectively.

Acknowledgments This chapter was partially overlapped with the paper published by our group in *The Journal of Biomedical Research* (Gu and Zhang 2015; 29: 356–361). Zhang laboratory is supported by grants from the Natural Sciences and Engineering Research Council of Canada (RGPIN-2016-06479), the Canadian Institutes of Health Research (PS 155994), a Grant-in-Aid from the Heart and Stroke Foundation of Canada, and Pfizer Canada (PI ASPIRECCRG W1218248).

References

1. Libby P (2002) Inflammation in atherosclerosis. Nature 420:868–874
2. Goldstein JL, Brown MS (2009) The LDL receptor. Arterioscler Thromb Vasc Biol 29:431–438
3. Tiwari RL, Singh V, Barthwal MK (2008) Macrophages: an elusive yet emerging therapeutic target of atherosclerosis. Med Res Rev 28:483–544
4. Platt N, Gordon S (2001) Is the class A macrophage scavenger receptor (SR-A) multifunctional? – The mouse's tale. J Clin Invest 108:649–654
5. Quiroga AD, Lehner R (2012) Liver triacylglycerol lipases. Biochim Biophys Acta 1821:762–769
6. Brown MS, Goldstein JL (2006) Biomedicine. Lowering LDL – not only how low, but how long? Science 311:1721–1723
7. Abifadel M, Varret M, Rabes JP, Allard D, Ouguerram K, Devillers M, Cruaud C, Benjannet S, Wickham L, Erlich D, Derre A, Villeger L, Farnier M, Beucler I, Bruckert E, Chambaz J, Chanu B, Lecerf JM, Luc G, Moulin P, Weissenbach J, Prat A, Krempf M, Junien C, Seidah NG, Boileau C (2003) Mutations in PCSK9 cause autosomal dominant hypercholesterolemia. Nat Genet 34:154–156
8. Leren TP (2004) Mutations in the PCSK9 gene in Norwegian subjects with autosomal dominant hypercholesterolemia. Clin Genet 65:419–422
9. Abifadel M, Rabes JP, Devillers M, Munnich A, Erlich D, Junien C, Varret M, Boileau C (2009) Mutations and polymorphisms in the proprotein convertase subtilisin kexin 9 (PCSK9) gene in cholesterol metabolism and disease. Hum Mutat 30:520–529
10. Gu HM, Zhang DW (2015) Hypercholesterolemia, low density lipoprotein receptor and proprotein convertase subtilisin/kexin-type 9. J Biomed Res 29:356–361
11. Cohen JC, Boerwinkle E, Mosley TH Jr, Hobbs HH (2006) Sequence variations in PCSK9, low LDL, and protection against coronary heart disease. N Engl J Med 354:1264–1272
12. Seidah NG, Khatib AM, Prat A (2006) The proprotein convertases and their implication in sterol and/or lipid metabolism. Biol Chem 387:871–877
13. Abifadel M, Rabes JP, Jambart S, Halaby G, Gannage-Yared MH, Sarkis A, Beaino G, Varret M, Salem N, Corbani S, Aydenian H, Junien C, Munnich A, Boileau C (2009) The molecular basis of familial hypercholesterolemia in Lebanon: spectrum of LDLR mutations and role of PCSK9 as a modifier gene. Hum Mutat 30:E682–E691
14. Seidah NG, Mayer G, Zaid A, Rousselet E, Nassoury N, Poirier S, Essalmani R, Prat A (2008) The activation and physiological functions of the proprotein convertases. Int J Biochem Cell Biol 40:1111–1125
15. Cunningham D, Danley DE, Geoghegan KF, Griffor MC, Hawkins JL, Subashi TA, Varghese AH, Ammirati MJ, Culp JS, Hoth LR, Mansour MN, McGrath KM, Seddon AP, Shenolikar S, Stutzman-Engwall KJ, Warren LC, Xia D, Qiu X (2007) Structural and biophysical studies of PCSK9 and its mutants linked to familial hypercholesterolemia. Nat Struct Mol Biol 14:413–419
16. Du F, Hui Y, Zhang M, Linton MF, Fazio S, Fan D (2011) Novel domain interaction regulates secretion of proprotein convertase subtilisin/kexin type 9 (PCSK9) protein. J Biol Chem 286:43054–43061
17. Saavedra YG, Day R, Seidah NG (2012) The M2 module of the Cys-His-rich domain (CHRD) of PCSK9 protein is needed for the extracellular low-density lipoprotein receptor (LDLR) degradation pathway. J Biol Chem 287:43492–43501

18. Hampton EN, Knuth MW, Li J, Harris JL, Lesley SA, Spraggon G (2007) The self-inhibited structure of full-length PCSK9 at 1.9 A reveals structural homology with resistin within the C-terminal domain. Proc Natl Acad Sci U S A 104:14604–14609
19. Piper DE, Jackson S, Liu Q, Romanow WG, Shetterly S, Thibault ST, Shan B, Walker NP (2007) The crystal structure of PCSK9: a regulator of plasma LDL-cholesterol. Structure 15:545–552
20. Kwon HJ, Lagace TA, McNutt MC, Horton JD, Deisenhofer J (2008) Molecular basis for LDL receptor recognition by PCSK9. Proc Natl Acad Sci U S A 105:1820–1825
21. Henrich S, Lindberg I, Bode W, Than ME (2005) Proprotein convertase models based on the crystal structures of furin and kexin: explanation of their specificity. Journal of molecular biology 345:211–227
22. Holyoak T, Wilson MA, Fenn TD, Kettner CA, Petsko GA, Fuller RS, Ringe D (2003) 2.4 A resolution crystal structure of the prototypical hormone processing protease Kex2 in complex with an Ala-Lys-Arg boronic acid inhibitor. Biochemistry 42:6709–6718
23. Steppan CM, Bailey ST, Bhat S, Brown EJ, Banerjee RR, Wright CM, Patel HR, Ahima RS, Lazar MA (2001) The hormone resistin links obesity to diabetes. Nature 409:307–312
24. Seidah NG, Benjannet S, Wickham L, Marcinkiewicz J, Jasmin SB, Stifani S, Basak A, Prat A, Chretien M (2003) The secretory proprotein convertase neural apoptosis-regulated convertase 1 (NARC-1): liver regeneration and neuronal differentiation. Proc Natl Acad Sci U S A 100:928–933
25. Zaid A, Roubtsova A, Essalmani R, Marcinkiewicz J, Chamberland A, Hamelin J, Tremblay M, Jacques H, Jin W, Davignon J, Seidah NG, Prat A (2008) Proprotein convertase subtilisin/kexin type 9 (PCSK9): hepatocyte-specific low-density lipoprotein receptor degradation and critical role in mouse liver regeneration. Hepatology 48:646–654
26. Benjannet S, Rhainds D, Essalmani R, Mayne J, Wickham L, Jin W, Asselin MC, Hamelin J, Varret M, Allard D, Trillard M, Abifadel M, Tebon A, Attie AD, Rader DJ, Boileau C, Brissette L, Chretien M, Prat A, Seidah NG (2004) NARC-1/PCSK9 and its natural mutants: zymogen cleavage and effects on the low density lipoprotein (LDL) receptor and LDL cholesterol. J Biol Chem 279:48865–48875
27. Cameron J, Holla OL, Laerdahl JK, Kulseth MA, Berge KE, Leren TP (2009) Mutation S462P in the PCSK9 gene reduces secretion of mutant PCSK9 without affecting the autocatalytic cleavage. Atherosclerosis 203:161–165
28. Benjannet S, Hamelin J, Chretien M, Seidah NG (2012) Loss- and gain-of-function PCSK9 variants: cleavage specificity, dominant negative effects, and low density lipoprotein receptor (LDLR) degradation. J Biol Chem 287:33745–33755
29. Chorba JS, Galvan AM, Shokat KM (2018) Stepwise processing analyses of the single-turnover PCSK9 protease reveal its substrate sequence specificity and link clinical genotype to lipid phenotype. J Biol Chem 293:1875–1886
30. Deng SJ, Shen Y, Gu HM, Guo S, Wu SR, Zhang DW (2020) The role of the C-terminal domain of PCSK9 and SEC24 isoforms in PCSK9 secretion. Biochim Biophys Acta Mol Cell Biol Lipids 1865:158660
31. Wieland FT, Gleason ML, Serafini TA, Rothman JE (1987) The rate of bulk flow from the endoplasmic reticulum to the cell surface. Cell 50:289–300
32. Martinez-Menarguez JA, Geuze HJ, Slot JW, Klumperman J (1999) Vesicular tubular clusters between the ER and Golgi mediate concentration of soluble secretory proteins by exclusion from COPI-coated vesicles. Cell 98:81–90
33. Dancourt J, Barlowe C (2010) Protein sorting receptors in the early secretory pathway. Annu Rev Biochem 79:777–802
34. Szul T, Sztul E (2011) COPII and COPI traffic at the ER-Golgi interface. Physiology (Bethesda) 26:348–364
35. Mizuno M, Singer SJ (1993) A soluble secretory protein is first concentrated in the endoplasmic reticulum before transfer to the Golgi apparatus. Proc Natl Acad Sci U S A 90:5732–5736
36. Zhang B, Cunningham MA, Nichols WC, Bernat JA, Seligsohn U, Pipe SW, McVey JH, Schulte-Overberg U, de Bosch NB, Ruiz-Saez A, White GC, Tuddenham EG, Kaufman RJ, Ginsburg D (2003) Bleeding due to disruption of a cargo-specific ER-to-Golgi transport complex. Nat Genet 34:220–225
37. Nyfeler B, Reiterer V, Wendeler MW, Stefan E, Zhang B, Michnick SW, Hauri HP (2008) Identification of ERGIC-53 as an intracellular transport receptor of alpha1-antitrypsin. J Cell Biol 180:705–712
38. Hara-Kuge S, Ohkura T, Ideo H, Shimada O, Atsumi S, Yamashita K (2002) Involvement of VIP36 in intracellular transport and secretion of glycoproteins in polarized Madin-Darby canine kidney (MDCK) cells. J Biol Chem 277:16332–16339
39. Strating JR, Hafmans TG, Martens GJ (2009) Functional diversity among p24 subfamily members. Biol Cell 101:207–219
40. Neve EP, Svensson K, Fuxe J, Pettersson RF (2003) VIPL, a VIP36-like membrane protein with a putative function in the export of glycoproteins from the endoplasmic reticulum. Exp Cell Res 288:70–83
41. Chen XW, Wang H, Bajaj K, Zhang P, Meng ZX, Ma D, Bai Y, Liu HH, Adams E, Baines A, Yu G, Sartor MA, Zhang B, Yi Z, Lin J, Young SG, Schekman R, Ginsburg D (2013) SEC24A deficiency lowers plasma cholesterol through reduced PCSK9 secretion. Elife 2:e00444

42. Emmer BT, Hesketh GG, Kotnik E, Tang VT, Lascuna PJ, Xiang J, Gingras AC, Chen XW, Ginsburg D (2018) The cargo receptor SURF4 promotes the efficient cellular secretion of PCSK9. Elife 7:e38839
43. Reeves JE, Fried M (1995) The surf-4 gene encodes a novel 30 kDa integral membrane protein. Mol Membr Biol 12:201–208
44. Shen Y, Wang B, Deng S, Zhai L, Gu HM, Alabi A, Xia X, Zhao Y, Chang X, Qin S, Zhang DW (2020) Surf4 regulates expression of proprotein convertase subtilisin/kexin type 9 (PCSK9) but is not required for PCSK9 secretion in cultured human hepatocytes. Biochim Biophys Acta Mol Cell Biol Lipids 1865:158555
45. Gustafsen C, Kjolby M, Nyegaard M, Mattheisen M, Lundhede J, Buttenschon H, Mors O, Bentzon JF, Madsen P, Nykjaer A, Glerup S (2014) The hypercholesterolemia-risk gene SORT1 facilitates PCSK9 secretion. Cell Metab 19:310–318
46. Butkinaree C, Canuel M, Essalmani R, Poirier S, Benjannet S, Asselin MC, Roubtsova A, Hamelin J, Marcinkiewicz J, Chamberland A, Guillemot J, Mayer G, Sisodia SS, Jacob Y, Prat A, Seidah NG (2015) Amyloid precursor-like protein 2 and sortilin do not regulate the PCSK9 convertase-mediated low density lipoprotein receptor degradation but interact with each other. J Biol Chem 290:18609–18620
47. Allard D, Amsellem S, Abifadel M, Trillard M, Devillers M, Luc G, Krempf M, Reznik Y, Girardet JP, Fredenrich A, Junien C, Varret M, Boileau C, Benlian P, Rabes JP (2005) Novel mutations of the PCSK9 gene cause variable phenotype of autosomal dominant hypercholesterolemia. Hum Mutat 26:497
48. Timms KM, Wagner S, Samuels ME, Forbey K, Goldfine H, Jammulapati S, Skolnick MH, Hopkins PN, Hunt SC, Shattuck DM (2004) A mutation in PCSK9 causing autosomal-dominant hypercholesterolemia in a Utah pedigree. Hum Genet 114:349–353
49. Cohen J, Pertsemlidis A, Kotowski IK, Graham R, Garcia CK, Hobbs HH (2005) Low LDL cholesterol in individuals of African descent resulting from frequent nonsense mutations in PCSK9. Nat Genet 37:161–165
50. Folsom AR, Peacock JM, Boerwinkle E (2009) Variation in PCSK9, low LDL cholesterol, and risk of peripheral arterial disease. Atherosclerosis 202:211–215
51. Kotowski IK, Pertsemlidis A, Luke A, Cooper RS, Vega GL, Cohen JC, Hobbs HH (2006) A spectrum of PCSK9 alleles contributes to plasma levels of low-density lipoprotein cholesterol. Am J Hum Genet 78:410–422
52. Hooper AJ, Marais AD, Tanyanyiwa DM, Burnett JR (2007) The C679X mutation in PCSK9 is present and lowers blood cholesterol in a Southern African population. Atherosclerosis 193:445–448
53. Zhao Z, Tuakli-Wosornu Y, Lagace TA, Kinch L, Grishin NV, Horton JD, Cohen JC, Hobbs HH (2006) Molecular characterization of loss-of-function mutations in PCSK9 and identification of a compound heterozygote. Am J Hum Genet 79:514–523
54. Maxwell KN, Breslow JL (2004) Adenoviral-mediated expression of Pcsk9 in mice results in a low-density lipoprotein receptor knockout phenotype. Proc Natl Acad Sci U S A 101:7100–7105
55. Park SW, Moon YA, Horton JD (2004) Post-transcriptional regulation of low density lipoprotein receptor protein by proprotein convertase subtilisin/kexin type 9a in mouse liver. J Biol Chem 279:50630–50638
56. Rashid S, Curtis DE, Garuti R, Anderson NN, Bashmakov Y, Ho YK, Hammer RE, Moon YA, Horton JD (2005) Decreased plasma cholesterol and hypersensitivity to statins in mice lacking Pcsk9. Proc Natl Acad Sci U S A 102:5374–5379
57. Frank-Kamenetsky M, Grefhorst A, Anderson NN, Racie TS, Bramlage B, Akinc A, Butler D, Charisse K, Dorkin R, Fan Y, Gamba-Vitalo C, Hadwiger P, Jayaraman M, John M, Jayaprakash KN, Maier M, Nechev L, Rajeev KG, Read T, Rohl I, Soutschek J, Tan P, Wong J, Wang G, Zimmermann T, de Fougerolles A, Vornlocher HP, Langer R, Anderson DG, Manoharan M, Koteliansky V, Horton JD, Fitzgerald K (2008) Therapeutic RNAi targeting PCSK9 acutely lowers plasma cholesterol in rodents and LDL cholesterol in nonhuman primates. Proc Natl Acad Sci U S A 105:11915–11920
58. Lagace TA, Curtis DE, Garuti R, McNutt MC, Park SW, Prather HB, Anderson NN, Ho YK, Hammer RE, Horton JD (2006) Secreted PCSK9 decreases the number of LDL receptors in hepatocytes and in livers of parabiotic mice. J Clin Invest 116:2995–3005
59. McNutt MC, Kwon HJ, Chen C, Chen JR, Horton JD, Lagace TA (2009) Antagonism of secreted PCSK9 increases low-density lipoprotein receptor expression in HEPG2 cells. J Biol Chem 284:10561–10570
60. Zhang DW, Lagace TA, Garuti R, Zhao Z, McDonald M, Horton JD, Cohen JC, Hobbs HH (2007) Binding of proprotein convertase subtilisin/kexin type 9 to epidermal growth factor-like repeat A of low density lipoprotein receptor decreases receptor recycling and increases degradation. J Biol Chem 282:18602–18612
61. Zhang DW, Garuti R, Tang WJ, Cohen JC, Hobbs HH (2008) Structural requirements for PCSK9-mediated degradation of the low-density lipoprotein receptor. Proc Natl Acad Sci U S A 105:13045–13050
62. Cameron J, Holla OL, Ranheim T, Kulseth MA, Berge KE, Leren TP (2006) Effect of mutations in the PCSK9 gene on the cell surface LDL receptors. Hum Mol Genet 15:1551–1558
63. Lalanne F, Lambert G, Amar MJ, Chetiveaux M, Zair Y, Jarnoux AL, Ouguerram K, Friburg J, Seidah NG, Brewer HB Jr, Krempf M, Costet P (2005) Wild-

type PCSK9 inhibits LDL clearance but does not affect apoB-containing lipoprotein production in mouse and cultured cells. J Lipid Res 46:1312–1319
64. Qian YW, Schmidt RJ, Zhang Y, Chu S, Lin A, Wang H, Wang X, Beyer TP, Bensch WR, Li W, Ehsani ME, Lu D, Konrad RJ, Eacho PI, Moller DE, Karathanasis SK, Cao G (2007) Secreted PCSK9 downregulates low density lipoprotein receptor through receptor-mediated endocytosis. J Lipid Res 48:1488–1498
65. Wang Y, Huang Y, Hobbs HH, Cohen JC (2012) Molecular characterization of proprotein convertase subtilisin/kexin type 9-mediated degradation of the LDLR. J Lipid Res 53:1932–1943
66. Fasano T, Sun XM, Patel DD, Soutar AK (2009) Degradation of LDLR protein mediated by 'gain of function' PCSK9 mutants in normal and ARH cells. Atherosclerosis 203:166–171
67. Maxwell KN, Fisher EA, Breslow JL (2005) Overexpression of PCSK9 accelerates the degradation of the LDLR in a post-endoplasmic reticulum compartment. Proc Natl Acad Sci U S A 102:2069–2074
68. Sanchez-Hernandez RM, Di Taranto MD, Benito-Vicente A, Uribe KB, Lamiquiz-Moneo I, Larrea-Sebal A, Jebari S, Galicia-Garcia U, Novoa FJ, Boronat M, Wagner AM, Civeira F, Martin C, Fortunato G (2019) The Arg499His gain-of-function mutation in the C-terminal domain of PCSK9. Atherosclerosis 289:162–172
69. Homer VM, Marais AD, Charlton F, Laurie AD, Hurndell N, Scott R, Mangili F, Sullivan DR, Barter PJ, Rye KA, George PM, Lambert G (2008) Identification and characterization of two non-secreted PCSK9 mutants associated with familial hypercholesterolemia in cohorts from New Zealand and South Africa. Atherosclerosis 196:659–666
70. Poirier S, Mayer G, Poupon V, McPherson PS, Desjardins R, Ly K, Asselin MC, Day R, Duclos FJ, Witmer M, Parker R, Prat A, Seidah NG (2009) Dissection of the endogenous cellular pathways of PCSK9-induced LDLR degradation: Evidence for an intracellular route. J Biol Chem 284:28856–28864
71. Grefhorst A, McNutt MC, Lagace TA, Horton JD (2008) Plasma PCSK9 preferentially reduces liver LDL receptors in mice. J Lipid Res 49:1303–1311
72. Luo Y, Warren L, Xia D, Jensen H, Sand T, Petras S, Qin W, Miller KS, Hawkins J (2009) Function and distribution of circulating human PCSK9 expressed extrahepatically in transgenic mice. J Lipid Res 50:1581–1588
73. Tavori H, Fan D, Blakemore JL, Yancey PG, Ding L, Linton MF, Fazio S (2013) Serum proprotein convertase subtilisin/kexin type 9 and cell surface low-density lipoprotein receptor: evidence for a reciprocal regulation. Circulation 127:2403–2413
74. Cariou B, Benoit I, Le May C (2014) Preserved adrenal function in fully PCSK9-deficient subject. Int J Cardiol 176:499–500
75. Gustafsen C, Olsen D, Vilstrup J, Lund S, Reinhardt A, Wellner N, Larsen T, Andersen CBF, Weyer K, Li JP, Seeberger PH, Thirup S, Madsen P, Glerup S (2017) Heparan sulfate proteoglycans present PCSK9 to the LDL receptor. Nat Commun 8:503
76. Maxwell KN, Breslow JL (2005) Proprotein convertase subtilisin kexin 9: the third locus implicated in autosomal dominant hypercholesterolemia. Curr Opin Lipidol 16:167–172
77. Nguyen MA, Kosenko T, Lagace TA (2014) Internalized PCSK9 dissociates from recycling LDL receptors in PCSK9-resistant SV-589 fibroblasts. J Lipid Res 55:266–275
78. McNutt MC, Lagace TA, Horton JD (2007) Catalytic activity is not required for secreted PCSK9 to reduce low density lipoprotein receptors in HepG2 cells. J Biol Chem 282:20799–20803
79. Rudenko G, Henry L, Henderson K, Ichtchenko K, Brown MS, Goldstein JL, Deisenhofer J (2002) Structure of the LDL receptor extracellular domain at endosomal pH. Science 298:2353–2358
80. Brown MS, Goldstein JL (1986) A receptor-mediated pathway for cholesterol homeostasis. Science 232:34–47
81. Goldstein JL, Hobbs HH, Brown MS (2001) Familial hypercholesterol. In: Scriver CR (ed) The metabolic & molecular bases of inherited disease, 8th edn. McGraw-Hill, New York, pp 2863–2913
82. Gu HM, Adijiang A, Mah M, Zhang DW (2013) Characterization of the role of EGF-A of low-density lipoprotein receptor in PCSK9 binding. J Lipid Res 54:3345–3357
83. Romagnuolo R, Scipione CA, Boffa MB, Marcovina SM, Seidah NG, Koschinsky ML (2015) Lipoprotein (a) catabolism is regulated by proprotein convertase subtilisin/kexin type 9 through the low density lipoprotein receptor. J Biol Chem 290:11649–11662
84. Nassoury N, Blasiole DA, Tebon Oler A, Benjannet S, Hamelin J, Poupon V, McPherson PS, Attie AD, Prat A, Seidah NG (2007) The cellular trafficking of the secretory proprotein convertase PCSK9 and its dependence on the LDLR. Traffic 8:718–732
85. Jang H-D, Lee SE, Yang J, Lee H-C, Shin D, Lee H, Lee J, Jin S, Kim S, Lee SJ, You J, Park H-W, Nam K-Y, Lee S-H, Park SW, Kim J-S, Kim S-Y, Kwon Y-W, Kwak SH, Yang H-M, Kim H-S (2020) Cyclase-associated protein 1 is a binding partner of proprotein convertase subtilisin/kexin type-9 and is required for the degradation of low-density lipoprotein receptors by proprotein convertase subtilisin/kexin type-9. Eur Heart J 41:239–252
86. Surdo PL, Bottomley MJ, Calzetta A, Settembre EC, Cirillo A, Pandit S, Ni YG, Hubbard B, Sitlani A, Carfi A (2011) Mechanistic implications for LDL receptor degradation from the PCSK9/LDLR structure at neutral pH. EMBO Rep 12:1300–1305
87. Tveten K, Holla OL, Cameron J, Strom TB, Berge KE, Laerdahl JK, Leren TP (2012) Interaction

between the ligand-binding domain of the LDL receptor and the C-terminal domain of PCSK9 is required for PCSK9 to remain bound to the LDL receptor during endosomal acidification. Hum Mol Genet 21:1402–1409
88. Holla OL, Cameron J, Tveten K, Strom TB, Berge KE, Laerdahl JK, Leren TP (2011) Role of the C-terminal domain of PCSK9 in degradation of the LDL receptors. J Lipid Res 52:1787–1794
89. Yamamoto T, Lu C, Ryan RO (2011) A two-step binding model of PCSK9 interaction with the low density lipoprotein receptor. J Biol Chem 286:5464–5470
90. Deng SJ, Alabi A, Gu HM, Adijiang A, Qin S, Zhang DW (2019) Identification of amino acid residues in the ligand binding repeats of LDLR important for PCSK9 binding. J Lipid Res 60:516–527
91. Strom TB, Holla OL, Tveten K, Cameron J, Berge KE, Leren TP (2010) Disrupted recycling of the low density lipoprotein receptor by PCSK9 is not mediated by residues of the cytoplasmic domain. Mol Genet Metab 101:76–80
92. Canuel M, Sun X, Asselin M-C, Paramithiotis E, Prat A, Seidah NG (2013) Proprotein convertase subtilisin/kexin type 9 (PCSK9) can mediate degradation of the low density lipoprotein receptor-related protein 1 (LRP-1). PloS one 8:e64145
93. DeVay RM, Shelton DL, Liang H (2013) Characterization of proprotein convertase subtilisin/kexin type 9 (PCSK9) trafficking reveals a novel lysosomal targeting mechanism via amyloid precursor-like protein 2 (APLP2). J Biol Chem 288:10805–10818
94. Fu T, Guan Y, Xu J, Wang Y (2017) APP, APLP2 and LRP1 interact with PCSK9 but are not required for PCSK9-mediated degradation of the LDLR in vivo. Biochim Biophys Acta 1862:883–889
95. Ly K, Essalmani R, Desjardins R, Seidah NG, Day R (2016) An unbiased mass spectrometry approach identifies Glypican-3 as an interactor of proprotein convertase subtilisin/kexin type 9 (PCSK9) and LDL receptor in hepatocellular carcinoma cells. J Biol Chem
96. Sun H, Krauss RM, Chang JT, Teng BB (2018) PCSK9 deficiency reduces atherosclerosis, apolipoprotein B secretion, and endothelial dysfunction. J Lipid Res 59:207–223
97. Huang M, Zhao Z, Cao Q, You X, Wei S, Zhao J, Bai M, Chen Y (2019) PAQR3 modulates blood cholesterol level by facilitating interaction between LDLR and PCSK9. Metabolism 94:88–95
98. Rashid S, Tavori H, Brown PE, Linton MF, He J, Giunzioni I, Fazio S (2014) Proprotein convertase subtilisin kexin type 9 promotes intestinal overproduction of triglyceride-rich apolipoprotein B lipoproteins through both low-density lipoprotein receptor-dependent and -independent mechanisms. Circulation 130:431–441
99. Le May C, Kourimate S, Langhi C, Chetiveaux M, Jarry A, Comera C, Collet X, Kuipers F, Krempf M, Cariou B, Costet P (2009) Proprotein convertase subtilisin kexin type 9 null mice are protected from postprandial triglyceridemia. Arterioscler Thromb Vasc Biol 29:684–690
100. Drouin-Chartier JP, Tremblay AJ, Hogue JC, Lemelin V, Lamarche B, Couture P (2018) Plasma PCSK9 correlates with apoB-48-containing triglyceride-rich lipoprotein production in men with insulin resistance. J Lipid Res 59:1501–1509
101. Clarke R, Peden JF, Hopewell JC, Kyriakou T, Goel A, Heath SC, Parish S, Barlera S, Franzosi MG, Rust S, Bennett D, Silveira A, Malarstig A, Green FR, Lathrop M, Gigante B, Leander K, de Faire U, Seedorf U, Hamsten A, Collins R, Watkins H, Farrall M, Consortium P (eds) (2009) Genetic variants associated with Lp(a) lipoprotein level and coronary disease. N Engl J Med 361:2518–2528
102. Emerging Risk Factors C, Erqou S, Kaptoge S, Perry PL, Di Angelantonio E, Thompson A, White IR, Marcovina SM, Collins R, Thompson SG, Danesh J (2009) Lipoprotein(a) concentration and the risk of coronary heart disease, stroke, and nonvascular mortality. JAMA 302:412–423
103. Kamstrup PR, Tybjaerg-Hansen A, Steffensen R, Nordestgaard BG (2009) Genetically elevated lipoprotein(a) and increased risk of myocardial infarction. JAMA 301:2331–2339
104. Gencer B, Rigamonti F, Nanchen D, Vuilleumier N, Kern I, Aghlmandi S, Klingenberg R, Raber L, Auer R, Carballo D, Carballo S, Heg D, Windecker S, Luscher TF, Matter CM, Rodondi N, Mach F (2019) Prognostic value of elevated lipoprotein(a) in patients with acute coronary syndromes. Eur J Clin Invest 49:e13117
105. Rader DJ, Mann WA, Cain W, Kraft HG, Usher D, Zech LA, Hoeg JM, Davignon J, Lupien P, Grossman M et al (1995) The low density lipoprotein receptor is not required for normal catabolism of Lp(a) in humans. J Clin Invest 95:1403–1408
106. Koschinsky ML, Marcovina SM (2004) Structure-function relationships in apolipoprotein(a): insights into lipoprotein(a) assembly and pathogenicity. Curr Opin Lipidol 15:167–174
107. Tavori H, Christian D, Minnier J, Plubell D, Shapiro MD, Yeang C, Giunzioni I, Croyal M, Duell PB, Lambert G, Tsimikas S, Fazio S (2016) PCSK9 association with lipoprotein(a). Circ Res 119:29–35
108. O'Donoghue ML, Fazio S, Giugliano RP, Stroes ESG, Kanevsky E, Gouni-Berthold I, Im K, Lira Pineda A, Wasserman SM, Ceska R, Ezhov MV, Jukema JW, Jensen HK, Tokgozoglu SL, Mach F, Huber K, Sever PS, Keech AC, Pedersen TR, Sabatine MS (2019) Lipoprotein(a), PCSK9 Inhibition, and Cardiovascular Risk. Circulation 139:1483–1492
109. Ray KK, Vallejo-Vaz AJ, Ginsberg HN, Davidson MH, Louie MJ, Bujas-Bobanovic M, Minini P, Eckel RH, Cannon CP (2019) Lipoprotein(a) reductions

from PCSK9 inhibition and major adverse cardiovascular events: Pooled analysis of alirocumab phase 3 trials. Atherosclerosis 288:194–202
110. Poirier S, Mayer G, Benjannet S, Bergeron E, Marcinkiewicz J, Nassoury N, Mayer H, Nimpf J, Prat A, Seidah NG (2008) The proprotein convertase PCSK9 induces the degradation of low density lipoprotein receptor (LDLR) and its closest family members VLDLR and ApoER2. J Biol Chem 283:2363–2372
111. Demers A, Samami S, Lauzier B, Des Rosiers C, Sock ET, Ong H, Mayer G (2015) PCSK9 induces CD36 degradation and affects long-chain fatty acid uptake and triglyceride metabolism in adipocytes and in mouse liver. Arterioscler Thromb Vasc Biol 35:2517–2525
112. Roubtsova A, Munkonda MN, Awan Z, Marcinkiewicz J, Chamberland A, Lazure C, Cianflone K, Seidah NG, Prat A (2011) Circulating proprotein convertase subtilisin/kexin 9 (PCSK9) regulates VLDLR protein and triglyceride accumulation in visceral adipose tissue. Arterioscler Thromb Vasc Biol 31:785–791
113. Liu M, Wu G, Baysarowich J, Kavana M, Addona GH, Bierilo KK, Mudgett JS, Pavlovic G, Sitlani A, Renger JJ, Hubbard BK, Fisher TS, Zerbinatti CV (2010) PCSK9 is not involved in the degradation of LDL receptors and BACE1 in the adult mouse brain. J Lipid Res 51:2611–2618
114. Kysenius K, Muggalla P, Matlik K, Arumae U, Huttunen HJ (2012) PCSK9 regulates neuronal apoptosis by adjusting ApoER2 levels and signaling. Cell Mol Life Sci 69:1903–1916
115. Wang L, Wang Z, Shi J, Jiang Q, Wang H, Li X, Hao D (2018) Inhibition of proprotein convertase subtilisin/kexin type 9 attenuates neuronal apoptosis following focal cerebral ischemia via apolipoprotein E receptor 2 downregulation in hyperlipidemic mice. Int J Mol Med 42:2098–2106
116. Apaijai N, Moisescu DM, Palee S, McSweeney CM, Saiyasit N, Maneechote C, Boonnag C, Chattipakorn N, Chattipakorn SC (2019) Pretreatment with PCSK9 inhibitor protects the brain against cardiac ischemia/reperfusion injury through a reduction of neuronal inflammation and amyloid beta aggregation. J Am Heart Assoc 8: e010838
117. Jonas MC, Costantini C, Puglielli L (2008) PCSK9 is required for the disposal of non-acetylated intermediates of the nascent membrane protein BACE1. EMBO Rep 9:916–922
118. Horton JD, Shah NA, Warrington JA, Anderson NN, Park SW, Brown MS, Goldstein JL (2003) Combined analysis of oligonucleotide microarray data from transgenic and knockout mice identifies direct SREBP target genes. Proc Natl Acad Sci U S A 100:12027–12032
119. Goldstein JL, Brown MS (2015) A century of cholesterol and coronaries: from plaques to genes to statins. Cell 161:161–172
120. DeBose-Boyd RA (2008) Feedback regulation of cholesterol synthesis: sterol-accelerated ubiquitination and degradation of HMG CoA reductase. Cell Res 18:609–621
121. Radhakrishnan A, Goldstein JL, McDonald JG, Brown MS (2008) Switch-like control of SREBP-2 transport triggered by small changes in ER cholesterol: a delicate balance. Cell Metab 8:512–521
122. Li H, Liu J (2012) The novel function of HINFP as a co-activator in sterol-regulated transcription of PCSK9 in HepG2 cells. Biochem J 443:757–768
123. Melone M, Wilsie L, Palyha O, Strack A, Rashid S (2012) Discovery of a new role of human resistin in hepatocyte low-density lipoprotein receptor suppression mediated in part by proprotein convertase subtilisin/kexin type 9. J Am Coll Cardiol 59:1697–1705
124. Lin Z, Pan X, Wu F, Ye D, Zhang Y, Wang Y, Jin L, Lian Q, Huang Y, Ding H, Triggle C, Wang K, Li X, Xu A (2015) Fibroblast growth factor 21 prevents atherosclerosis by suppression of hepatic sterol regulatory element-binding protein-2 and induction of adiponectin in mice. Circulation 131:1861–1871
125. Li H, Dong B, Park SW, Lee HS, Chen W, Liu J (2009) Hepatocyte nuclear factor 1alpha plays a critical role in PCSK9 gene transcription and regulation by the natural hypocholesterolemic compound berberine. J Biol Chem 284:28885–28895
126. Dong B, Singh AB, Shende VR, Liu J (2017) Hepatic HNF1 transcription factors control the induction of PCSK9 mediated by rosuvastatin in normolipidemic hamsters. Int J Mol Med 39:749–756
127. Ai D, Chen C, Han S, Ganda A, Murphy AJ, Haeusler R, Thorp E, Accili D, Horton JD, Tall AR (2012) Regulation of hepatic LDL receptors by mTORC1 and PCSK9 in mice. J Clin Invest 122:1262–1270
128. Lai Q, Giralt A, Le May C, Zhang L, Cariou B, Denechaud PD, Fajas L (2017) E2F1 inhibits circulating cholesterol clearance by regulating Pcsk9 expression in the liver. JCI Insight 2
129. Tao R, Xiong X, DePinho RA, Deng CX, Dong XC (2013) FoxO3 transcription factor and Sirt6 deacetylase regulate low density lipoprotein (LDL)-cholesterol homeostasis via control of the proprotein convertase subtilisin/kexin type 9 (Pcsk9) gene expression. J Biol Chem 288:29252–29259
130. Naeli P, Mirzadeh Azad F, Malakootian M, Seidah NG, Mowla SJ (2017) Post-transcriptional regulation of PCSK9 by miR-191, miR-222, and miR-224. Front Genet 8:189
131. Benjannet S, Rhainds D, Hamelin J, Nassoury N, Seidah NG (2006) The proprotein convertase (PC) PCSK9 is inactivated by furin and/or PC5/6A: functional consequences of natural mutations and post-translational modifications. J Biol Chem 281:30561–30572

132. Dewpura T, Raymond A, Hamelin J, Seidah NG, Mbikay M, Chretien M, Mayne J (2008) PCSK9 is phosphorylated by a Golgi casein kinase-like kinase ex vivo and circulates as a phosphoprotein in humans. The FEBS journal 275:3480–3493
133. Ben Djoudi Ouadda A, Gauthier MS, Susan-Resiga-D, Girard E, Essalmani R, Black M, Marcinkiewicz J, Forget D, Hamelin J, Evagelidis A, Ly K, Day R, Galarneau L, Corbin F, Coulombe B, Caku A, Tagliabracci VS, Seidah NG (2019) Ser-phosphorylation of PCSK9 (proprotein convertase subtilisin-kexin 9) by Fam20C (family with sequence similarity 20, member C) kinase enhances its ability to degrade the LDLR (low-density lipoprotein receptor). Arterioscler Thromb Vasc Biol 39:1996–2013
134. Han B, Eacho PI, Knierman MD, Troutt JS, Konrad RJ, Yu X, Schroeder KM (2014) Isolation and characterization of the circulating truncated form of PCSK9. J Lipid Res 55:1505–1514
135. Kosenko T, Golder M, Leblond G, Weng W, Lagace TA (2013) Low density lipoprotein binds to proprotein convertase subtilisin/kexin type-9 (PCSK9) in human plasma and inhibits PCSK9-mediated low density lipoprotein receptor degradation. J Biol Chem 288:8279–8288
136. Fisher TS, Lo Surdo P, Pandit S, Mattu M, Santoro JC, Wisniewski D, Cummings RT, Calzetta A, Cubbon RM, Fischer PA, Tarachandani A, De Francesco R, Wright SD, Sparrow CP, Carfi A, Sitlani A (2007) Effects of pH and low density lipoprotein (LDL) on PCSK9-dependent LDL receptor regulation. J Biol Chem 282:20502–20512
137. Galvan AM, Chorba JS (2019) Cell-associated heparin-like molecules modulate the ability of LDL to regulate PCSK9 uptake. J Lipid Res 60:71–84
138. Poirier S, Mamarbachi M, Chen WT, Lee AS, Mayer G (2015) GRP94 regulates circulating cholesterol levels through blockade of PCSK9-induced LDLR degradation. Cell Rep 13:2064–2071
139. Mayer G, Poirier S, Seidah NG (2008) Annexin A2 is a C-terminal PCSK9 binding protein that regulates endogenous LDL receptor levels. J Biol Chem 283:31791-31780
140. Wang X, Berry E, Hernandez-Anzaldo S, Sun D, Adijiang A, Li L, Zhang D, Fernandez-Patron C (2015) MMP-2 inhibits PCSK9-induced degradation of the LDL receptor in Hepa1-c1c7 cells. FEBS Lett 589:490–496
141. Spolitu S, Okamoto H, Dai W, Zadroga JA, Wittchen ES, Gromada J, Ozcan L (2019) Hepatic glucagon signaling regulates PCSK9 and low-density lipoprotein cholesterol. Circ Res 124:38–51
142. Fitchett DH, Hegele RA, Verma S (2015) Cardiology patient page. Statin intolerance. Circulation 131: e389–e391
143. Naoumova RP, Tosi I, Patel D, Neuwirth C, Horswell SD, Marais AD, van Heyningen C, Soutar AK (2005) Severe hypercholesterolemia in four British families with the D374Y mutation in the PCSK9 gene: long-term follow-up and treatment response. Arterioscler Thromb Vasc Biol 25:2654–2660
144. Benn M, Tybjaerg-Hansen A, Nordestgaard BG (2019) Low LDL cholesterol by PCSK9 variation reduces cardiovascular mortality. J Am Coll Cardiol 73:3102–3114
145. Leander K, Malarstig A, Van't Hooft FM, Hyde C, Hellenius ML, Troutt JS, Konrad RJ, Ohrvik J, Hamsten A, de Faire U (2016) Circulating proprotein convertase subtilisin/kexin type 9 (PCSK9) predicts future risk of cardiovascular events independently of established risk factors. Circulation 133:1230–1239
146. Denis M, Marcinkiewicz J, Zaid A, Gauthier D, Poirier S, Lazure C, Seidah NG, Prat A (2012) Gene inactivation of proprotein convertase subtilisin/kexin type 9 reduces atherosclerosis in mice. Circulation 125:894–901
147. Robinson JG, Farnier M, Krempf M, Bergeron J, Luc G, Averna M, Stroes ES, Langslet G, Raal FJ, El Shahawy M, Koren MJ, Lepor NE, Lorenzato C, Pordy R, Chaudhari U, Kastelein JJ, Investigators OLT (2015) Efficacy and safety of alirocumab in reducing lipids and cardiovascular events. N Engl J Med 372:1489–1499
148. Schwartz GG, Steg PG, Szarek M, Bhatt DL, Bittner VA, Diaz R, Edelberg JM, Goodman SG, Hanotin C, Harrington RA, Jukema JW, Lecorps G, Mahaffey KW, Moryusef A, Pordy R, Quintero K, Roe MT, Sasiela WJ, Tamby JF, Tricoci P, White HD, Zeiher AM, Committees OO, Investigators (2018) Alirocumab and cardiovascular outcomes after acute coronary syndrome. N Engl J Med 379:2097–2107
149. Sabatine MS, De Ferrari GM, Giugliano RP, Huber K, Lewis BS, Ferreira J, Kuder JF, Murphy SA, Wiviott SD, Kurtz CE, Honarpour N, Keech AC, Sever PS, Pedersen TR (2018) Clinical benefit of evolocumab by severity and extent of coronary artery disease: an analysis from FOURIER. Circulation 138:756–766
150. Ray KK, Stoekenbroek RM, Kallend D, Leiter LA, Landmesser U, Wright RS, Wijngaard P, Kastelein JJP (2018) Effect of an siRNA therapeutic targeting PCSK9 on atherogenic lipoproteins. Circulation 138:1304–1316
151. Ray KK, Landmesser U, Leiter LA, Kallend D, Dufour R, Karakas M, Hall T, Troquay RP, Turner T, Visseren FL, Wijngaard P, Wright RS, Kastelein JJ (2017) Inclisiran in patients at high cardiovascular risk with elevated LDL cholesterol. N Engl J Med 376:1430–1440
152. Sabatine MS, Giugliano RP, Wiviott SD, Raal FJ, Blom DJ, Robinson J, Ballantyne CM, Somaratne R, Legg J, Wasserman SM, Scott R, Koren MJ, Stein EA, Open-Label Study of Long-Term Evaluation against, L. D. L. C. I (2015) Efficacy and safety of evolocumab in reducing lipids and cardiovascular events. N Engl J Med 372:1500–1509

153. Cannon CP, Cariou B, Blom D, McKenney JM, Lorenzato C, Pordy R, Chaudhari U, Colhoun HM, for the, O. C. I. I. I (2015) Efficacy and safety of alirocumab in high cardiovascular risk patients with inadequately controlled hypercholesterolaemia on maximally tolerated doses of statins: the ODYSSEY COMBO II randomized controlled trial. Eur Heart J
154. Kazi DS, Moran AE, Coxson PG, Penko J, Ollendorf DA, Pearson SD, Tice JA, Guzman D, Bibbins-Domingo K (2016) Cost-effectiveness of PCSK9 inhibitor therapy in patients with heterozygous familial hypercholesterolemia or atherosclerotic cardiovascular disease. JAMA 316:743–753
155. Jackson AL, Bartz SR, Schelter J, Kobayashi SV, Burchard J, Mao M, Li B, Cavet G, Linsley PS (2003) Expression profiling reveals off-target gene regulation by RNAi. Nat Biotechnol 21:635–637
156. Pecot CV, Calin GA, Coleman RL, Lopez-Berestein-G, Sood AK (2011) RNA interference in the clinic: challenges and future directions. Nat Rev Cancer 11:59–67
157. Adorni MP, Ruscica M, Ferri N, Bernini F, Zimetti F (2019) Proprotein convertase subtilisin/kexin type 9, brain cholesterol homeostasis and potential implication for alzheimer's disease. Front Aging Neurosci 11:120

LDL and HDL Oxidative Modification and Atherosclerosis

10

Shucun Qin

Abstract

Low-density lipoprotein (LDL) and high-density lipoprotein (HDL) are two kinds of common lipoproteins in plasma. The level of LDL cholesterol in plasma is positively correlated with atherosclerosis (AS), which is related to the complex macromolecular components, especially the easy oxygenation of protein and lipid components. However, the plasma HDL cholesterol level is negatively correlated with AS, but the results of recent studies show that the oxidative modified HDL in pathological state will not reduce and may aggravate the occurrence and development of AS. Therefore, the oxidative modification of lipoproteins is closely related to vascular homeostasis, which has become a hot research area for a long time.

Keywords

Lipoproteins · LDL oxidative modification · HDL oxidative modification · Atherosclerosis · Cardiovascular diseases

10.1 LDL Oxidative Modification

LDL is a compound particle of lipid and protein in human blood circulation. The mature LDL particles are composed of hydrophobic core and hydrophilic outer layer [1]. The lipid hydrophobic core of LDL consists of triglycerides and cholesterol esters, mainly cholesterol esters. The surface lipid phospholipid and free cholesterol are amphiphilic. The main protein component of LDL is apoB-100, which contains only one apoB-100 in each lipoprotein particle [2]. ApoB-100 has a binding region with LDL receptor, which binds to apoB-100 through the ligand-binding domain of LDL receptor on cell membrane, recognizes and absorbs lipids. LDL granules can contain a little apolipoprotein E (ApoE) and paraoxonase (PON). The paraoxonase in lipoproteins is an antioxidant, which can resist LDL lipid peroxidation. LDL granule is also the main carrier of lipophilic antioxidant vitamin E, carrying k-tocopherol and a small amount of q-tocopherol [3]. However, LDL granules in plasma have weak antioxidation and anti-inflammatory ability and are prone to oxidative modification. This oxidative modification causes lipid deposition in the vascular intima and triggers the pathological process of AS.

S. Qin (✉)
Institute of Atherosclerosis, Shandong First Medical University, Taian, China

X.-C. Jiang (ed.), *Lipid Transfer in Lipoprotein Metabolism and Cardiovascular Disease*, Advances in Experimental Medicine and Biology 1276, https://doi.org/10.1007/978-981-15-6082-8_10

10.1.1 LDL Oxidative Modification Types

LDL oxidative modification types may include nonenzyme-mediated modification, such as free radical, proteoglycan, glycosylation, repair of immune complex; enzyme-mediated modification, such as lipase, oxidase, MPO, etc. In addition, according to the different modified components, it can also be divided into lipid component modification and protein component modification. The physical structure, chemical properties and biological activity of LDL particles will be changed after nonenzyme modification and enzyme modification [4]. The results showed that endothelial cells, macrophages, and smooth muscle cell (SMC) could also modify LDL. LDL is a complex particle with different oxidation sensitivity. LDL oxidation is a gradual process, leading to the formation of a continuous oxidation LDL from mild to extensive, containing various potentially toxic components of oxidized lipids and oxidized proteins in different proportions, i.e., the composition, metabolism and biological characteristics of oxidized LDL are heterogeneous. Lipid aldehyde and sterol and lipid peroxide were oxidized and existed in different proportion. All kinds of bioactive lipids of oxidized LDL interact with cell molecular targets through various mechanisms and play physiological or pathological roles. So far, many mechanisms remain unclear [5].

10.1.2 LDL Oxidative Modification Degree

LDL oxidative modification can occur during fetal growth, and lipoprotein oxidative modification can be observed in human fetal artery samples [6]. LDL oxidation is a gradual process from minimum to mild to severe extensive oxidation. The smallest oxidized LDL and mild oxidized LDL mainly changed in lipid, while the extensive oxidized LDL showed lipid oxidation and aldehyde carbohydrate modification [7]. Obviously, LDL oxidation process can be divided into two stages. The first is the initial stage of LDL oxidation, at which the consumption of lipophilic antioxidants occurs in LDL particles, and then the oxidation of polyunsaturated fatty acids of phospholipids. In this stage of oxidation, LDL particles with low levels of lipid oxide products and relatively complete apoB-100 are considered as the minimal modified LDL particles (MM-LDL) [8], which can also be recognized by LDL receptors. In the stage of severe modification, the lipid and protein components of LDL will be further modified by oxidation, and a large number of lipid components will be modified to generate aldehyde lipid peroxides. ApoB-100 will also be modified by oxidation. In this stage, the oxidized LDL will lose the ability to recognize LDL receptor, but it can combine with scavenger receptor (SR) of macrophages infiltrating into the subcutaneous, and then it will be swallowed by cells, which will lead to macrophages moving to foam cell transformation.

10.1.3 LDL Oxidative Modified Site

It is still believed that the oxidative modification of LDL may occur in the arterial wall, specifically in the subendothelial layer, rather than in the circulating blood. Endothelial cells retain LDL in the subendothelial layer through endocytosis, vesicular transport and particle exocytosis. The fluorescent labeled LDL was transported to the subendothelial layer in a scavenger receptor-b1 dependent manner. Small dense LDL (sdLDL) is more likely to bind to proteoglycan, trapped in extracellular matrix, where it is susceptible to oxidative modification. LDL oxidation is unlikely to occur in plasma because of the high concentration of antioxidants and proteins chelating metal ions. Although in vitro experiments show that transition metal ions can oxidize LDL, it is unlikely that there are a large number of free metal ions in vivo, because they are mostly combined with proteins to form ceruloplasmin or transferrin [9].

10.1.4 LDL Lipid Oxidative Modification

10.1.4.1 Nonenzymatic (Free Radical) Mechanism of Lipid Peroxidation

The surface lipids of LDL, including phospholipids and their derivatives and free cholesterol, can undergo enzymatic or nonenzymatic reactions, and the surface lipids are oxidized to form modified lipids. Lipid peroxidation may be caused by oxidants in the vascular system. The new chemical commonality of as plaque in vivo is the formation of oxidized LDL [10] through the mechanism involving free radicals or lipoxygenase. LDL lipid peroxidation is a free radical-mediated process involving the peroxidation modification of polyunsaturated fatty acids (PUFAs) in phospholipids [11]. The phospholipid components of LDL cause lipid peroxidation under the action of hydroxyl radicals, which generate active hydroxyl radicals in the cell system and react with adjacent lipids as soon as they are produced. Polyunsaturated fatty acids in lipids can react with free radicals. The lipid radicals can react with the neighboring lipid molecules to form lipid peroxides and lipid hydroperoxides, which lead to more and more lipid peroxidation. Lipid hydroperoxides can split under the action of transition metals and break into various oxygen-containing products, such as alkanes, olefins, aldehydes, ketones, and other products. The oxidation reaction is terminated by the depletion of substrate or the formation of stable lipid peroxide derivatives (alkanes, alcohols, ketones, aldehydes, carboxylic acids) [5].

10.1.4.2 Enzyme-Mediated Lipid Peroxidation of Lipoproteins

LDL can be directly modified by various enzymes, such as phospholipase, sphingomyelinase, and lipoxygenase. The cells of arterial wall contain lipoprotein lipase, cholesterol esterase, phospholipase A1, phospholipase A2, phospholipase C and phospholipase D, lipoxygenase, and cholesterol ester oxidase. Enzyme modification of LDL lipids may occur in the arterial wall [4].

10.1.5 LDL Lipid Peroxidation Products

10.1.5.1 Lysophosphatidylcholine (LPC) and Lysophosphatidic Acid (LPA)

The first component of oxidized LDL is LPC, which also exists in normal LDL, but the concentration is low. A large amount of LPC is continuously produced in plasma through the action of sn-2 position of LCAT hydrolysis phospholipid. LPC is the chemokine of monocyte, which can stimulate the production of superoxide and inflammatory cytokines, and stimulate the proliferation of lymphocytes. LPA is produced from LPC by lysophosphatidy esterase D, which is a known mitogen and plays a role through a specific G protein coupled receptor.

10.1.5.2 Sn-2 Short-Chain Phosphatide

The position of sn-2 of LPC in LDL may be changed, and the derivatives obtained by this change include 1-palmitoyl-2-(5-oxomethyl)-sn-glycerol-3-phosphocholine (povpc), 1-palmitoyl-2-glutaryl-sn-glycerol-3-phosphocholine (pgpc), 1-palmitoyl-2-(9oxononyl)-sn-glycerol-3-phosphocholine (ponpc), and palmitoyl-2-arachidonol-Sn phosphatidylcholine (PAPC) [12, 13].

10.1.5.3 Other Lipid Peroxidation Products

Sn-2 epoxide is generally considered as one of the most reliable markers of oxidative stress in vivo [12]; sphingosine-1-phosphate (s-1-p), the metabolite of ceramide, can induce the proliferation of vascular smooth muscle cells, platelet activation, and endothelial cell stimulation [13]; FFA and its metabolites 15-hete, 9-hode, and 13-hode can be further oxidized to form a large number of complex isoprostaglandin (ox-ce) products [12]; arachidonic acid and linoleic acid, as well as esterified fatty acids, are oxidized to hydroperoxides derivatives, 15-hete, 9-hode, and 13-hode. These modified FFA have been proved to be ligands of PPARa and PPARG [14]; in lipid peroxidation products, 27 hydroxycholesterol, 7-ketcholesterol and 5 α, 6 α-epoxide, 5b, 6b-epoxide, and cholesterol-3b, 5A, 6b-triol are the most abundant oxidized sterols in plasma and as lesions [15]; CE hydroperoxides and hydroxides are the main lipid oxidation products found in human as lesions [14]; 4-hydroxynonene aldehyde and malondialdehyde (HNE and MDA), HNE, and MDA are carbonyl compounds and also the most abundant ones in LDL lipid peroxidation abundant α, β-unsaturated hydroxyene [4].

In short, LDL produces a large number of different types of lipid peroxidation bioactive molecules, which promote the occurrence of local inflammation of endothelium, stimulate the migration and infiltration of chemotactic inflammatory cells and smooth muscle cells through different mechanisms and molecular pathways, and constantly promote the slow occurrence and development of AS.

10.1.6 LDL Protein Oxidative Modification

10.1.6.1 Protein Modification Caused by Lipid Peroxidation Products

LDL lipid peroxidation products can react with apoB-100 amino acid residues. There are 357 lysine residues in apoB-100, of which a considerable part (225 lysine residues) are exposed on the surface of LDL, and the remaining 132 are embedded in the lipid part of LDL [16]. Polyunsaturated fatty acids in LDL oxidize aldehydes, such as HNE and MDA, which can react with lysine and other amino acid residues.

10.1.6.2 Modification of ApoB-100 by Enzyme Mechanism

Exposure of LDL to reagents or enzyme-catalyzed hypochlorite (HOCl) results in the oxidation of amino acid residues of apoB-100, which transforms LDL into the high uptake form of macrophages [16]. Myeloperoxidase is an enzyme related to inflammation and oxidative stress. It can catalyze H2O2 and chloride to form a powerful cytotoxin, HOCl, and then react with the tyrosine residue of apoB-100. Chlorinated biomolecules such as 3-chlorotyrosine are considered to be specific markers of MPO catalyzed oxidation [17]. The modification of LDL by the active nitrogen produced by MPO of monocyte transformed lipoproteins into NO2-LDL with high uptake and promoted the lipid loading of macrophages and the formation of foam cells through CD36 pathway [18].

10.1.7 LDL Acetylation and Oxidation

Deacetylation may be the first step in the chain of as changes caused by LDL particles in the blood of as patients. LDL is deacetylated firstly, followed by the loss of free cholesterol and cholesterol ester, phospholipid and triglyceride, the increase of particle density, and the decrease of particle size; secondly, the increase of particle negative charge leads to the formation of electronegative LDL part and the appearance of misfolded apolipoprotein B in large quantities; in the later stage, the increase of LDL particle oxidant and the decrease of antioxidant, and the heavily modified LDL particle can produce autoantibody [19].

10.1.8 LDL Glycation and Oxidation

LDL glycosylation is a nonenzymatic reaction between the carbonyl group of reducing sugar and the amino group of L-lysine residue of apoB-100. It can also occur in the phospholipid component of LDL, leading to functional changes in LDL and increased susceptibility to oxidative modification [20]. The body can also have nonenzymatic glycosylation reaction at the normal blood glucose level, and the carbonyl metabolites can be eliminated by the body's enzymes, which will not cause harm to the human body [21]. But in diabetes and hyperglycemia, the concentration of sdLDL increased. It has been reported that sdLDL apoB isolated from individuals without diabetes mellitus is more widely glycosylated than general LDL granules, and more than 90% of glycosylated apolipoprotein B in plasma is present in sdLDL granules. SdLDL may be more susceptible to glycosylation [22]. The glycosylation and oxidation of LDL are not mutually exclusive modification of LDL, because the glycosylation itself will produce free radicals. Even in vitro, glycosylation is accompanied by some degree of oxidation when molecular oxygen and oxygen free radical generation processes do not exist [23].

10.1.9 Cellular Mechanism of LDL Oxidation on AS

10.1.9.1 Effect of Oxidized LDL on Endothelial Cells

Endothelial cells play an important role in maintaining vascular homeostasis. They can synthesize and secrete a large number of enzymes and cytokines to maintain the balance between vasodilation and contraction, inhibition and stimulation of smooth muscle cell proliferation and migration, thrombosis, and fibrinolysis [24]. Cardiovascular risk factors such as hyperlipidemia, diabetes, hypertension, obesity, smoking, and chronic mental stress can lead to endothelial dysfunction and oxidative stress reaction of endothelial cells, and initially aggravate the pathological process of AS [25]. Endothelial dysfunction can be manifested as endothelial activation, which eventually leads to the transformation of arterial endothelial cells from a resting phenotype to an inflammatory phenotype involving host defense response [26]. Under the action of endothelial cells, or because of its own characteristics, sdLDL in the circulatory system stays in the subendothelial space or the extracellular matrix of the arterial wall. Under the action of free radicals and enzymes, LDL components are modified. A large number of LDL-oxidized lipid components stimulate scavenger receptor (SR), toll-like receptor, and other receptors of activated endothelial cells, which lead to NF-kB activation, and activate a variety of target genes related to vascular wall pathophysiology, including cytokines, chemokines, and leukocyte adhesion molecules, as well as genes regulating cell proliferation and cell survival [27]. Various lipid components of oxidized LDL participate in the activation of NF-kB. 13-hpode, oxidized phospholipid of MM-LDL, and LPC can activate NF-kB and induce the expression of VCAM-1, ICAM-1, and MCP-1. However, some bioactive components of oxidized LDL can inhibit NF-kB activation, indicating that oxidized LDL has biphasic effect on NF-kB [26].

10.1.9.2 Effect of Oxidized LDL on Macrophages

In the early pathological changes of AS, oxidized LDL promotes the activation of endothelial cells and the activation of inflammatory pathway, which results in the increase of the expression of inflammatory factors; under the chemotaxis of MCP-1 and oxidized lipid components, blood monocytes enter into the endothelium, derive into macrophages, and develop into foam cells through SR recognition and phagocytosis of oxidized LDL. Oxidative LDL stimulates macrophages to express a large number of SR, CD36, LOX-1, SR-A, SR-B1, CD68, etc. and to recognize and ingest specific oxidative LDL components. Under normal conditions, the phagocytized lipids form an endocytosome, and cholesterol esters (CE) and TG related to lysosomal fusion lipoproteins are hydrolyzed by cholesterol esterase with high activity in the acidic pH of lysosomal lumen [28]. In order to prevent the potential cytotoxicity caused by excessive accumulation of free cholesterol (FC), FC can be re-esterified by ACAT to form CE on the endoplasmic reticulum, which is stored in the cytoplasm as lipid droplets. Cholesterol esterification is considered as a protective defense mechanism, which can avoid excessive accumulation of cytotoxic FC. Under the background of serious lipoproteins oxidation, lipid uptake, and cholesterol esterification increase, while cholesterol outflow is insufficient. The final result is excessive accumulation of CE in macrophages, forming foam cells. The formation of foam cells depends on the balance of three main related biological processes, including fat uptake, cholesterol esterification, and cholesterol efflux. A large amount of oxidized LDL also promotes macrophages to absorb modified lipids without restriction, and it destroys the pathway of cholesterol outflow, promotes cholesterol storage, and then the cholesterol esterification mechanism is also destroyed [29].

In addition, macrophages absorb a lot of oxidized LDL through SR, which destroys the normal lipid outflow pathway, and more and more oxidized lipid and protein components are trapped in the cytoplasm. These components interfere with the function of ER modified folding proteins, resulting in the accumulation of misfolded proteins in ER and ER stress. In the case of ERS, the ability of protein folding must be restored rapidly to meet the needs of protein folding. In the presence of high levels of misfolded proteins in the endoplasmic reticulum, an intracellular signaling pathway called UPR induces a series of transcription and translation events to restore the homeostasis of the endoplasmic reticulum. Macrophage-derived foam cells engulf a large number of oxidized lipids, and the FC esterification in the cytoplasm is blocked. A large number of FC is trapped in the cytoplasm, which reflects its cytotoxicity and starts the process of apoptosis. In the early pathological changes of AS, the apoptotic foam cells can be phagocytized by local macrophages, and then be cleared. This effect is called exocytosis, which can maintain the stability of early pathological plaques and reduce the extracellular disintegration of foam cells, thus causing lipid accumulation under the intima [30]. However, in the middle and late stage of AS, excessive ERS will aggravate the lipid phagocytosis and even apoptosis of macrophages, resulting in more subintimal lipid accumulation, forming a typical atheroma [31].

10.1.9.3 Effect of Ox LDL on Vascular Smooth Muscle Cells

In early pathological changes of AS, with the infiltration of LDL and the function of entering into the vascular wall, a variety of lipid active components are activated to diffuse and act on smooth muscle cells, NADPH oxidase is activated to produce a large number of oxygen free radicals; the polarity of smooth muscle cells changes, from contractile to synmorphic [32]; and matrix metalloproteinases activated, secreted, and degraded the matrix components around cells, making them smooth. Under the action of ox LDL, smooth muscle cells located in the middle membrane of blood vessels pass through the inner elastic layer, migrate into the inner membrane, proliferate [33], synthesize, and secrete a large number of extracellular matrix components, and form fiber caps. In the early pathological process of AS, the formation of fibrous cap is helpful to reduce plaque rupture and prevent the occurrence of vascular embolism. However, when the disease entered the progressive stage, under the stimulation of ox LDL, smooth muscle cells expressed LOX-1 and other scavenger receptors, and phagocytized lipids through scavenger receptors. In the late stage of AS, the foam cells derived from smooth muscle cells secrete a large number of matrix metalloproteinases to degrade the collagen fibers of the fibrous cap, resulting in the thinning of the fibrous cap. Under the effect of the blood flow shear force in the vascular cavity, it is easy to locate in the upstream and downstream of the plaque on the lumen surface, namely, the shoulder rupture, leading to lipid outflow of plaque, and then to thrombosis, acute clinical event of vascular stenosis [34].

10.2 HDL Oxidative Modification

10.2.1 Introduction of HDL

High-density lipoprotein (HDL) is a kind of small, dense, and rich in a variety of lipid and protein macromolecular components in the blood. The average size is 8–10 nm, and the density is 1.063–1.21 g/ml [35]. HDL mainly contains polar lipids and apolipoproteins, in addition to many other proteins, including enzymes and acute phase proteins, and may contain a small amount of nonpolar lipids. HDL can be isomers with different macromolecule components, which have different structure, chemical and biological characteristics. HDL has strong antioxidant modification ability in physiological state, but it also has various modifications in pathological state [36], such as oxidation [37]. The lipids of HDL are mainly the surface phospholipids and the internal cholesterol esters and triglycerides. Phospholipids are mainly phosphatidylcholine, accounting for 32–35 mol% of total lipids in HDL. Another important phospholipid of HDL is lysophosphatidylcholine, which accounts for 1.4–8.1 mol% of total lipids. Sphingomyelin on HDL is a kind of structural lipid, which can enhance the rigidity of surface lipid. It is also the main sphingolipid in blood circulation, accounting for 5.6–6.6 mol% of the total lipid. It is mainly derived from triglyceride rich lipoproteins, only to a small extent from new HDL. The cholesterol ester (CE) on HDL is the result of lecithin cholesterol acyltransferase (LCAT) catalyzed transesterification of phospholipids with cholesterol. These high hydrophobic lipids form the lipid core of HDL, accounting for 36 mol% of the total lipids of HDL. A small amount of free sterol is located in the lipid monolayer on the surface of HDL particles, which regulates its fluidity.

HDL carries a large number of different proteins, which can be divided into apolipoproteins, enzymes, lipid transfer proteins, acute phase response proteins, complement components, protease inhibitors, and other protein components. Apolipoproteins and lipases are widely considered as the key functional components of HDL, while the secondary protein components mainly play the role of complement regulation, infection prevention and acute phase response. ApoAI is the main structural and functional component of HDL, accounting for 70% of HDL protein. Almost all HDL particles are believed to contain apoAI. The main functions of apoAI include the interaction with cell receptor, activation of LCAT, and multiple anti-atherosclerosis (as) activities of HDL. ApoA II is the second largest HDL apolipoprotein, accounting for 15–20% of the total HDL protein. About half of the HDL particles may contain ApoA II [35].

10.2.2 Clinical Evidence and Pathological Effect of Oxidative Modification of HDL

10.2.2.1 Evidence of Oxidative Modification of HDL In Vivo

There is oxidized HDL (ox HDL) in the human body [38]. With the specific antibody of Cu2 + oxidized HDL, the presence of ox HDL was detected in the intima and endothelial cells of human abdominal aortic atheroma plaque by immunohistochemistry. The enzyme-linked immunosorbent assay (ELISA) based on monoclonal antibody can detect ox HDL in hemorrhagic plasma sensitively and has reliable specificity. In addition, ox HDL also exists in plasma of patients with endogenous hypertriglycerides. At present, the oxidation mechanism of HDL is not clear. In vitro, HDL can be oxidized by different media, such as metal ions Cu2 +, Fe2 +, Mn2 +, etc., among which Cu2 + oxidation is the most commonly used method in vitro. Hypochlorite (HOCl) can also cause oxidative modification of HDL, but HOCl and Cu2 + mediated oxidative modification of HDL are different in properties and kinetics. In 2004, it was found for the first time that tyrosine can be nitrated and chlorinated by myeloperoxidase (MPO) in plasma and plaque of patients with coronary heart disease, reducing the cholesterol reverse transport capacity of HDL granules. HDL in the intima of aortic atherosclerotic plaque contains 3-chlorotyrosine, which is the product of HOCl oxidation, and its content is much higher than that of HDL in blood [39]; it is also found that MPO is a component of HDL in plaque, and MPO can produce HOCl, so it is speculated that MPO can mediate HDL oxidation through HOCl. MPO is the only source of 3-chlorotyrosine [40], which proves that MPO can oxidize HDL in vivo. Paraoxonase-1 (PON-1) in HDL granules is negatively related to the oxidative susceptibility of HDL, which can inhibit the oxidative modification of HDL mediated by Cu2 +. PON-1 can inhibit the oxidative modification of HDL in a dose-dependent manner, and the ability of oxidative modified HDL to obtain and stabilize PON-1 from hepatocytes decreases. The decrease of PON-1 content makes HDL easier to be oxidized. In patients with coronary heart disease, tyrosine at 166 and 192 sites can be nitrated and chlorinated by MPO, and the modified content is inversely proportional to the reverse transport capacity of cholesterol [41]. In vitro studies have shown that MPO can modify several amino acid residues of human ApoA1 by producing nitrite and hypochlorite, such as methionine residues at 86, 112, and 148, tryptophan residues at 8, 50, 72, and 108, and tyrosine residues at 192, 236, etc., which can be modified by nitration or chlorination. The mutation of tryptophan in ApoA1 to phenylalanine can not only protect, it can keep HDL normal function and avoid oxidative modification of HDL. The oxidation of specific areas of ApoA1 was measured by tandem mass spectrometry with selective reaction monitoring mode. It was found that 192 tyrosine residues of ApoA1 were the main chlorination sites, and 18 tyrosine residues were the main nitration sites in human as plaque. 192 tyrosine residues of ApoA1 in healthy human blood circulation were both the main chlorination sites and the main nitration sites [42]. Trp72 is a site of ApoA1 oxidation, and its main mechanism is mpo-h2o2-cl-system. Trp72 can resist the oxidative modification and functional degradation of HDL induced by mpo-h2o2-cl-system. Tyrosine 166 is a nitration site of ApoA1, which accounts for 8% of human atherosclerotic plaque, and its function is damaged compared with normal HDL [43].

10.2.2.2 Functional Abnormality After Oxidative Modification of HDL

In vitro study shows that HDL oxidized by plasma and MPO hypochlorite system in patients with coronary heart disease has significantly reduced reverse transport capacity of cholesterol and its ability to activate LCAT. Other important functional molecules in HDL, such as ApoA1, PON-1, CETP, and so on, are oxidized and modified to change the structure, which also causes the reverse transport of cholesterol to be blocked. For example, ApoA1 as a ligand mediates the binding of HDL with ATP-binding cassette transporter (ABCA1) on foam cell membrane, and ABCA1 becomes one of the main pathways for cholesterol transfer to HDL in foam cell. The combination of ApoA1 and ABCA1 is the initial link of cholesterol reverse transport in AS plaque, but the change of structure of ApoA1 cannot combine with ABCA1, which results in the obstruction of cholesterol outflow in foam cells. It was also found that the antioxidation and anti-inflammatory ability of HDL decreased significantly in the plasma of patients with psoriasis, which may have an impact on the pathogenesis of psoriasis [44].

10.2.3 Cellular Mechanism of HDL Oxidative Modification Impairing Anti-AS Function

10.2.3.1 Effect of HDL Oxidative Modification on Endothelial Cells

Vascular endothelial cells (EC) cover the smooth intima on the surface of blood vessels and maintain the state of blood flow. Meanwhile, endothelium is the largest endocrine organ of the body. It can secrete a variety of bioactive substances, including vasodilator factor and vasoconstrictor factor, which are in balance under physiological state. For vascular endothelial cells, the steady state of holding cycle plays a very important role. HDL oxidized by MPO in vitro significantly reduced the migration ability of endothelial cells. In the model of electrical injury of carotid artery, HDL modified in vitro decreased the endothelial repair ability [45]. Vascular endothelial cell injury and dysfunction are the early links of AS, which are manifested in the decrease of endothelial nitric oxide synthase (eNOS) activity and no production. HDL has the functions of activating eNOS, promoting no production and anti-endothelial apoptosis. As a gas signal molecule, NO plays an important role in maintaining normal vasodilation, inhibiting platelet aggregation and proliferation of arterial smooth muscle cells, and inhibiting monocyte and endothelial adhesion. In addition, NO is also an oxygen free radical scavenger in vivo, which can inhibit the oxidation of lipoproteins. ENOS is the key enzyme of NO synthesis. Its activity and function directly regulate the production and biological function of NO. The oxidative modification of HDL can improve the endothelial function and reduce the ability of anti-endothelial apoptosis [46]. At the same time, ox HDL can promote the release of endothelin-1 (ET-1), which can promote the proliferation of smooth muscle cells (SMC), constrict blood vessels and raise blood pressure, thus aggravating the injury of EC and promoting the development of AS.

10.2.3.2 Effect of HDL Oxidative Modification on Macrophages

As a main feature of advanced atherosclerotic plaques, macrophage apoptosis promotes enlargement of the necrotic cores and plaque rupture, and then leads to cardiovascular complications [47]. Ox HDL, like ox LDL, also plays a crucial role in macrophage-derived foam cell formation and apoptosis. It has been reported that ox HDL exerts a cytotoxic effect on macrophages and accelerates atherosclerosis progression [48, 49]. It has been found that ox HDL prepared in vitro and HDL isolated from patients with metabolic syndrome (MS) activated ER stress-CHOP-mediated apoptotic pathway in macrophages, which could be blocked by oxidative stress inhibitors, toll-like receptor 4 (TLR4)-specific small interfering RNA (siRNA), and TLR4 antibody [50]. HDL exposure to hyperglycemic conditions could contribute to the acceleration of atherosclerosis in DM patients. Glycated HDL may induce macrophage apoptosis through activating ER stress-CHOP pathway, and ER stress mediates glycated HDL-induced autophagy, which in turn protects macrophages against apoptosis by alleviating CHOP pathway [51].

10.2.3.3 Effect of HDL Oxidation on Other Cells

Smooth muscle cell (SMC) is the main cell component in as plaque, and its proliferation plays an important role in the formation of as. As early as the twentieth century, it has been reported that ox HDL can promote the proliferation of SMC. In addition, platelets are also affected by HDL oxidative modification. Under physiological condition, HDL can inhibit platelet aggregation and prevent as. The effect of ox HDL on platelets in pathological state is concerned, although there are inconsistent reports. For example, HOCl oxidized HDL can cause inflammation and coagulation by binding to CD36 on platelets [52]. CD36 belongs to class B scavenger receptor family and is the receptor of ox LDL on macrophages. When CD36 helps to absorb ox HDL, it will increase foam cell formation [53]. At the same time, ox HDL will reduce the expression of CD36 mRNA and protein in human peripheral macrophages in vitro. CD36 can selectively ingest lipids in Cu2 + oxidized HDL, but not in ordinary HDL or LDL [54], which may lead to AS. MPO or Cu2 + oxidized HDL can bind SR-BI receptor on platelets, inhibit platelet aggregation, and produce antithrombotic effect [55]. Adipocyte differentiation is also affected by oxidative modification of HDL. Ox HDL changes the number and size of adipocytes through several unknown mechanisms.

10.2.4 Effect of HDL Oxidative Modification and Intervention on Its Anti-AS Function

10.2.4.1 Oxidative Modification of HDL Protein Components and Its Effect on Anti-AS Function

There are more than 80 protein components in HDL, and the modification of some protein components will also affect the anti-AS function. The oxidative modification of HDL occurs on the methionine and aromatic amino acid residues of apoAI, which leads to the separation of apoAI from HDL and the decrease of lipid content in HDL. After oxidative modification, the structure and function of apoAI changed [56], resulting in the inability of apoAI to combine with ABCAI, the loss of the ability to activate LCAT, the failure of cholesterol esterification, and the obstruction of cholesterol transfer to LDL, thus affecting the whole reverse cholesterol transport process. The oxidative modification of apoAI by HOCl resulted in the cleavage of apoAI and apoAI, which reduced the anti-AS function of HDL. HDL glycosylation may be a nonenzymatic glycosylation of protein, which mainly occurs on the lysine of apoAI, thus affecting the cholesterol outflow, antioxidant, anti-inflammatory, endothelial protection, and other functions of HDL. PON1 in HDL can resist oxidation, but it can also be oxidized. PON1 itself is oxidized and ox HDL can inactivate PON1 [57]. PON1 activity is negatively correlated with age, and its mechanism may be related to the decrease of free sulfhydryl group on the 284th cysteine related to the active site of lipid peroxidation [58]. The activity of PON1 in glycosylated HDL decreased, and the degree of decline was positively correlated with glucose concentration and incubation time [59]. Therefore, this modification of PON1 can affect its activity, thus leading to the decline of antioxidant function of HDL.

10.2.4.2 Oxidative Modification of HDL Reduces the Protection of LDL and Promotes AS

Normal HDL has antioxidant capacity, which can inhibit the oxidative modification of LDL by macrophages, endothelial cells, and smooth muscle cells, but the oxidized HDL loses the ability to inhibit the oxidative modification of LDL. When macrophages sense that LDL changes into ox LDL, they enter the endothelium and phagocytize ox LDL. These macrophages come from plasma monocytes. Stimulated by chemical factors such as plasma monocyte chemoattractant protein-1 (MCP-1), monocytes infiltrate into the vascular wall, recognize, and phagocytize ox LDL specifically by scavenger receptor, then become macrophages, and further form foam cells. The accumulation of a large number of foam cells in the arterial wall promotes the formation of lipid striation and early pathological changes of as. Foam cells are the early signs of the formation of as lipid striation. It is believed that the inhibition of LDL oxidation by ox HDL loss is related to the decrease of PON1 activity in HDL. PON1 is synthesized by hepatocytes, and the ability of obtaining and stabilizing PON1 from hepatocytes after HDL oxidative modification decreases, resulting in the decrease of HDL antioxidant capacity. There is a negative correlation between PON1 activity and HDL oxidation in obese patients [60], which may be related to the increased risk of cardiovascular disease in obese patients. In addition, when Cu2 + oxidizes HDL in vitro, the activity of PON1 in HDL decreases significantly, and the inactivation mode is inconsistent with that of PON1 mediated by Cu2 +, which indicates that ox HDL promotes PON1 inactivation [61].

10.2.4.3 HDL Modification Intervention and Its Effect on Anti-AS Function

HDL function is the result of multiple protein synergies, and any abnormal component will affect its function. Therefore, the intervention of HDL components and its modification may be an important measure for as prevention and treatment. It was found that antioxidants can inhibit the oxidative modification of HDL in vitro and in vivo, and enhance its anti-AS effect. Improving the level of apoAI can resist the toxic effect of inflammation. ApoAI mimic peptide d4f can significantly reduce the level of oxLDL in serum of mice fed with high-fat diet, reduce the total area of aortic root lesions, the percentage of lipid positive areas, macrophage aggregation and apoptosis rate [62]. Omega-3 polyunsaturated fatty acids modify the lipoproteins containing the apoAI proteome. These protein changes can improve the function of HDL. After eating foods with high omega-3 polyunsaturated fatty acids content, PON1 and apoAI in HDL increase, thus enhancing its antioxidation and anti-inflammatory ability [63]. The combination of statins and niacin can improve the protein function of HDL in patients with coronary heart disease, increase the expression of cholesterol reverse transporter, and promote the anti-AS effect [64]. Animal experiments and population experiments show that the new small molecule antioxidant hydrogen molecule has a clear role in promoting HDL antioxidation and anti-inflammatory and inhibiting as [65]. In addition, the anti-AS function of HDL can also be improved by proper regular aerobic exercise [66].

In conclusion, the complexity of HDL structure and function can maintain its vascular homeostasis under physiological conditions. However, in some pathological conditions, the main component modification changed its anti-AS function and vascular homeostasis. Further study on the molecular mechanism of component modification is beneficial to the precise location of intervention target and the restoration of biological characteristics of HDL against AS.

References

1. Prassl R, Laggner P (2009) Molecular structure of low density lipoprotein: current status and future challenges. Eur Biophys J 38:145–158
2. Obama T, Kato R, Masuda Y, Takahashi K, Aiuchi T, Itabe H (2007) Analysis of modified apolipoprotein B-100 structures formed in oxidized low-density lipoprotein using LC-MS/MS. Proteomics 7:2132–2141
3. Singh U, Devaraj S, Jialal I (2005) Vitamin E, oxidative stress, and inflammation. Annu Rev Nutr 25:151–174
4. Aviram M (1993) Modified forms of low density lipoprotein and atherosclerosis. Atherosclerosis 98:1–9
5. Salvayre R, Auge N, Benoist H, Negre-Salvayre A (2002) Oxidized low-density lipoprotein-induced apoptosis. Biochim Biophys Acta 1585:213–221
6. Back M, Yurdagul A Jr, Tabas I, Oorni K, Kovanen PT (2019) Inflammation and its resolution in atherosclerosis: mediators and therapeutic opportunities. Nat Rev Cardiol 16:389–406
7. Negre-Salvayre A, Garoby-Salom S, Swiader A, Rouahi M, Pucelle M, Salvayre R (2017) Proatherogenic effects of 4-hydroxynonenal. Free Radic Biol Med 111:127–139
8. Cox DA, Cohen ML (1996) Effects of oxidized low-density lipoprotein on vascular contraction and relaxation: clinical and pharmacological implications in atherosclerosis. Pharmacol Rev 48:3–19
9. Yoshida H, Kisugi R (2010) Mechanisms of LDL oxidation. Clin Chim Acta 411:1875–1882
10. Esterbauer H, Gebicki J, Puhl H, Jurgens G (1992) The role of lipid peroxidation and antioxidants in oxidative modification of LDL. Free Radic Biol Med 13:341–390
11. Jialal I (1998) Evolving lipoprotein risk factors: lipoprotein(a) and oxidized low-density lipoprotein. Clin Chem 44:1827–1832
12. Choi SH, Sviridov D, Miller YI (2017) Oxidized cholesteryl esters and inflammation. Biochim Biophys Acta Mol Cell Biol Lipids 1862:393–397
13. Najafi M, Roustazadeh A, Alipoor B (2011) Ox-LDL particles: modified components, cellular uptake, biological roles and clinical assessments. Cardiovasc Hematol Disord Drug Targets 11:119–128
14. Levitan I, Volkov S, Subbaiah PV (2010) Oxidized LDL: diversity, patterns of recognition, and pathophysiology. Antioxid Redox Signal 13:39–75
15. Gargiulo S, Testa G, Gamba P, Staurenghi E, Poli G, Leonarduzzi G (2017) Oxysterols and 4-hydroxy-2-nonenal contribute to atherosclerotic plaque destabilization. Free Radic Biol Med 111:140–150
16. Yang CY, Gu ZW, Yang HX, Yang M, Gotto AM Jr, Smith CV (1997) Oxidative modifications of apoB-100 by exposure of low density lipoproteins to HOCL in vitro. Free Radic Biol Med 23:82–89
17. Berliner JA, Heinecke JW (1996) The role of oxidized lipoproteins in atherogenesis. Free Radic Biol Med 20:707–727
18. Podrez EA, Febbraio M, Sheibani N, Schmitt D, Silverstein RL, Hajjar DP, Cohen PA, Frazier WA, Hoff HF, Hazen SL (2000) Macrophage scavenger receptor CD36 is the major receptor for LDL modified by monocyte-generated reactive nitrogen species. J Clin Invest 105:1095–1108
19. Nikiforov NG, Zakiev ER, Elizova NV, Sukhorukov VN, Orekhov AN (2017) Multiple-modified low-density lipoprotein as atherogenic factor of patients' blood: development of therapeutic approaches to reduce blood atherogenicity. Curr Pharm Des 23:932–936
20. Abidi M, Khan MS, Ahmad S, Kausar T, Nayeem SM, Islam S, Ali A, Alam K, Moinuddin (2018) Biophysical and biochemical studies on glycoxidatively modified human low density lipoprotein. Arch Biochem Biophys 645:87–99
21. Graier WF, Kostner GM (1997) Glycated low-density lipoprotein and atherogenesis: the missing link between diabetes mellitus and hypercholesterolaemia? Eur J Clin Investig 27:457–459
22. Younis N, Sharma R, Soran H, Charlton-Menys V, Elseweidy M, Durrington PN (2008) Glycation as an atherogenic modification of LDL. Curr Opin Lipidol 19:378–384
23. Soran H, Durrington PN (2011) Susceptibility of LDL and its subfractions to glycation. Curr Opin Lipidol 22:254–261
24. Lyons TJ, Jenkins AJ (1997) Lipoprotein glycation and its metabolic consequences. Curr Opin Lipidol 8:174–180
25. Davignon J, Ganz P (2004) Role of endothelial dysfunction in atherosclerosis. Circulation 109:III27–III32
26. Huang PL (2009) eNOS, metabolic syndrome and cardiovascular disease. Trends Endocrinol Metab 20:295–302
27. Deanfield JE, Halcox JP, Rabelink TJ (2007) Endothelial function and dysfunction: testing and clinical relevance. Circulation 115:1285–1295
28. Ouimet M, Marcel YL (2012) Regulation of lipid droplet cholesterol efflux from macrophage foam cells. Arterioscler Thromb Vasc Biol 32:575–581

29. Chistiakov DA, Bobryshev YV, Orekhov AN (2016) Macrophage-mediated cholesterol handling in atherosclerosis. J Cell Mol Med 20:17–28
30. Hetz C, Papa FR (2018) The unfolded protein response and cell fate control. Mol Cell 69:169–181
31. Yao S, Miao C, Tian H, Sang H, Yang N, Jiao P, Han J, Zong C, Qin S (2014) Endoplasmic reticulum stress promotes macrophage-derived foam cell formation by up-regulating cluster of differentiation 36 (CD36) expression. J Biol Chem 289:4032–4042
32. Lee SJ, Seo KW, Yun MR, Bae SS, Lee WS, Hong KW, Kim CD (2008) 4-Hydroxynonenal enhances MMP-2 production in vascular smooth muscle cells via mitochondrial ROS-mediated activation of the Akt/NF-kappaB signaling pathways. Free Radic Biol Med 45:1487–1492
33. Newby AC (2005) Dual role of matrix metalloproteinases (matrixins) in intimal thickening and atherosclerotic plaque rupture. Physiol Rev 85:1–31
34. Xue XH, Zhou XM, Wei W, Chen T, Su QP, Tao J, Chen LD (2016) Alisol A 24-acetate, a triterpenoid derived from alisma orientale, inhibits ox-LDL-induced phenotypic transformation and migration of rat vascular smooth muscle cells through suppressing ERK1/2 signaling. J Vasc Res 53:291–300
35. Camont L, Chapman MJ, Kontush A (2011) Biological activities of HDL subpopulations and their relevance to cardiovascular disease. Trends Mol Med 17:594–603
36. Norata GD, Pirillo A, Catapano AL (2006) Modified HDL: biological and physiopathological consequences. Nutr Metab Cardiovasc Dis 16:371–386
37. Nakajima T, Origuchi N, Matsunaga T, Kawai S, Hokari S, Nakamura H, Inoue I, Katayama S, Nagata A, Komoda T (2000) Localization of oxidized HDL in atheromatous plaques and oxidized HDL binding sites on human aortic endothelial cells. Ann Clin Biochem 37(Pt 2):179–186
38. Nakano T, Nagata A (2003) Immunochemical detection of circulating oxidized high-density lipoprotein with antioxidized apolipoprotein A-I monoclonal antibody. J Lab Clin Med 141(6):378–384
39. Bergt C, Pennathur S, Fu X, Byun J, O'Brien K, McDonald TO, Singh P, Anantharamaiah GM, Chait A, Brunzell J, Geary RL, Oram JF, Heinecke JW (2004) The myeloperoxidase product hypochlorous acid oxidizes HDL in the human artery wall and impairs ABCA1-dependent cholesterol transport. Proc Natl Acad Sci U S A 101(35):13032–13037
40. Shao B, Pennathur S, Heinecke JW (2012) Myeloperoxidase targets apolipoprotein A-I, the major high density lipoprotein protein, for site-specific oxidation in human atherosclerotic lesions. J Biol Chem 287:6375–6386
41. Zheng L, Settle M, Brubaker G, Schmitt D, Hazen SL, Smith JD, Kinter M (2005) Localization of nitration and chlorination sites on apolipoprotein A-I catalyzed by myeloperoxidase in human atheroma and associated oxidative impairment in ABCA1-dependent cholesterol efflux from macrophages. J Biol Chem 280 (1):38–47
42. Huang Y, DiDonato JA, Levison BS, Schmitt D, Li L, Wu Y, Buffa J, Kim T, Gerstenecker GS, Gu X, Kadiyala CS, Wang Z, Culley MK, Hazen JE, Didonato AJ, Fu X, Berisha SZ, Peng D, Nguyen TT, Liang S, Chuang CC, Cho L, Plow EF, Fox PL, Gogonea V, Tang WH, Parks JS, Fisher EA, Smith JD, Hazen SL (2014) An abundant dysfunctional apolipoprotein A1 in human atheroma. Nat Med 20 (2):193–203
43. DiDonato JA, Aulak K, Huang Y, Wagner M, Gerstenecker G, Topbas C, Gogonea V, DiDonato AJ, Tang WH, Mehl RA, Fox PL, Plow EF, Smith JD, Fisher EA, Hazen SL (2014) Site-specific nitration of apolipoprotein A-I at tyrosine 166 is both abundant within human atherosclerotic plaque and dysfunctional. J Biol Chem 289(15):10276–10292
44. He L, Qin S, Dang L, Song G, Yao S, Yang N, Li Y (2014) Psoriasis decreases the anti-oxidation and anti-inflammation properties of high-density lipoprotein. Biochim Biophys Acta 1841(12):1709–1715
45. Pan B, Yu B, Ren H, Willard B, Pan L, Zu L, Shen X, Ma Y, Li X, Niu C, Kong J, Kang S, Eugene Chen Y, Pennathur S, Zheng L (2013) High-density lipoprotein nitration and chlorination catalyzed by myeloperoxidase impair its effect of promoting endothelial repair. Free Radic Biol Med 60:272–281
46. Undurti A, Huang Y, Lupica JA, Smith JD, DiDonato JA, Hazen SL (2009) Modification of high density lipoprotein by myeloperoxidase generates a pro-inflammatory particle. J Biol Chem 284 (45):30825–30835
47. Moore KJ, Tabas I (2011) Macrophages in the pathogenesis of atherosclerosis. Cell 145:341–355
48. Soumyarani VS, Jayakumari N (2014) Oxidized HDL induces cytotoxic effects: implications for atherogenic mechanism. J Biochem Mol Toxicol 28:481–489
49. Ru D, Zhiqing H, Lin Z, Feng W, Feng Z, Jiayou Z, Yusheng R, Min F, Chun L, Zonggui W (2015) Oxidized high-density lipoprotein accelerates atherosclerosis progression by inducing the imbalance between treg and teff in LDLR knockout mice. APMIS 123:410–421
50. Yao S, Tian H, Zhao L, Li J, Yang L, Yue F, Li Y, Jiao P, Yang N, Wang Y, Zhang X, Qin S (2017) Oxidized high density lipoprotein induces macrophage apoptosis via toll-like receptor 4-dependent CHOP pathway. J Lipid Res 58:164–177
51. Tian H, Li Y, Kang P, Wang Z, Yue F, Jiao P, Yang N, Qin S, Yao S (2019) Endoplasmic reticulum stress-dependent autophagy inhibits glycated high-density lipoprotein-induced macrophage apoptosis by inhibiting CHOP pathway. J Cell Mol Med 23 (4):2954–2969
52. Assinger A, Koller F, Schmid W, Zellner M, Babeluk R, Koller E, Volf I (2010) Specific binding of hypochlorite-oxidized HDL to platelet CD36 triggers proinflammatory and procoagulant effects. Atherosclerosis 212(1):153–160

53. Ren J, Jin W, Chen H (2010) oxHDL decreases the expression of CD36 on human macrophages through PPARgamma and p38 MAP kinase dependent mechanisms. Mol Cell Biochem 342(1–2):171–181
54. Thorne RF, Mhaidat NM, Ralston KJ, Burns GF (2007) CD36 is a receptor for oxidized high density lipoprotein: implications for the development of atherosclerosis. FEBS Lett 581(6):1227–1232
55. Valiyaveettil M, Kar N, Ashraf MZ, Byzova TV, Febbraio M, Podrez EA (2008) Oxidized high-density lipoprotein inhibits platelet activation and aggregation via scavenger receptor BI. Blood 111(4):1962–1971
56. Shao B, Tang C, Heinecke JW, Oram JF (2010) Oxidation of apolipoprotein A-I by myeloperoxidase impairs the initial interactions with ABCA1 required for signaling and cholesterol export. J Lipid Res 51:1849–1858
57. Matsuo Y, Oberbach A, Till H, Inge TH, Wabitsch M, Moss A, Jehmlich N, Volker U, Muller U, Siegfried W, Kanesawa N, Kurabayashi M, Schuler G, Linke A, Adams V (2013) Impaired HDL function in obese adolescents: impact of lifestyle intervention and bariatric surgery. Obesity (Silver Spring) 21:E687–E695
58. Henning MF, Herlax V, Bakas L (2011) Contribution of the C-terminal end of apolipoprotein AI to neutralization of lipopolysaccharide endotoxic effect. Innate Immun 17:327–337
59. Brown BE, Nobecourt E, Zeng J, Jenkins AJ, Rye KA, Davies MJ (2013) Apolipoprotein A-I glycation by glucose and reactive aldehydes alters phospholipid affinity but not cholesterol export from lipid-laden macrophages. PLoS One 8:e65430
60. Ferretti G, Bacchetti T, Moroni C, Savino S, Liuzzi A, Balzola F, Bicchiega V (2005) Paraoxonase activity in high-density lipoproteins: a comparison between healthy and obese females. J Clin Endocrinol Metab 90(3):1728–1733
61. Nguyen SD, Kim JR, Kim MR, Jung TS, Soka DE (2004) Copper ions and hypochlorite are mainly responsible for oxidative inactivation of paraoxon-hydrolyzing activity in human high density lipoprotein. Toxicol Lett 147(3):201–208
62. Qin S, Kamanna VS, Lai JH, Liu T, Ganji SH, Zhang L, Bachovchin WW, Kashyap ML (2012) Reverse D4F, an apolipoprotein-AI mimetic peptide, inhibits atherosclerosis in ApoE-null mice. J Cardiovasc Pharmacol Ther 17(3):334–343
63. Chahal N, Manlhiot C, Wong H, McCrindle BW (2014) Effectiveness of omega-3 polysaturated fatty acids (fish oil) supplementation for treating hypertriglyceridemia in children and adolescents. Clin Pediatr (Phila) 53(7):645–651
64. Green PS, Vaisar T, Pennathur S, Kulstad JJ, Moore AB, Marcovina S, Brunzell J, Knopp RH, Zhao XQ, Heinecke JW (2008) Combined statin and niacin therapy remodels the high-density lipoprotein proteome. Circulation 118(12):1259–1267
65. Sang H, Yao S, Zhang L, Li X, Yang N, Zhao J, Zhao L, Si Y, Zhang Y, Lv X, Xue Y, Qin S (2015) Walk-run training improves the anti-inflammation properties of high-density lipoprotein in patients with metabolic syndrome. J Clin Endocrinol Metab 100 (3):870–879
66. Song G, Lin Q, Zhao H, Liu M, Ye F, Sun Y, Yu Y, Guo S, Jiao P, Wu Y, Ding G, Xiao Q, Qin S (2015) Hydrogen activates ATP-binding cassette transporter A1-dependent efflux ex vivo and improves high-density lipoprotein function in patients with hypercholesterolemia: a double-blinded, randomized, and placebo-controlled trial. J Clin Endocrinol Metab 100 (7):2724–2273

Further Reading

67. Kontush A, Chapman MJ (2012) High-density lipoproteins: structure, metabolism, function and therapeutics. Wiley, New York
68. Zheng L, Nukuna B, Brennan ML et al (2004) Apolipoprotein A-I is a selective target for myeloperoxidase-catalyzed oxidation and functional impairment in subjects with cardiovascular disease. J Clin Invest 114(4):529–541
69. Shao B, Pennathur S, Heinecke JW (2012) Myeloperoxidase targets apolipoprotein ai, the major high density lipoprotein protein, for site-specific oxidation in human atherosclerotic lesions. J Biol Chem 287(9):6375–6386
70. Shao B, Cavigiolio G, Brot N et al (2008) Methionine oxidation impairs reverse cholesterol transport by apolipoprotein A-I. Proc Natl Acad Sci U S A 105 (34):12224–12229
71. Peng DQ, Brubaker G, Wu Z et al (2008) Apolipoprotein A-I tryptophan substitution leads to resistance to myeloperoxidase-mediated loss of function. Arterioscler Thromb Vasc Biol 28(11):2063–2070
72. Yao S, Tian H, Zhao L, Li J, Yang L, Yue F, Li Y, Jiao P, Yang N, Wang Y, Zhang X, Qin S (2017) Oxidized high-density lipoprotein induces macrophage apoptosis via toll-like receptor 4-dependent CHOP pathway. J Lipid Res 58(1):164–177
73. Shao B, Tang C, Heinecke JW et al (2010) Oxidation of apolipoprotein AI by myeloperoxidase impairs the initial interactions with ABCAl required for signaling and cholesterol export. J Lipid Res 51(7):1849–1858
74. Matsuo Y, Oberbach A, Till H et al (2013) Impaired HDL function inobese adolescents: impact of lifestyle intervention and bariatric surgery. Obesity (Silver Spring) 21(12):687–695
75. Henning MF, Herlax V, Bakas L (2011) Contribution of the Cterminalend of apolipoprotein Al to neutralization of lipopolysaccharide endotoxic effect. Innate Immunity 17(3):327–337
76. Brown BE, Nobecourt E, Zeng J et al (2013) Apolipoprotein AI glycationby glucose and reactive aldehydes alters phospholipid affinity but notcholesterol export from lipidladen macrophages. PLoS One 8(5):65430

Rare Diseases Related with Lipoprotein Metabolism

11

Hongwen Zhou, Yingyun Gong, Qinyi Wu, Xuan Ye, Baowen Yu, Chenyan Lu, Wanzi Jiang, Jingya Ye, and Zhenzhen Fu

Abstract

Rare diseases are gathering increasing attention in last few years, not only for its effects on innovation scientific research, but also for its propounding influence on common diseases. One of the most famous milestones made by Michael Brown and Joseph Goldstein in metabolism field is the discovery of the defective gene in familial hypercholesterolemia, a rare human genetic disease manifested with extreme high level of serum cholesterol (Goldstein JL, Brown MS, Proc Natl Acad Sci USA 70:2804–2808, 1973; Brown MS, Dana SE, Goldstein JL, J Biol Chem 249:789–796, 1974). Follow-up work including decoding the gene function, mapping-related pathways, and screening therapeutic targets are all based on the primary finding (Goldstein JL, Brown MS Arterioscler Thromb Vasc Biol 29:431–438, 2009). A series of succession win the two brilliant scientists the 1985 Nobel Prize, and bring about statins widely used for lipid management and decreasing cardiovascular disease risks. Translating the clinical extreme phenotypes into laboratory bench work has turned out to be the first important step in the paradigm conducting translational and precise medical research. Here we review the main categories of rare disorders related with lipoprotein metabolism, aiming to strengthen the notion that human rare inheritable genetic diseases would be the window to know ourselves better, to treat someone more efficiently, and to lead a healthy life longer. Few rare diseases related with lipoprotein metabolism were clustered into six sections based on changes in lipid profile, namely, hyper- or hypocholesterolemia, hypo- or hyperalphalipoproteinemia, abetalipoproteinemia, hypobetalipoproteinemia, and sphingolipid metabolism diseases. Each section consists of a brief introduction, followed by a summary of well-known disease-causing genes in one table, and supplemented with one or two diseases as example for detailed description. Here we aimed to raise more attention on rare lipoprotein metabolism diseases, calling for more work from basic research and clinical trials.

Keywords

Hypercholesterolemia · Hypocholesterolemia · Hypoalphalipoproteinemia · Hyperalphalipoproteinemia · Abetalipoproteinemia ·

H. Zhou (✉) · Y. Gong · Q. Wu · X. Ye · B. Yu · C. Lu · W. Jiang · J. Ye · Z. Fu
Department of Endocrinology and Metabolism, The First Affiliated Hospital of Nanjing Medical University, Nanjing, China
e-mail: drhongwenzhou@njmu.edu.cn

X.-C. Jiang (ed.), *Lipid Transfer in Lipoprotein Metabolism and Cardiovascular Disease*, Advances in Experimental Medicine and Biology 1276, https://doi.org/10.1007/978-981-15-6082-8_11

Hypobetalipoproteinemia · Sphingolipid metabolism diseases · Rare disease

Abbreviations

7-DHC	7-dehydrocholesterol
ABCA1	ATP-binding cassette subfamily A member 1
ABCG5	ATP-binding cassette subfamily G member 5
ABCG8	ATP-binding cassette subfamily G member 8
ABL	abetalipoproteinemia
ANGPTL3	angiopoietin-like protein 3
apoB	apolipoprotein B
apoC3	apolipoprotein C3
apoE	apolipoprotein E
ASCVD	arteriosclerotic cardiovascular disease
ASM	acid sphingomyelinase
CDL	calvarial doughnut lesions
CESD	cholesteryl ester storage disease
CETP	cholesteryl ester transfer protein
CMRD	chylomicron retention disease
CYP27A1	cytochrome P450 family 27 subfamily A member 1
CYP7A1	cytochrome P450 family 7 subfamily A member 1
DHCR7	7-dehydrocholesterol reductase
FH	familial hypercholesterolemia
FHBL	familial hypobetalipoproteinemia
FLD	familial LCAT deficiency
HA	hypoalphalipoproteinemia
HALP	hyperalphalipoproteinemia
HBL	hypobetalipoproteinemia
HeFH	heterozygous familial hypercholesterolemia
HoFH	homozygous familial hypercholesterolemia
LALD	lysosomal acid lipase deficiency
LCAT	lecithin-cholesterol acyltransferase
LDLR	low-density lipoprotein receptor
LDLRAP1	low-density lipoprotein receptor adapter protein 1
LIMA1	LIM domain and actin-binding 1
LIPA	lysosomal acid lipase A
LIPC	lipase C hepatic type
MTTP	microsomal triglyceride transfer protein
NPA	Niemann-Pick disease type A
NPB	Niemann-Pick disease type B
NPC1L1	Niemann-Pick C1-like 1
NPD	Niemann-Pick disease
PCSK9	proprotein convertase subtilisin/kexin type 9
PLTP	phospholipid transfer protein
PNPLA5	patatin-like phospholipase domain-containing 5
SAR1B	secretion associated ras-related GTPase 1B
SGMS2	sphingomyelin synthase 2
SLOS	Smith-Lemli-Opitz syndrome
STAP1	signal transducing adaptor protein family 1

11.1 Hypercholesterolemia

Hypercholesterolemia is a disease characterized by high level of cholesterol in blood [4], and it could be segregated into two subtypes: (1) polygenic hypercholesterolemia, with plasma levels of LDL-C over 130 mg/dl and high risk of arteriosclerotic cardiovascular disease (ASCVD) [5], and (2) monogenic hypercholesterolemia, familial hypercholesterolemia (FH) is the most common, but severe even life-threatening monogenic hypercholesterolemia [6]. According to the genetic inherence, FH can be characterized into two types: (1) heterozygous familial hypercholesterolemia (HeFH), in which the loss in low-density lipoprotein receptor (LDLR) activity is up to 50%, leading to plasma LDL-C levels 2 to 3 times above average(>190 mg/dl), development of xanthomas, and high risk of ASCVD, and (2) homozygous familial hypercholesterolemia (HoFH), in which the LDLR pathway is almost nonfunctional or markedly defective, leading to plasma LDL-C levels 4 to 8 times above average (>500 mg/dl), development of xanthomas and high risk of ASCVD [5, 7].

The prevalence of HeFH was traditionally thought to be 1 in 500 [8]. However, the estimated overall prevalence of HeFH from the data of 19 studies including 2,458,456 individuals is about 1 in 250 [9].The prevalence of HoFH has also been revised upward to 1 in 300,000 (ranging from 1 in 160,000 to 1 in 1000,000) [8]. FH is mainly caused by rare dysfunctional mutations affecting either LDLR (>95%), apolipoprotein B (apoB) (2 ~ 11%), or proprotein convertase subtilisin/kexin type 9 (PCSK9) (<1%) [10].

Additionally, low-density lipoprotein receptor adapter protein 1 (LDLRAP1), cytochrome P450 family 7 subfamily A member 1 (CYP7A1), lysosomal acid lipase A (LIPA), apolipoprotein E (APOE), cytochrome P450 family 27 subfamily A member 1 (CYP27A1), signal transducing adaptor protein family 1 (STAP1), patatin-like phospholipase domain-containing 5 (PNPLA5), and so on could also result in hypercholesterolemia [8, 11, 12]. Mutations affecting ATP-binding cassette subfamily G member 5 or 8 (ABCG5 or ABCG8) leading to sitosterolemia, a rare monogenic condition affecting 1 in 2000,00 individuals in the population and is manifested with increased levels of plant sterols, such as sitosterol [13].

We summarize information of these pathogenic genes listed as Table 11.1 below.

Given that hypercholesterolemia caused by mutations in LDLR, APOB, and PCSK9 have been thoroughly studied and reviewed elsewhere [23–26], we are going to discuss some other types of hypercholesterolemia induced by rare monogenic mutations in details.

11.1.1 Lysosomal Acid Lipase Deficiency

Lysosomal acid lipase deficiency (LALD) is a rare autosomal recessive lysosomal storage disease caused by mutations in the LIPA gene [17]. The LIPA gene encodes lipase A, the lysosomal acid lipase, also known as cholesterol ester hydrolase. It functions in the lysosome to catalyze the hydrolysis of cholesteryl esters and triglycerides.

LIPA mutations can result in two distinct diseases depending on the extent of deficiency: the severe one, called early-onset Wolman disease, and the less severe one known as cholesteryl ester storage disease (CESD) [27]. Wolman disease, with 1% or less than 1% of residual LAL activity, often accompanied with hepatosplenomegaly, adrenal cortical insufficiency, vomiting, and malnutrition in the first month of life [14]. The less severe CESD, with 1% to approximatively 12% of residual LAL activity, is often accompanied with increased total cholesterol, increased LDL-C, decreased HDL-C, progressive liver steatosis, and a high risk of contracting coronary heart disease [28, 29].

Liver biopsy, LAL activity, and molecular sequencing of the LIPA mutations, as well as the serum transaminase activities and lipid levels are used to diagnose LALD.

The hypercholesterolemic phenotype and lipid accumulated in lysosome in patients with LIPA gene deficiency can be attenuated by using HMG-CoA reductase inhibitors such as statins [27]. Among hypolipidemic agents, fibrates, cholestyramine, and ezetimibe can be used for treatment. In 2015, the US Food and Drug Administration approved Kanuma (sebelipase alfa), a recombinant human lysosomal acid lipase, as the first treatment for LALD [30]. Additionally, the enzyme replacement therapy was shown to be successful in animal models, but the clinical trials for CESD are now underway [31].

11.1.2 Cholesterol-7-Alpha-Hydroxylase Deficiency

Pullinger et al. firstly reported cases with hypercholesterolemia that existed mutations in CYP7A1 gene [16]. The CYP7A1 gene encodes the enzyme cholesterol 7α-hydroxylase, which catalyzes the first reaction in cholesterol catabolism and classic bile acid synthesis. Deficiency of CYP7A1 could decrease bile acid production and may lead to accumulation of cholesterol in the liver, thus downregulating LDLR expression and then developing hypercholesterolemia.

Table 11.1 Summary of genetic causes and characteristics of monogenic hypercholesterolemia

Gene	Disease	Inheritance	Lipid profile	Prevalence	Clinical features
LDLR	Familial hypercholesterolemia (Type I)	AD	Elevated plasma levels of cholesterol, specifically LDL-C [14]	1/500 in HeFH; 1/1000,000 in HoFH [6]	Tendon and skin xanthomas, arcus corneae and/or cardiovascular deposits and leads to increased risk in ASCVD and mortality [14]
APOB (gain of function)	Familial hypercholesterolemia (Type II); familial defective apolipoprotein B	AD	Elevated plasma LDL-C levels with normal triglyceride levels [6]	1/1000 in HeFH; 1/4,000,000 in HoFH [6]	Tendon xanthomas and premature atherosclerosis [6]
PCSK9 (gain of function)	Familial hypercholesterolemia (Type III)	AD	Elevated plasma LDL-C levels [14]	<1/2500 [6]	Accelerated atherosclerosis and premature coronary heart disease [14]
LDLRAP1	Familial hypercholesterolemia (Type IV); ARH	AR	Defective LDLR activity, elevated plasma LDL-C levels [6]	<1/ 5,000,000 [6]	The presence of planar, tuberous, tendon xanthomas, and coronary artery disease [15]
CYP7A1	Cholesterol 7alpha-hydroxylase deficiency	ACD	Elevated LDL-C and triglyceride levels [16]	N/A	Hepatitis, increased risk of cardiovascular and gallstone disease [16]
LIPA	Lysosomal acid lipase deficiency; cholesterol ester storage disease	AR	Massive accumulation of cholesteryl ester and triglycerides [17]	1/ 40,000 ~ 1/ 300,000 [18]	Elevation of cholesterol, nausea, vomiting, global developmental delay, hepatomegaly, hypertriglyceridemia, and abdominal pain [11]
APOE	Hyperlipoproteinemia (Type III)	AD	Increased LDL and triglycerides levels	1 ~ 7/5000 [19]	Xanthomas and increased risk of cardiovascular diseases [20]
ABCG5 or ABCG8	Sitosterolemia	AR	Elevated plasma levels of LDL-C and defect in sterol efflux from cells [6]	<1/ 5,000,000 [6]	Xanthomatosis, hypercholesterolemia and increased risk of aortic stenosis and premature coronary atherosclerotic disease [6]
CYP27A1	Cerebrotendinous xanthomatosis	AR	Elevated plasma levels of cholesterol [21]	<1/20,000 [22]	Xanthelasma, cholesterol deposition in the brain and other tissues [21]

AD autosomal dominant, *AR* autosomal recessive, *ACD* autosomal codominant

The clinical phenotypes of homozygotes are prominent with significantly increased total cholesterol and LDL-C levels, evaluated triglycerides, as well as more than 90% reduction in fecal bile acid excretion in that patient reported by Pullinger et al. Consequently, individuals with CYP7A1 gene variants are associated with increased risk of myocardial infarction and gallstone disease [32].

Patients with CYP7A1 mutation may be resistant to lipid lowering treatments. Sustained combination of large doses of statins and niacin could help to bring their plasma cholesterol levels under control [16]. In addition, it is suggested that increasing intestinal reclamation of bile acids may help compensate the decreased biosynthesis [33].

11.2 Hypocholesterolemia

Previous work mainly focused on hypercholesterolemia for its significant association with increased risk of cardiovascular diseases, but little is known about hypocholesterolemia. In related studies, it was defined as plasma cholesterol lower than the 5th percentile of the distribution in the population adjusted for age and gender [34].

Sparing no efforts working on hypocholesterolemia is of great importance. On the one hand, hypocholesterolemia due to cholesterol biosynthesis defect may manifest as multiple malformation and developmental disabilities, so it is of huge necessity to find out therapeutic regimens for those patients. On the other hand, some natural occurring mutations causing hypocholesterolemia may become novel therapeutic target candidates for treating hypercholesterolemia.

As we all know, PSCK9 inhibitor and ezetimibe targeting Niemann-Pick C1-like 1 (NPC1L1) have been widely used in clinical practice, thus we will review 7-dehydrocholesterol reductase (DHCR7) associated with congenital malformation and the newly found LIM domain and actin-binding 1 (LIMA1) in the following parts.

Table 11.2 shows the pathogenic gene, lipid profiles, prevalence, and clinical features of major types of monogenic hypocholesterolemia.

11.2.1 Smith-Lemli-Opitz Syndrome

Smith-Lemli-Opitz syndrome (SLOS) is an autosomal recessive multiple malformation syndrome that was firstly described in 1964 by Smith et al. [48]. This congenital disease is resulted from deficiency of the DHCR7 gene, which encodes the enzyme 7-dehydrocholesterol (7-DHC) reductase and mediates the conversion of 7-DHC to cholesterol. It is reported that SLOS is more common among northern and central European population relatively and the prevalence was estimated to be 1/70,000 ~ 1/30,000 [44].

Most patients with SLOS have varied and complicated clinical manifestations. Microcephaly, cleft palate, 2–3 toe syndactyly, growth failure, intellectual disability, and mental and behavior abnormalities are frequent in SLOS patients [44, 49]. Moreover, congenital defects, including holoprosencephaly, atrial, and ventricular septal defects, intestinal malrotation and renal hypoplasia or agenesis were also reported [44, 50–52]. Due to the deficiency of 7-DHC reductase, cholesterol precursor 7-DHC would accumulate in the blood and tissues, and total cholesterol could be decreased, which is an important basis for diagnosis [49]. If the biochemical indicators are ambiguous, the diagnosis could be confirmed by testing DHCR7 mutations [44].

Surgical operation could help correct cardiovascular, renal, craniofacial, and other anomalies occurred in SLOS patients [53]. Additionally, supplementation of cholesterol to some extent may relieve some symptoms [53]. Statins is also suggested as possible treatment for SLOS to reduce abnormally elevated 7-DHC, which was considered to be potentially toxic, and clinical studies demonstrated that it could have positive effects on patients suffered from SLOS [53–56].

11.2.2 LIMA1 Variant with Hypocholesterolemia Phenotype

In 2018, Song et.al made great progress in finding a rare frameshift mutation (K306fs) in LIMA1 encoding LIM domain and actin-binding protein 1 in a Chinese Kazakh family with low plasma LDL-C and decreased intestinal cholesterol absorption [47].

LIMA1-K306fs carriers seem to have healthy phenotype and reduced risk of ASCVD. Their plasma total cholesterol and LDL-C levels were remarkably lower than those of wild-type individuals, while triglyceride, HDL-C and fasting glucose levels were similar between the two groups. In addition, the campesterol: lathosterol ratio (Ca: L ratio) was significantly lower in LIMA1-K306fs carriers, indicating that they have reduced intestinal cholesterol absorption [47].

Moreover, they confirmed the phenotypes in both intestine-specific *Lima1*-deficient and whole-body Lima1 knockout mice. Then they investigated the potential mechanism that LIMA1 binds with NPC1L1 and myosin Vb to

Table 11.2 Summary of genetic causes and characteristics of monogenic hypocholesterolemia

Gene	Disease	Inheritance	Lipid profile	Prevalence	Clinical features
PCSK9 (loss of function)	–	–	Reduction in LDL-C levels [35]	2–2.6% (African Americans) [35]	Normal health, protective Decreased prevalence of ASCVD [35]
APOB (loss of function)	Hypobetalipoproteinemia	ACD	Decreased LDL-C and apoB levels [36]	1/1000 ~ 1/3000 (heterozygotes) [36]	Rod-cone dystrophy, ataxia, reduced tendon reflexes, acanthocytosis, retinal degeneration, and hepatic steatosis [37]
MTP	Abetalipoproteinemia	AR	Low plasma triglycerides, cholesterol and undetectable *levels* of LDL-C and apoB [38].	<1/1000,000 [38, 39]	Malabsorption, ataxia, muscular hypotonia, abnormality of retinal pigmentation, and reduced tendon reflexes [37, 38]
ANGPTL3	Familial combined hypolipidemia	AR	Pan-hypolipidemia [40]	About 1 / 382,000 [41]	Normal health, protective Decreased prevalence of ASCVD
SAR1B	Chylomicron retention disease/Anderson disease	AR	Reduced LDL-C and HDL-C Normal triglycerides [42]	<60 cases	Failure to thrive, diarrhea, steatorrhea, fat malabsorption, retinopathy, peripheral neuropathy [42, 43]
DHCR7	Smith-Lemli-Opitz syndrome	AR	Abnormally elevated cholesterol precursor 7-DHC Reduced cholesterol [44]	1/30,000 ~ 1/70,000 [44]	Intellectual disability, global developmental delay, microcephaly, and malformations of different organs [44]
NPC1L1 (loss of function)	–	–	Low total cholesterol and LDL-C levels [45]	About 1/650 (heterozygotes) [46]	Normal health, protective Decreased prevalence of ASCVD [46]
LIMA1	Hypocholesterolemia phenotype	–	Significantly lower total cholesterol and LDL-C levels [47]	First identified in a Chinese Kazakh family [47]	Normal health, protective Decreases intestinal cholesterol absorption [47]

AR autosomal recessive, *ACD* autosomal codominant

regulate the transportation of NPC1L1 from the endocytic recycling compartment to the plasma membrane. Deficiency of LIMA1 may block the process, leading to decreased intestinal cholesterol absorption [47].

Just as the application of statins, ezetimibe, and PCSK9 inhibitors, LIMA1 may have the potential to be another emerging target for LDL-C lowering therapy.

11.3 Hypoalphalipoproteinemia

Hypoalphalipoproteinemia (HA) is a group of disorders of HDL deficiency (Table 11.3). HA can be caused by many genetic defects including APOAI, ABCA1, lecithin-cholesterol acyltransferase (LCAT), and some other genes. Diagnosis concludes both clinical and biochemical evaluation after management designed to correct known secondary causes of low HDL. Low

Table 11.3 Summary of genetic causes and characteristics of hypoalphalipoproteinemia

Gene	Disease	Inheritance	Lipid profile	Prevalence	Clinical features
APOAI LCAT	Fish-eye disease	AR	Decreased HDL levels	<1/1000000 (worldwide)	Corneal clouding, splenomegaly, hepatomegaly, visual impairment
APOAI LCAT	Familial LCAT deficiency	AR	Elevated TG, VLDL-C and LDL levels, marked HDL deficiency [59]	<1/1000000 (worldwide) ~ 70 families have been reported worldwide [59]	Corneal opacities, hemolytic anemia, kidney failure [60]
ABCA1	Tangier disease	AR	HDL < 5 mg/dl, apoA-I < 5 mg/dl, low total plasma cholesterol (<150 mg/dl), normal or high plasma triglycerides	<1/1000000 (worldwide)	Dry skin, risk of ASCVD; very large, yellow-orange tonsils, enlarged liver, spleen and lymph nodes; hypertriglyceridemia, neuropathy, corneal clouding, type 2 diabetes.

AR autosomal recessive

serum HDL level (< 40 mg/dl or <1.0 mmol/L in men and <50 mg/dl or <1.3 mmol/L in women) is a well-known independent risk factor for ASCVD and is also common in patients with hypertriglyceridemia, insulin resistance, obesity, and diabetes. Patients with significant deficiency of HDL-C (<20 mg/dl or <0.5 mmol/L) and free from secondary causes (<1% of the population) is grouped as HA [57]. To date, HA is defined according to the following criteria: (1) low HDL-C level with normal VLDL and LDL-C levels; (2) without secondary causes to hypoalphalipoproteinemia; and (3) share a similar lipoprotein pattern with at least one first-degree relative [58].

We conclude the major characteristics of HA interns of pathogenic genes, related disease, inheritance, lipid profile, prevalence, and clinical features. Some other genes are mentioned in the section Hypocholesterolemia.

11.3.1 Familial LCAT Deficiency

Familial LCAT deficiency (FLD) is a rare autosomal recessive disease caused by mutation in the LCAT gene. About 70 families have been reported worldwide [59]. LCAT activities can be detected in different lipoproteins which include alpha-LCAT activity and beta-LCAT activity. These two types of LCAT activity are two functional aspects of the same protein. There are two clinical syndromes that arise out of mutations in LCAT gene, named FLD and fish-eye disease (FED).

The main genetic defects of FLD are the mutations on LCAT gene. Recent studies identified many rare mutations associated with FLD, we enumerate some of them as follows: V333M and M404V mutations [61], P274S LCAT mutation [62], missense variation c.301 G > A in exon 2 [63], and LCAT G30S mutation [64]. There are still a lot of genetic loci on LCAT that needs to be discovered.

The main clinical features include progressive corneal opacity, mild hemolytic anemia, multiple impaired lipid-related traits, deterioration in kidney function, and mild thrombocytopenia. FLD is also associated with an increased prevalence of ASCVD [62]. FLD patients have very low plasma HDL-C levels accompanied by some other lipid metabolism disorders. The diagnosis often relies on clinical and biochemical parameters, clinical evaluation, and urine examination. The measurement of LCAT activity and genetic testing can also be helpful in identifying the underlying mutations [59].

There is no precise treatment or cure for familial LCAT deficiency so far, but some new treatment targets are under evaluation. Even though, we can find some ways to manage the clinical symptoms.

To prevent renal disease in patients with FLD, recombinant human LCAT infusion may be an effective therapy as recommended in recent studies [65, 66]. Both LCAT gene replacement and enzyme replacement are under development [66].

11.3.2 Tangier Disease

Tangier disease, caused by mutations in ABCA1 gene, is one of the most severe subtype of familial HDL deficiency. Classic manifestations are severe plasma deficiency or even absence of major HDL apolipoprotein (apoA-I) and HDL particle, thus causing the accumulation of cholesteryl esters in multiple tissues including tonsils, the liver, peripheral nerves, intestinal mucosa, skin, cornea, and immune organs [67]. Multiple and diverse mutations in ATP-binding cassette subfamily A member 1 (ABCA1) are linked to Tangier disease, for example, c.1824delG, c.1881C > G, and c.4121C > T are notable ABCA1 pathogenic variants [68–70].

The major clinical signs of this disease are not limited to hyperplastic yellow-orange tonsils, corneal opacities, neuropathy, hepatosplenomegaly, thrombocytopenia, anemia, and stomatocytosis. For individuals, one or few signs would be presented [67, 69]. The major biochemical signs of this condition are very low plasma HDL-C concentration, typically <5 mg/dl (0.125 mmol/L), rarely 5–10 mg/dl; very low or absent apoA-I concentration, usually <30 mg/dl (typically <5 mg/dl); small or absent alpha band on lipoprotein electrophoresis [71, 72]. Till now, there is no precise regimen for treating Tangier disease. Even though cholesteryl ester transfer protein (CETP) inhibition raises HDL levels, but it has not been shown to be effective in patients with Tangier disease [57]. In the future, the possible goal for a Tangier disease therapeutic strategy would be aimed to obtain a selective increase in mature HDL level to restore cholesterol efflux capacity. To this end, the first attempt could be applying reconstituted HDL as replacement, until the access to reliable gene therapy [69].

11.4 Hyperalphalipoproteinemia

Hyperalphalipoproteinemia (HALP) is a condition of elevated HDL-C level attributed to both genetic and environmental factors, which is related coronary stenosis [73].

Familial HALP often coexists with longevity and that higher HDL-C levels are found among healthy elderly. The most significant cause of primary HALP is a genetic deficiency of CETP, which has been reported mainly in Japanese [74] and is mainly related with PPAR signaling pathway. Some studies [75] have shown that hetero- and homozygotes for CETP gene mutations are associated with increased risks for ASCVD.

The prevalence of HALP was traditionally thought to be heterozygous mutation present in 5–7% of the Japanese population [76]. HALP is mainly caused by dysfunctional rare mutations affecting the CETP. Additionally, minor genes such as apolipoprotein C3 (APOC3) [77], lipase C hepatic type (LIPC) [78], LCAT [79], and so on could also result in HALP. Moreover, some low-frequency gene variants have already been identified in several studies, such as phospholipid transfer protein (PLTP) [80].

We summarize information of these pathogenic genes listed as Table 11.4 below, including related disease, inheritance, lipid profile, prevalence, and clinical features.

11.5 Abetalipoproteinemia and Hypobetalipoproteinemia

Abetalipoproteinemia [ABL; Online Mendelian Inheritance in Man (OMIM) 200100] and hypobetalipoproteinemia (HBL) are inherited lipoprotein disorders defined as absence or low levels (below the 5th percentile of sex- and

Table 11.4 Summary of genetic causes and characteristics of hyperalphalipoproteinemia

Gene	Related disease	Inheritance	Lipid profile	Prevalence	Clinical features
CETP	Familial hepatic lipase deficiency, Huntington disease-like, familial hyperlipoproteinemia, type iii [73]	AD	Hypercholesterolemia, markedly increased levels of HDL cholesterol, apolipoprotein A-I, decreased levels of low-density lipoprotein cholesterol and apolipoprotein B [73]	Heterozygous mutation present in 5–7% of the Japanese population	Hypercholesterolemia markedly increased levels of HDL-C, atherosclerotic, coronary heart disease
APOC3 (loss of function)	Apolipoprotein C-iii deficiency	–	Low plasma apo C-III concentrations and atypically large HDL [81]	Two cases reported [77, 81]	Coronary artery disease
PLTP	–	–	High TC, LDL-C and HDL-TG, and lower VLDL-TG, and VLDL-C [82]	One case reported [82]	Coronary artery disease
LIPC	Hepatic lipase deficiency	AR	Elevated plasma HDL-C as well as triglyceride-rich beta VLDL, LDL, and HDL lipoproteins [74]	3%	Hyperlipidemia
LCAT	Hyperalphalipoproteinemia	AR	Hypertriglyceridemia and reduced LDL apoprotein B concentration [83]	4.8%	Corneal opacities, microalbuminuria, hypertriglyceridemia, and reduced LDL apoprotein B concentration, anemia [83]

AD autosomal dominant, *AR* autosomal recessive

age-matched individuals in the population) of apoB and LDL-C in the plasma. HBL represents a heterogeneous group of diseases, and familial hypobetalipoproteinemia (FHBL) is the most frequent form, which contains FHBL1 (OMIM 615558) and FHBL2 (OMIM 605019). Less frequently, HBL also includes chylomicron retention disease (CMRD; OMIM 246700) and specific dyslipidemia for mutations in PCSK9.

The exact prevalence of ABL and HBL is unknown, but it is reported by Lee and Hegele [84] that the incidences of both are less than 1 in one million. There are unknown gender, racial, or ethnic preferences for ABL and HBL, but the disorder is more prevalent in consanguineous marriages.

ABL is caused by homozygous or compound heterozygous mutation in microsomal triglyceride transfer protein (*MTTP*, alias *MTP*) gene on chromosome locus 4q22–24 encoding the protein. MTTP forms a heterodimer with protein disulfide isomerase, which facilitates the transfer of triglyceride from cytosol to rough endoplasmic reticulum containing nascent apo B during the assembly of chylomicron in enterocytes and VLDL in hepatocytes [85, 86]. Therefore, mutations in *MTTP* may disrupt MTTP formation, affect its ability to transfer lipids [87] and ultimately impair apoB-containing lipoproteins processing and secretion.

FHBL1 occurs with an autosomal codominant mode resulted from mutations in *APOB* gene on chromosome locus 2p23–24. ApoB is the major protein constituent of chylomicron (apoB-48), VLDL (apoB-100), and LDL (apoB-100). The gene mutations abolish or interfere with the translation of full-length apoB and usually give rise to a truncated apoB protein [88–90], consequently impairing apoB-containing particles processing and secretion.

Mutations in angiopoietin-like protein 3 (*ANGPTL3*) gene on chromosome 1p31 are associated with FHBL2, a recessive disorder characterized by global reduction of plasma lipoproteins [91, 92]. The function of ANGPTL3 appears to be the reversible inhibition of lipoprotein lipase activity, and the disruption of ANGPTL3 production increases lipolysis and causes familial combined hypolipidemia [93, 94].

CMRD is inherited in an autosomal recessive pattern and derives from two mutations in secretion associated ras-related GTPase 1B (*SAR1B*) gene, the gene product of which is critical for the fusion of the intestine-specific pre-chylomicron transport vesicle to the Golgi apparatus [42, 95, 96]. Mutations result in dysfunction of intracellular trafficking of chylomicron particles.

HBL can also be caused by mutations in *PCSK9* gene, which encodes a serine protease that binds to LDLR and targets for lysosomal degradation within hepatocytes [97]. Heterozygous carriers of *PCSK9* mutations increase the number of LDLRs on the cell surface, promoting the catabolic rate of LDL-C [98] and reducing circulating LDL and apoB levels.

Clinical manifestations of ABL and HBL usually become apparent in the neonatal period and malabsorption is the central feature. Affected individuals present fatty, foul-smelling stools, vomiting, abdominal distension, and failure to grow at the expected rate. Later in life, this condition may progress to neuro-ophthalmological dysfunction [86]. Typically, neuromuscular signs present in the first or second decade of life, and both the central and the peripheral nervous system are involved. Neurological manifestations include declined muscle coordination and inability to maintain balance and movement, loss of sensation in the extremities, muscle weakness or other involuntary movements. Ophthalmological findings tend to be variable, with many patients being asymptomatic until adulthood. Loss of night or color vision tends to occur at early stage in the disease but can progress to a phase in which the light-sensitive layer breaks down and consequently results in atypical retinitis pigmentosa. Additionally, individuals may have anemia with acanthocytes, hemolysis, and increased international normalized ratio. Liver involvement includes hepatic steatosis and abnormal levels of transaminase, with a high prevalence of severe fibrosis which can potentially progress to cirrhosis or liver carcinoma [99, 100].

Till now, no formal clinical diagnostic criteria for ABL and HBL have been published. The diagnoses of the conditions are established in typical clinical symptoms, lab examinations and most importantly, the pathogenic variants identified by molecular genetic testing. If two mutations in alleles are identified, testing of the proband's parents is recommended to investigate whether the variations originate from two different chromosomes.

A framework for clinical management of ABL and homozygous or compound heterozygous HBL has been proposed by Lee J and Hegele RA in 2014 [84], focusing on monitoring growth in children and preventing complications in all affected subjects. While for heterozygous HBL, although there seems to be no obviously adverse clinical outcomes, several reports of complications caused by vitamin deficiency and hepatic injuries due to fat accumulation over a long period of time suggest that follow-up assessments and appropriate interventions are also indispensable [86, 101].

11.6 Sphingolipid Metabolism Diseases

Sphingolipids are a class of lipids derived from the aliphatic amino alcohol sphingosine. Ceramides, sphingomyelins, and glycosphingolipids are three main types of sphingolipids, differing in the substituents on head groups. Sphingolipid metabolism diseases, termed as disorders of sphingolipid metabolism, are a class of diseases mainly occur in neural or even systemic systems, including Niemann-Pick disease (NPD), Fabry disease, Gaucher disease, Krabbe disease, Tay-Sachs disease, metachromatic leukodystrophy, and calvarial doughnut lesions (CDL) with bone fragility [102, 103]. Most sphingolipid metabolism diseases are inherited in an autosomal recessive pattern. Sphingolipidoses, caused by accumulation of lysosomal sphingolipid storage, occupy the largest part of sphingolipid metabolism diseases. In general, sphingolipidoses have a prevalence of approximately 1/10,000 [104], but it is actually higher in certain ethnics such as Ashkenazi Jewish [105]. Enzyme replacement therapy is accessible for people with Fabry disease [106] and Gaucher disease and may help them live well into adulthood. Unfortunately, the other types of sphingolipidoses are generally fatal in infantile stage, but the progression may be mild if the diseases are onset in juvenile or adult.

We here summarize the table of sphingolipid metabolism diseases considering gene and related characteristics as following (Table 11.5).

Generally, sphingolipidoses such as NPD, Fabry disease, Gaucher disease, Farber disease, and Krabbe disease are well-known to us caused by accumulation of different sphingolipids in lysosomes because of their degradation dysfunction [108]. However, as important as sphingolipidoses, disease caused by aberrant sphingolipid synthesis in cells like CDL is still elusive. Take sphingomyelin metabolism for an example, we give a brief summary to two sphingolipid disorders, NPD and CDL, which caused by the mutations in enzymes from opposite biological reactions.

11.6.1 Niemann-Pick Disease (NPD) (Type A and Type B)

11.6.1.1 Introduction and Genetic Defects

NPD is a classic kind of autosomal recessive inherited lipid storage diseases. In 1961, Crocker [109] divided NPD into four subtypes, and the majority is NPD type A and B (NPA, MIM# 257200 and NPB, MIM# 607616), both caused by mutations of SMPD1 gene (MIM# 607608), which located on chromosome 11 (11p15.1–11p15.4), encoding a 631 amino-acid protein named acid sphingomyelinase (ASM) in the lysosome [9], a kind of enzyme catalyzing the hydrolysis of sphingomyelin to ceramide and phosphorylcholine [110]. So far, approximately 1200 cases worldwide have been reported as NPA or NPB. NPA is also known as the intermediate protracted neurovisceral or the classic infantile form, and NPB is the visceral form [111].

Table 11.5 Summary of genetic causes and characteristics of sphingolipid metabolism diseases

Gene	Disease	Inheritance	Lipid profile/ dysfunction	Prevalence	Clinical features
GBA	Gaucher disease	AR	Deficiency of β-glucocerebrosidase, glucocerebroside accumulation, especially in the bone marrow, spleen, and liver	1–9/100000 (Europe) 1–9/100000 (Sweden)	Hepatosplenomegaly, anemia, thrombocytopenia, lung disease, bone abnormalities, hepatosplenomegaly, hematological defects
SMPD1/ 3	Niemann-pick disease, type A/ Niemann-pick disease, type B	AR	Little or no acid sphingomyelinase (ASM) (type A 1% or less, type B 10%); sphingomyelin accumulation in the nerve system, spleen and liver	1/250,000 (Ashkenazi Jewish)	Vomiting, diarrhea, hepatosplenomegaly, hypotrophy, pain, respiratory disorder
NPC1/2	NPD, type C1/ NPD, type C2	AR	Dysfunction of NPC intracellular cholesterol transporter 1/2, cholesterol (LDL-C, HDL-C) accumulation in the nerve system	1/150,000 (C1 95%, C2 5%, Spanish-American; type D, Nova Scotia)	Characteristic vertical supranuclear gaze palsy (VSGP), psychiatric disturbances
SCARB2			Deficiency of lysosomal integral membrane protein type 2 (LIMP-2) [107], LDL-C accumulation in nerve system		
GLA	Fabry disease	X-linked recessive	Dysfunction of α-galactosidase A, intracellular accumulation of globotriaosylceramide (GL-3), particularly in the vascular tree, nerve system, kidney, and heart	1–9/100000 (Sweden)	Episodes of pain, especially in hands and feet, clusters of small, angiokeratomas, hypohidrosis, corneal opacity, progressive kidney damage, heart attacks, and strokes
SGMS 1/2	Calvarial doughnut lesions with bone fragility (CDL)	AD	Dysfunction of sphingomyelin synthase 2, increased phosphatidylcholine, and decreased ceramide in the bone and nerve system	<1/1000000 (worldwide)	Cranial sclerosis and spondylometaphyseal dysplasia (CDLSMD)
HEXA	Tay-Sachs disease	AR	Deficiency of hexosaminidase α, GM2 ganglioside accumulation, in the nerve system (brain and spinal cord), testes, and eye	1–9/1000000 (worldwide)	Macrocephaly, seizures, tremor, and back pain
ARSA	Metachromatic leukodystrophy	AR	Deficiency of arylsulfatase A, sphingolipids (sulfatides)	–	Intellectual disability and seizures

(continued)

Table 11.5 (continued)

Gene	Disease	Inheritance	Lipid profile/ dysfunction	Prevalence	Clinical features
			accumulation in the brain, bone, and bone marrow		
LASS	Autosomal recessive congenital ichthyosis (ARCI)	AR	Deficiency of ceramide synthase (CerS3), decreased ceramide in the skin, eye, and testes	1–9/1000000 (Europe)	Ichthyosis (abnormal skin scaling over the whole body)
SPTLC	Neuropathy, hereditary sensory, and autonomic, Type 1	AD	Decreased serine palmitoyltransferase (SPT), glucosylceramides accumulation in nerve system	–	Hearing loss, dementia, peripheral neuropathy
GLB1	GM1 gangliosidosis	AR	Dysfunction of galactosidase β1, GM1 ganglioside accumulation in the brain, spinal cord, and bone	1–9/1000000 (Europe), 1–5/ 10000 (Malta), 1–9/100000 (Brazil), 1–9/ 1000000 (Sweden).	Exaggerated startle reactions Hepatosplenomegaly, skeletal abnormalities, seizures, intellectual disability, and the cherry-red spot
GALC	Krabbe disease (globoid cell leukodystrophy, GLD)	AR	Deficiency of galactosylceramidase, psychosine accumulation in the brain, bone, and bone marrow	1–9/1000000 (United States, France); 1–9/100000 (Europe, Sweden)	Irritability, muscle weakness, feeding difficulties, fever, stiff posture, delayed mental and physical development, weak muscles
CERKL	Retinitis pigmentosa	AR	Deficiency of ceramide kinase, decreased N-acetylcysteine in the eye, retina, and testes	–	Rod-cone dystrophy and optic disc pallor
ACER3	Leukodystrophy, progressive, early childhood-onset (PLDECO)	AR	Deficiency of alkaline ceramidase 3, abnormal production of myelin in the brain	–	Low-set ears and intellectual disability
ASAH1	Farber lipogranulomatosis (FRBRL)	AR	Dysfunction of N-acylsphingosine amidohydrolase 1 (ceramidase), ceramide accumulation in the liver, skin, and spleen	<1/1000000 (worldwide), <1/1000000 (Europe)	Hoarse voice, lipogranulomas, swollen, painful joints; difficulty breathing, hepatosplenomegaly

AD autosomal dominant, *AR* autosomal recessive, *ACD* autosomal codominant

11.6.1.2 Clinical Features, Diagnosis, and Treatment

Symptoms of NPD are related to the organs in which sphingomyelin accumulates, including the peripheral symptoms (hepatosplenomegaly) and the central symptoms (dementia, dysarthria, dysphagia, and ataxia). NPA is different from NPB in the aspect of decrease of neurological function [112]. Most NPA patients show severe neuropathic symptoms and shorter lifespan (between 1.5 and 3 years) than NPB [111, 113]. Besides, NPA eventually presents classic cherry-red spots of the macula of the retina in all affected children and fatally interstitial lung diseases, which are characteristics for the differential diagnosis [113].

The diagnosis is established by demonstrating the deficiency of acid sphingomyelinase activity in white blood cells or cultured skin fibroblasts. Prenatal diagnosis is possible by measurement of sphingomyelinase activity or the neonatal screening panel based on the gene sequencing technic on uncultured or cultured chorionic villus sampling, or cultured amniocytes [114].

Liver transplantation is more efficient to NPB patients with severe liver and pulmonary dysfunction than NPA [115]. Meanwhile, enzyme replacement therapy, especially in NPB patients, has been tested worked in clinical trials [116]. Although nowadays no curable treatment is available for NPA, recent study [117] has revealed that cerebellomedullary cistern injection of adeno-associated viral vector serotype 9 encoding human ASM can effectively recover the ASM activity in ASM knockout mice, which indicate possibility of genetic therapy in clinical trials in NPA as well as other lysosomal storage brain disorders. Further clinic studies using enzyme replacement therapy or gene therapy might be promising in the foreseeable future.

11.6.2 Calvarial Doughnut Lesions with Bone Fragility (CDL)

CDL (OMIM# 126550), also known as familial doughnut lesions of skull, is a rare autosomal dominant disease and occurs in less than 1/1000,000 of the population [118]. Till now, 9 familial cases and 29 sporadic cases have been reported worldwide [118–123].

Pathogenic variants of sphingomyelin synthase 2 (SGMS2 OMIM* 611574) is the main cause of CDL, which located in chromosome 4, encoding a key terminal limiting enzyme in control of sphingomyelin synthesis, presenting plasma membrane-bound sphingomyelin metabolism dysfunction in skeletal homeostasis [118] and recurrent facial nerve palsy [119].

Several characteristics such as bone mass, bone microarchitecture are negatively affected by the SGMS2 mutations [124]. The main characteristic phenotypes are low bone mineral density, primary sclerotic doughnut lesions in distinctive X-ray translucencies of the skull, increased spinal and peripheral fractures, carious teeth, and other related phenotypes [118, 122]. However, different from patients' phenotypes, SGMS2 defective mice were protected from insulin resistance in diet-induced obesity, but no evidence of any overt bone defect [125, 126], which indicates that a bone phenotype in SMS2 knockout mice may have been overlooked. Neurotoxicity induced by aberrant sphingomyelin metabolism is similar to neuronal damage in ASM deficiency NPD [127].

Bisphosphonate treatment in patients with SGMS2 mutation brought notable improvement in back pain and quality of life [124], but long-term effects still remain unclear. Molecular and biochemical mechanism exploration may identify novel therapeutic targets to enhance bone strength. Moreover, S1P, generated through ceramide deacylation as first step and sphingosine kinases phosphorylation as second step, is recognized as a fundamental role in bone metabolism, especially in coupling osteoblasts and osteoclasts [128]. Thus, S1P lyase is likely to be a potential target for osteoporosis therapy [129].

References

1. Goldstein JL, Brown MS (1973) Familial hypercholesterolemia: identification of a defect in the regulation of 3-hydroxy-3-methylglutaryl coenzyme A reductase activity associated with overproduction of cholesterol. Proc Natl Acad Sci U S A 70:2804–2808
2. Brown MS, Dana SE, Goldstein JL (1974) Regulation of 3-hydroxy-3-methylglutaryl coenzyme A reductase activity in cultured human fibroblasts. Comparison of cells from a normal subject and from a patient with homozygous familial hypercholesterolemia. J Biol Chem 249:789–796
3. Goldstein JL, Brown MS (2009) The LDL receptor. Arterioscler Thromb Vasc Biol 29:431–438
4. Bhatnagar D, Soran H, Durrington PN (2008) Hypercholesterolaemia and its management. BMJ 337:a993
5. Ahangari N, Ghayour Mobarhan M, Sahebkar A, Pasdar A (2018) Molecular aspects of hypercholesterolemia treatment: current perspectives and hopes. Ann Med 50:303–311
6. Rader DJ, Cohen J, Hobbs HH (2003) Monogenic hypercholesterolemia: new insights in pathogenesis and treatment. J Clin Invest 111:1795–1803
7. Sniderman AD, Tsimikas S, Fazio S (2014) The severe hypercholesterolemia phenotype: clinical

diagnosis, management, and emerging therapies. J Am Coll Cardiol 63:1935–1947
8. Berberich AJ, Hegele RA (2019) The complex molecular genetics of familial hypercholesterolaemia. Nat Rev Cardiol 16:9–20
9. Akioyamen LE et al (2017) Estimating the prevalence of heterozygous familial hypercholesterolaemia: a systematic review and meta-analysis. BMJ Open 7:e016461
10. Benn M, Watts GF, Tybjærg-Hansen A, Nordestgaard BG (2016) Mutations causative of familial hypercholesterolaemia: screening of 98 098 individuals from the Copenhagen General Population Study estimated a prevalence of 1 in 217. Eur Heart J 37:1384–1394
11. Salen G, Steiner RD (2017) Epidemiology, diagnosis, and treatment of cerebrotendinous xanthomatosis (CTX). J Inherit Metab Dis 40:771–781
12. Sharifi M, Futema M, Nair D, Humphries SE (2017) Genetic architecture of familial Hypercholesterolaemia. Curr Cardiol Rep 19:44
13. Tada H et al (2018) Sitosterolemia, hypercholesterolemia, and coronary artery disease. J Atheroscler Thromb 25:783–789
14. Gu H, Zhang D (2015) Hypercholesterolemia, low density lipoprotein receptor and proprotein convertase subtilisin/kexin-type 9. J Biomed Res 29:356–361
15. Fellin R et al (2003) Clinical and biochemical characterisation of patients with autosomal recessive hypercholesterolemia (ARH). Nutr Metab Cardiovasc Dis 13:278–286
16. Pullinger CR et al (2002) Human cholesterol 7alpha-hydroxylase (CYP7A1) deficiency has a hypercholesterolemic phenotype. J Clin Invest 110:109–117
17. Reiner Ž et al (2014) Lysosomal acid lipase deficiency--an under-recognized cause of dyslipidaemia and liver dysfunction. Atherosclerosis 235:21–30
18. Desai NK, Wilson DP (2000) Lysosomal acid lipase deficiency. In: Feingold KR et al (eds) Endotext. MDText.com, Inc, Dartmouth
19. Henneman P et al (2009) The expression of type III hyperlipoproteinemia: involvement of lipolysis genes. Eur J Hum Genet 17:620–628
20. Sakuma N et al (1995) Coexisting type III hyperlipoproteinemia and familial hypercholesterolemia: a case report. Metab Clin Exp 44:460–465
21. Freeman LA, Remaley AT (2016) Chapter 6 – discovery of high-density lipoprotein gene targets from classical genetics to Genome-Wide Association Studies. In: Rodriguez-Oquendo A (ed) Translational cardiometabolic genomic medicine. Academic, pp 119–159. https://doi.org/10.1016/B978-0-12-799961-6.00006-8
22. Nie S, Chen G, Cao X, Zhang Y (2014) Cerebrotendinous xanthomatosis: a comprehensive review of pathogenesis, clinical manifestations, diagnosis, and management. Orphanet J Rare Dis 9:179
23. Defesche JC et al (2017) Familial hypercholesterolaemia. Nat Rev Dis Primers 3:17093
24. Abifadel M et al (2003) Mutations in PCSK9 cause autosomal dominant hypercholesterolemia. Nat Genet 34:154–156
25. Young SG (1990) Recent progress in understanding apolipoprotein B. Circulation 82:1574–1594
26. Norata GD, Tibolla G, Catapano AL (2014) Targeting PCSK9 for hypercholesterolemia. Annu Rev Pharmacol Toxicol 54:273–293
27. Fouchier SW, Defesche JC (2013) Lysosomal acid lipase A and the hypercholesterolaemic phenotype. Curr Opin Lipidol 24:332–338
28. Poinsot P et al (2017) Childhood/adult-onset lysosomal acid lipase deficiency: a serious metabolic and vascular phenotype beyond liver disease-four new pediatric cases. J Clin Lipidol 11:167–177.e3
29. Vinje T, Wierød L, Leren TP, Strøm TB (2018) Prevalence of cholesteryl ester storage disease among hypercholesterolemic subjects and functional characterization of mutations in the lysosomal acid lipase gene. Mol Genet Metab 123:169–176
30. Cholesteryl Ester Storage Disease – NORD (National Organization for Rare Disorders). https://rarediseases.org/rare-diseases/cholesteryl-ester-storage-disease/
31. Freudenberg F, Bufler P, Ensenauer R, Lohse P, Koletzko S (2013) Cholesteryl ester storage disease: an easily missed diagnosis in oligosymptomatic children. Z Gastroenterol 51:1184–1187
32. Qayyum F et al (2018) Genetic variants in CYP7A1 and risk of myocardial infarction and symptomatic gallstone disease. Eur Heart J 39:2106–2116
33. Beigneux A, Hofmann AF, Young SG (2002) Human CYP7A1 deficiency: progress and enigmas. J Clin Invest 110:29–31
34. Moutzouri E, Elisaf M, Liberopoulos EN (2011) Hypocholesterolemia. Curr Vasc Pharmacol 9:200–212
35. Zhao Z et al (2006) Molecular characterization of loss-of-function mutations in PCSK9 and identification of a compound heterozygote. Am J Hum Genet 79:514–523
36. Burnett JR, Bell DA, Hooper AJ, Hegele RA (2015) Clinical utility gene card for: familial hypobetalipoproteinaemia (APOB) – update 2014. Eur J Hum Genet 23
37. Hooper AJ, van Bockxmeer FM, Burnett JR (2005) Monogenic hypocholesterolaemic lipid disorders and apolipoprotein B metabolism. Crit Rev Clin Lab Sci 42:515–545
38. Zamel R, Khan R, Pollex RL, Hegele RA (2008) Abetalipoproteinemia: two case reports and literature review. Orphanet J Rare Dis 3:19
39. Junaid Z, Patel K (2020) Abetalipoproteinemia. In: StatPearls. StatPearls Publishing, Treasure Island
40. Minicocci I et al (2012) Mutations in the *ANGPTL3* gene and familial combined hypolipidemia: a clinical

and biochemical characterization. J Clin Endocrinol Metabol 97:E1266–E1275
41. Genest J (2017) ANGPTL3: a gene, a protein, a new target? Aye, there's the rub! J Am Coll Cardiol 69:2064–2066
42. Peretti N et al (2010) Guidelines for the diagnosis and management of chylomicron retention disease based on a review of the literature and the experience of two centers. Orphanet J Rare Dis 5:24
43. Levy E (2015) Insights from human congenital disorders of intestinal lipid metabolism. J Lipid Res 56:945–962
44. Nowaczyk MJM, Irons MB (2012) Smith-Lemli-Opitz syndrome: phenotype, natural history, and epidemiology. Am J Med Genet C Semin Med Genet 160C:250–262
45. Cohen JC et al (2006) Multiple rare variants in NPC1L1 associated with reduced sterol absorption and plasma low-density lipoprotein levels. Proc Natl Acad Sci U S A 103:1810–1815
46. Myocardial Infarction Genetics Consortium Investigators et al (2014) Inactivating mutations in NPC1L1 and protection from coronary heart disease. N Engl J Med 371:2072–2082
47. Zhang Y-Y et al (2018) A LIMA1 variant promotes low plasma LDL cholesterol and decreases intestinal cholesterol absorption. Science 360:1087–1092
48. Smith DW, Lemli L, Opitz JM (1964) A newly recognized syndrome of multiple congenital anomalies. J Pediatr 64:210–217
49. Porter FD (2008) Smith-Lemli-Opitz syndrome: pathogenesis, diagnosis and management. Eur J Hum Genet 16:535–541
50. Ryan AK et al (1998) Smith-Lemli-Opitz syndrome: a variable clinical and biochemical phenotype. J Med Genet 35:558–565
51. Weaver DD, Solomon BD, Akin-Samson K, Kelley RI, Muenke M (2010) Cyclopia (synophthalmia) in Smith-Lemli-Opitz syndrome: first reported case and consideration of mechanism. Am J Med Genet C Semin Med Genet 154C:142–145
52. Lin AE, Ardinger HH, Ardinger RH, Cunniff C, Kelley RI (1997) Cardiovascular malformations in Smith-Lemli-Opitz syndrome. Am J Med Genet 68:270–278
53. Svoboda MD, Christie JM, Eroglu Y, Freeman KA, Steiner RD (2012) Treatment of Smith-Lemli-Opitz syndrome and other sterol disorders. Am J Med Genet C Semin Med Genet 160C:285–294
54. Wassif CA et al (2017) A placebo-controlled trial of simvastatin therapy in Smith-Lemli-Opitz syndrome. Genet Med 19:297–305
55. Haas D et al (2007) Effects of cholesterol and simvastatin treatment in patients with Smith-Lemli-Opitz syndrome (SLOS). J Inherit Metab Dis 30:375–387
56. Jira PE et al (2000) Simvastatin. A new therapeutic approach for Smith-Lemli-Opitz syndrome. J Lipid Res 41:1339–1346
57. Schaefer EJ, Anthanont P, Diffenderfer MR, Polisecki E, Asztalos BF (2016) Diagnosis and treatment of high density lipoprotein deficiency. Prog Cardiovasc Dis 59:97–106
58. Dioguardi N (1983) Familial hypoalphalipoproteinemia. Vergani's disease. Minerva Med 74:2659–2664
59. Mahapatra HS, Ramanarayanan S, Gupta A, Bhardwaj M (2015) Co-existence of classic familial lecithin-cholesterol acyl transferase deficiency and fish eye disease in the same family. Indian J Nephrol 25:362–365
60. Balwani MR et al (2017) An unusual presentation of LCAT deficiency as nephrotic syndrome with normal serum HDL-C level. J Nephropharmacol 6:23–26
61. Tobar HE et al (2019) Identification and functional analysis of missense mutations in the lecithin cholesterol acyltransferase gene in a Chilean patient with hypoalphalipoproteinemia. Lipids Health Dis 18
62. Fountoulakis N et al (2019) The P274S mutation of lecithin-cholesterol acyltransferase (LCAT) and its clinical manifestations in a large kindred. Am J Kidney Dis 74:510–522
63. Oliaei F, Batebi B, Tabaripour R, Akhavan Niaki H (2018) Finding a very rare mutation in non-Caucasian LCAT patients from Southwest Asia for the first time. J Cell Biochem. https://doi.org/10.1002/jcb.27981
64. Rosset J, Wang J, Wolfe BM, Dolphin PJ, Hegele RA (2001) Lecithin: cholesterol acyl transferase G30S: association with atherosclerosis, hypoalphalipoproteinemia and reduced in vivo enzyme activity. Clin Biochem 34:381–386
65. Lucca F, Ossoli A, Boscutti G, Franceschini G, Calabresi L (2016) Lecithin: cholesterol acyltransferase deficiency, from genes to therapy. G Ital Nefrol 33
66. Shamburek RD et al (2016) Familial lecithin: cholesterol acyltransferase deficiency: first-in-human treatment with enzyme replacement. J Clin Lipidol 10:356–367
67. Burnett JR, Hooper AJ, McCormick SP, Hegele RA (1993) Tangier disease. In: Adam MP et al (eds) GeneReviews®. University of Washington, Seattle
68. Bodzioch M et al (1999) The gene encoding ATP-binding cassette transporter 1 is mutated in Tangier disease. Nat Genet 22:347–351
69. Puntoni M, Sbrana F, Bigazzi F, Sampietro T (2012) Tangier disease: epidemiology, pathophysiology, and management. Am J Cardiovasc Drugs 12:303–311
70. Rust S et al (1999) Tangier disease is caused by mutations in the gene encoding ATP-binding cassette transporter 1. Nat Genet 22:352–355
71. Mercan M, Yayla V, Altinay S, Seyhan S (2018) Peripheral neuropathy in Tangier disease: a literature review and assessment. J Peripher Nerv Syst 23:88–98
72. Oram JF (2002) Molecular basis of cholesterol homeostasis: lessons from Tangier disease and ABCA1. Trends Mol Med 8:168–173

73. Inazu A et al (1990) Increased high-density lipoprotein levels caused by a common cholesteryl-ester transfer protein gene mutation. N Engl J Med 323:1234–1238
74. Koizumi J et al (1985) Deficiency of serum cholesteryl-ester transfer activity in patients with familial hyperalphalipoproteinaemia. Atherosclerosis 58:175–186
75. Brown ML et al (1989) Molecular basis of lipid transfer protein deficiency in a family with increased high-density lipoproteins. Nature 342:448–451
76. Nagano M et al (2001) Point mutation (-69 G-->A) in the promoter region of cholesteryl ester transfer protein gene in Japanese hyperalphalipoproteinemic subjects. Arterioscler Thromb Vasc Biol 21:985–990
77. Assmann G, von Eckardstein A, Funke H (1993) High density lipoproteins, reverse transport of cholesterol, and coronary artery disease. Insights from mutations. Circulation 87:III28–III34
78. Hirano K et al (1995) Atherosclerotic disease in marked hyperalphalipoproteinemia. Combined reduction of cholesteryl ester transfer protein and hepatic triglyceride lipase. Arterioscler Thromb Vasc Biol 15:1849–1856
79. Oliveira HC et al (1997) Cholesteryl ester transfer protein activity enhances plasma cholesteryl ester formation. Studies in CETP transgenic mice and human genetic CETP deficiency. Arterioscler Thromb Vasc Biol 17:1045–1052
80. Oka T et al (2002) Distribution of human plasma PLTP mass and activity in hypo- and hyperalphalipoproteinemia. J Lipid Res 43:1236–1243
81. von Eckardstein A et al (1991) Apolipoprotein C-III (Lys58----Glu). Identification of an apolipoprotein C-III variant in a family with hyperalphalipoproteinemia. J Clin Invest 87:1724–1731
82. Aouizerat BE et al (2006) Genetic variation of PLTP modulates lipoprotein profiles in hypoalphalipoproteinemia. J Lipid Res 47:787–793
83. Miettinen HE et al (1998) Molecular genetic study of Finns with hypoalphalipoproteinemia and hyperalphalipoproteinemia: a novel Gly230 Arg mutation (LCAT[Fin]) of lecithin: cholesterol acyltransferase (LCAT) accounts for 5% of cases with very low serum HDL cholesterol levels. Arterioscler Thromb Vasc Biol 18:591–598
84. Lee J, Hegele RA (2014) Abetalipoproteinemia and homozygous hypobetalipoproteinemia: a framework for diagnosis and management. J Inherit Metab Dis 37:333–339
85. Hussain MM, Rava P, Walsh M, Rana M, Iqbal J (2012) Multiple functions of microsomal triglyceride transfer protein. Nutr Metab (Lond) 9:14
86. Berriot-Varoqueaux N, Aggerbeck LP, Samson-Bouma M, Wetterau JR (2000) The role of the microsomal triglygeride transfer protein in abetalipoproteinemia. Annu Rev Nutr 20:663–697
87. Ohashi K et al (2000) Novel mutations in the microsomal triglyceride transfer protein gene causing abetalipoproteinemia. J Lipid Res 41:1199–1204
88. Schonfeld G, Lin X, Yue P (2005) Familial hypobetalipoproteinemia: genetics and metabolism. Cell Mol Life Sci 62:1372–1378
89. Tarugi P, Averna M (2011) Hypobetalipoproteinemia: genetics, biochemistry, and clinical spectrum. Adv Clin Chem 54:81–107
90. Whitfield AJ, Barrett PHR, van Bockxmeer FM, Burnett JR (2004) Lipid disorders and mutations in the APOB gene. Clin Chem 50:1725–1732
91. Hooper AJ, Burnett JR (2013) Recent developments in the genetics of LDL deficiency. Curr Opin Lipidol 24:111–115
92. Musunuru K et al (2010) Exome sequencing, ANGPTL3 mutations, and familial combined hypolipidemia. N Engl J Med 363:2220–2227
93. Nakajima K et al (2010) Association of angiopoietin-like protein 3 with hepatic triglyceride lipase and lipoprotein lipase activities in human plasma. Ann Clin Biochem 47:423–431
94. Shan L et al (2009) The angiopoietin-like proteins ANGPTL3 and ANGPTL4 inhibit lipoprotein lipase activity through distinct mechanisms. J Biol Chem 284:1419–1424
95. Jones B et al (2003) Mutations in a Sar1 GTPase of COPII vesicles are associated with lipid absorption disorders. Nat Genet 34:29–31
96. Siddiqi SA, Gorelick FS, Mahan JT, Mansbach CM (2003) COPII proteins are required for Golgi fusion but not for endoplasmic reticulum budding of the pre-chylomicron transport vesicle. J Cell Sci 116:415–427
97. Lambert G, Sjouke B, Choque B, Kastelein JJP, Hovingh GK (2012) The PCSK9 decade. J Lipid Res 53:2515–2524
98. Cariou B et al (2009) PCSK9 dominant negative mutant results in increased LDL catabolic rate and familial hypobetalipoproteinemia. Arterioscler Thromb Vasc Biol 29:2191–2197
99. Cefalù AB et al (2013) A novel APOB mutation identified by exome sequencing cosegregates with steatosis, liver cancer, and hypocholesterolemia. Arterioscler Thromb Vasc Biol 33:2021–2025
100. Di Filippo M et al (2014) Homozygous MTTP and APOB mutations may lead to hepatic steatosis and fibrosis despite metabolic differences in congenital hypocholesterolemia. J Hepatol 61:891–902
101. Minicocci I et al (2013) Functional and morphological vascular changes in subjects with familial combined hypolipidemia: an exploratory analysis. Int J Cardiol 168:4375–4378
102. Ferreira CR, Gahl WA (2017) Lysosomal storage diseases. TRD 2:1–71
103. Kolter T, Sandhoff K (2006) Sphingolipid metabolism diseases. Biochim Biophys Acta Biomembr 1758:2057–2079

104. Meikle PJ (1999) Prevalence of lysosomal storage disorders. JAMA 281:249
105. Sphingolipids, sphingolipidoses and allied disorders: Proceedings of the symposium on sphingolipidoses and allied disorders held in Brooklyn, New York, October 25–27, 1971, vol 19. Springer US (1972)
106. McCafferty EH, Scott LJ (2019) Migalastat: a review in Fabry disease. Drugs 79:543–554
107. Heybrock S et al (2019) Lysosomal integral membrane protein-2 (LIMP-2/SCARB2) is involved in lysosomal cholesterol export. Nat Commun 10:3521
108. Kolter T, Sandhoff K (2005) Principles of lysosomal membrane digestion: stimulation of sphingolipid degradation by sphingolipid activator proteins and anionic lysosomal lipids. Annu Rev Cell Dev Biol 21:81–103
109. Crocker AC, Mays VB (1961) Sphingomyelin synthesis in Niemann-Pick disease. Am J Clin Nutr 9:63–67
110. Jenkins RW, Canals D, Hannun YA (2009) Roles and regulation of secretory and lysosomal acid sphingomyelinase. Cell Signal 21:836–846
111. Schuchman EH, Wasserstein MP (2015) Types A and B Niemann-Pick disease. Best Pract Res Clin Endocrinol Metab 29:237–247
112. Baskfield A et al (2019) An induced pluripotent stem cell line (TRNDi009-C) from a Niemann-Pick disease type A patient carrying a heterozygous p.L302P (c.905 T>C) mutation in the SMPD1 gene. Stem Cell Res 38:101461
113. Wasserstein MP, Schuchman EH (1993) Acid sphingomyelinase deficiency. In: Adam MP et al (eds) GeneReviews®. University of Washington, Seattle
114. Irun P et al (2013) Identification of seven novel *SMPD1* mutations causing Niemann-Pick disease types A and B: identification of mutations causing NPA and NPB. Clin Genet 84:356–361
115. Liu Y et al (2019) The effects of liver transplantation in children with Niemann-Pick Disease Type B. Liver Transpl lt.25457. https://doi.org/10.1002/lt.25457
116. Schuchman EH, Desnick RJ (2017) Types A and B Niemann-Pick disease. Mol Genet Metab 120:27–33
117. Samaranch L et al (2019) Adeno-associated viral vector serotype 9–based gene therapy for Niemann-Pick disease type A. Sci Transl Med 11:eaat3738
118. Pekkinen M et al (2019) Osteoporosis and skeletal dysplasia caused by pathogenic variants in SGMS2. JCI Insight 4:e126180
119. Bartlett JE, Kishore PRS (1976) Familial "doughnut" lesions of the skull: a benign, hereditary dysplasia. Radiology 119:385–387
120. Baumgartner D et al (2001) Calvarial 'doughnut lesions': clinical spectrum of the syndrome, report on a case, and review of the literature. Am J Med Genet 99:238–243
121. Colavita N, Kozlowski K, Vecchia GL, Fileni A, Ricci R (1984) Calvarial doughnut lesions with osteoporosis, multiple fractures, dentinogenesis imperfecta and tumorous changes in the jaws (Report of a case). Australas Radiol 28:226–231
122. Jaakkola E et al (2009) Calvarial doughnut lesions and osteoporosis: a new three-generation family and review. Am J Med Genet 149A:2371–2377
123. Nishimura G et al (1996) Fragile bone syndrome associated with craniognathic fibro-osseous lesions and abnormal modeling of the tubular bones: report of two cases and review of the literature. Skelet Radiol 25:717–722
124. Wagermaier W, Klaushofer K, Fratzl P (2015) Fragility of bone material controlled by internal interfaces. Calcif Tissue Int 97:201–212
125. Li Z et al (2011) Reducing plasma membrane sphingomyelin increases insulin sensitivity. Mol Cell Biol 31:4205–4218
126. Mitsutake S et al (2011) Dynamic modification of sphingomyelin in lipid microdomains controls development of obesity, fatty liver, and type 2 diabetes. J Biol Chem 286:28544–28555
127. McGovern MM, Avetisyan R, Sanson B-J, Lidove O (2017) Disease manifestations and burden of illness in patients with acid sphingomyelinase deficiency (ASMD). Orphanet J Rare Dis 12:41
128. Maceyka M, Harikumar KB, Milstien S, Spiegel S (2012) Sphingosine-1-phosphate signaling and its role in disease. Trends Cell Biol 22:50–60
129. Weske S et al (2018) Targeting sphingosine-1-phosphate lyase as an anabolic therapy for bone loss. Nat Med 24:667–678

Preβ1-High-Density Lipoprotein in Cardiovascular Diseases

12

Yunqin Chen and Jibin Dong

Abstract

Serum preβ1-high-density lipoprotein (preβ1-HDL) was defined by two-dimensional non-denaturing linear gel electrophoresis and apolipoprotein A-I immunoblotting. However, there are still debatable questions for its quantification and coronary artery disease (CAD) relevance. We have established a novel and simple method for human serum preβ1-HDL quantification. We found that human lower preβ1-HDL is an independent predictor for severer coronary artery stenosis. In this chapter, we will discuss all these.

Keywords

Preβ1-HDL · Preβ1-HDL measurement · HDL · LDL · Cardiovascular diseases

12.1 Regulation of Blood Lipids and Residual Cardiovascular Risks

Atherosclerotic cardiovascular disease is a serious threat to human health [1], but coronary heart disease can be prevented and controlled by regulating blood lipids with basic strategies of reducing the pathogenic effect of risk factors and enhancing the anti-atherosclerotic effect of protective factors. For the primary risk factor of low-density lipoprotein cholesterol (LDL-C), statins and PCSK9 inhibitors can significantly reduce LDL-C levels and stabilize plaque and the incidence of cardiovascular events [2]. However, even if LDL-C is reduced, it is difficult to completely inhibit the progress of atherosclerotic plaque lesions, and the residual risk of cardiovascular disease remains a challenging problem worldwide currently [3].

As we know, reverse cholesterol transport (RCT) is the key mechanism for plaque regression. As a protective factor against atherosclerosis, high-density lipoprotein (HDL) transports excess cholesterol from peripheral cells to the liver for metabolism [4]. Unfortunately, all previous clinical trials of drugs that attempted to increase HDL levels, including cholesteryl ester transfer protein (CETP) inhibitors, did not achieve the expected beneficial results [5]. CETP inhibitors, such as evacetrapib, can significantly increase large HDL particles but fail to effectively reduce cardiovascular endpoint events [6]. Therefore, it is necessary to reconsider the composition and function of HDL [7, 8]. As an extracellular receptor, small particles of HDL, especially nascent HDL (nHDL), promote cellular cholesterol efflux [9], which is an important focus of research in this field.

Y. Chen (✉)
Department of Cardiology, Zhongshan Hospital, Fudan University, Shanghai, China

J. Dong
School of Pharmacy, Fudan University, Shanghai, China
e-mail: jbdong@shmu.edu

X.-C. Jiang (ed.), *Lipid Transfer in Lipoprotein Metabolism and Cardiovascular Disease*, Advances in Experimental Medicine and Biology 1276, https://doi.org/10.1007/978-981-15-6082-8_12

12.2 Particle Characteristics of nHDL

Owing to differences in the physical and chemical properties of their constituents, HDL particles possess a complex heterogeneity [10]. Free apolipoprotein A-I (apo A-I) receives cholesterol from cells and initially produces nHDL [11]. This type of particle has the smallest diameter of 5 nm. Because the particle shows preβ1 mobility during gel electrophoresis, it is also known as preβ1-HDL. The preβ1-HDL mainly comprises apo A-I and trace lipids (phospholipids and free cholesterol) at an extremely high density, reaching 1.210 g/mL; in fact this preβ1-HDL particle may be defined as a kind of very high-density lipoprotein (VHDL) [12]. Notably, the lipoprotein particles comprise apolipoprotein and lipid, with a hydrophilic phospholipid-coated surface and a hydrophobic core of lipids such as cholesterol esters and triglycerides. apo A-I exists mainly in the form of lipoprotein particles with lipids, as well as in lipid-free forms, including precursors [13], monomers, and polymers [14]. Although lipid-free apo A-I and preβ1-HDL both exhibit electrophoretic characteristics of preβ1 migration, there is a fundamental conceptual difference between them.

12.3 Detection Methods for nHDL

Based on the physical and chemical characteristics of the particles, the main separation techniques for preβ1-HDL include gel electrophoresis, ultrafiltration, and chromatography. The quantitative methods mainly rely on protein immunity, but quantitative methods of lipid staining have also been reported recently.

The first detection method is protein immunoassay. In 1985, Kunitake [15] obtained lipoproteins containing apo A-I through immunoadsorption. After separation using agarose gel electrophoresis, he first reported that the lipoproteins containing apo A-I exhibited preβ electrophoretic migration characteristics. In 1988, Castro [16] established a method to separation and quantitative detection of apo A-I-containing particles using two-dimensional gel electrophoresis and Western blotting and reported that this two-dimensional polyacrylamide gel electrophoresis could distinguish lipid-free apo A-I (possibly a precursor or monomer) from preβ1-HDL. In 1993, Asztalos [17] improved the two-dimensional electrophoretic separation and used a radioactive 125I-antibody to quantify preβ1-HDL, which has since been widely used for a long time (Boston Heart HDL Map®). In addition, other methods that take advantage of the difference in the particle size of the lipoproteins have been reported in the literature, including ultrafiltration (membrane pore size 100 kDa) [18] and high-performance size-exclusion chromatography [19] for separation and purification of small particles of lipoproteins containing apo A-I (preβ1-HDL or Sm Lp-AI) and quantification of lipoprotein composition by apo A-I antibody immunoassay. In 2000, Miyazaki [20] reported the use of a specific apo A-I monoclonal antibody (MAb55201) to detect plasma preβ1-HDL via ELISA. But in another study in 2014 [21], he found no difference between the electrophoretic migration of preβ1-HDL purified from plasma with MAb55201 and lipid-free apo A-I. Based on the observation that no phospholipid or cholesterol was detected by chemical analysis, this study confirmed that the preβ1-HDL detected in the plasma by immunoblotting was a lipid-free apo A-I monomer. Therefore, to avoid interference from lipid-free apo A-I, the prerequisite for quantitative assessment of preβ1-HDL by protein immunoassay is to completely distinguish the lipid-free apo A-I (including monomers and polymers) or selective recognition through specific antibodies.

The second detection method is lipid staining. Recognition of lipoproteins by lipophilic dyes such as Sudan Black B can completely avoid the interference of lipid-free apolipoprotein. In a previous study, we built the MEDLiPO system and set three layers of polyacrylamide gels at concentrations of 3.0%, 3.6%, and 7.0% as the electrophoretic medium. The serum lipoproteins were prestained by Sudan Black B for electrophoretic separation, and the HDL with the fastest electrophoretic migration showed relatively an isolated staining band [12]. After analyzing the particle size, density, charge, and chemical composition, it was proved that this isolated fast-moving lipoprotein was the so-called preβ1-HDL. In the study, Sudan Black B was dissolved in a mixed solvent of isopropyl alcohol and ethylene glycol in a volume ratio of 4:1, which improved the stability of lipoprotein staining. Using BeneScan-1000 scanner customized by BENEFI and MICROTEK Technology Co., Ltd (Shanghai, China), the separated gel images and gray scale of the lipid staining were acquired and quantitatively analyzed by measuring the optical density. Then, the absolute content and percentage of total lipid staining of preβ1-HDL were quantified. After repeated experiments, the results showed that the intra- and inter-assay coefficients of variation of serum preβ1-HDL were <5%. The MEDLiPO system was easy to operate and could meet the actual requirements of clinical testing with a unique performance.

12.4 nHDL and Coronary Heart Disease

Most clinical studies have reported an increase in plasma preβ1-HDL levels in patients with coronary heart disease, with a significantly positive correlation [22]. Guey et al. [23] reported that preβ1-HDL was an independent predictor of myocardial infarction. Sethi et al. [24] showed that preβ1-HDL levels in patients with ischemic heart disease (IHD) were twice as high as those in the control group, and the high preβ1- HDL and low activity of lecithin cholesterol acyltransferase (LCAT) were considered risk factors for IHD, independent of HDL-C. Because preβ1-HDL exerts a protective effect by promoting cholesterol efflux from peripheral cells, the main reason for the increase in preβ1-HDL content in patients with coronary heart disease is that the accumulated preβ1-HDL cannot be converted into large particles of mature HDL, leading to reverse cholesterol transport disorders. Under this circumstance, preβ1-HDL accumulates and increases in patients with Tangier disease [25]. The pathogenesis of Tangier disease is owing to a defect in the ATP-binding cassette (ABC) transporter gene *ABCA1* and impaired efflux of cellular cholesterol. Further, ABCA1 deficiency results in the inability of lipid-free apo A-I to receive cholesterol from cells, and theoretically, no preβ1-HDL is formed in this situation. However, immunoquantitative testing showed a controversial result. The MEDLiPO system demonstrated that preβ1-HDL and HDL were missing in the samples obtained from patients with Tangier disease [12]. Very few studies have reported a decrease in plasma preβ-HDL [26] or preβ1-HDL levels in patients with coronary heart disease [12, 27]. The reasons are more related to the detection methods and the differences observed in the included cases.

Using the MEDLiPO system can effectively avoid the interference of lipid-free apo A-I and accurately detect preβ1-HDL by quantitative determination of lipid staining. In 2016, we reported that the MEDLiPO system detected a decrease in serum preβ1- HDL levels in patients with coronary heart disease and the decrease was independently negatively correlated to the degree of coronary stenosis [12]. At the beginning of 2018, we fortunately obtained some blood samples from the ACCENTUATE clinical trial [28]. And using the MEDLiPO system, we found that plasma preβ1-HDL was significantly reduced after treatment with the CETP inhibitor evacetrapib [29]. These results were completely contrary to those of previous reports [30]. The preβ1-HDL reduction could give a clue to understand the failure of CETP inhibitors on cardiovascular outcomes.

12.5 nHDL Particle Reconstruction and Hypothesis

The ABC transporter family mediates free cholesterol efflux from cells [31]. As an acceptor, activated lipid-free apo A-I accepts cell membrane phospholipids and free cholesterol by ABCA1 to form a nascent type of preβ1-HDL particle. This process is the initiation of the reverse cholesterol transport mechanism. ABCA1-dependent cellular cholesterol efflux is

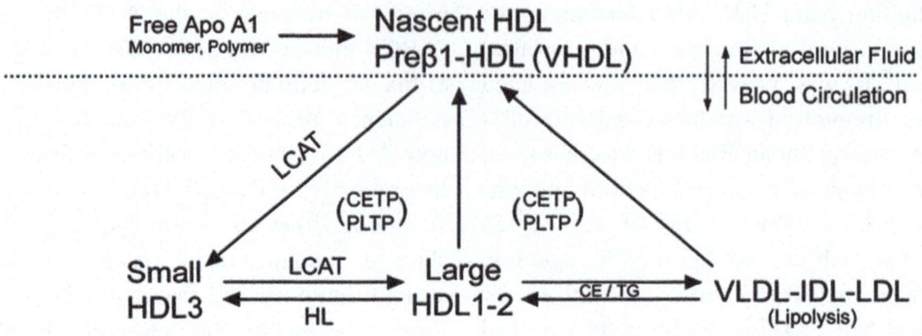

Fig. 12.1 *Generation and reconstruction of nHDL particles.* Extracellular activated apo A-I receives phospholipids and free cholesterol under ABCA1-mediated production of preβ1-HDL. The reconstruction of preβ1-HDL in blood circulation is in a state of dynamic equilibrium. The tissue barrier results in a difference in the composition and metabolism of extracellular fluid and plasma lipoproteins

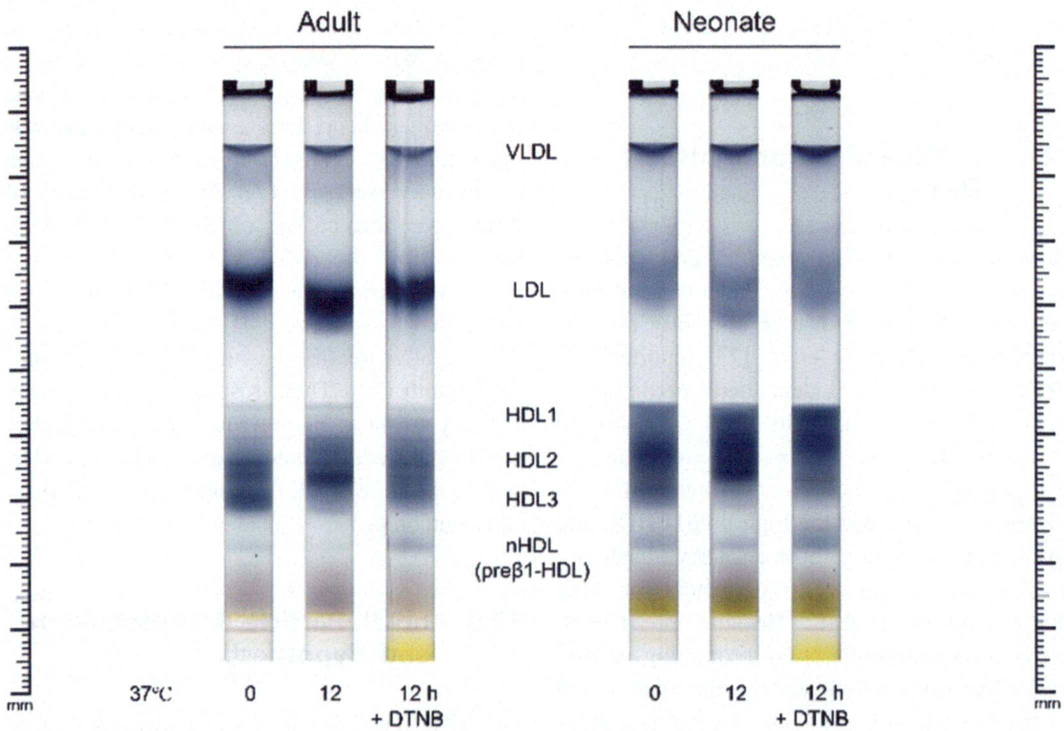

Fig. 12.2 *MEDLiPO system for lipoprotein detection.* Gel separation and quantification of blood lipoproteins through staining with Sudan Black B. 2-nitrobenzoic acid (DTNB), an LCAT inhibitor, inhibits the conversion of nHDL to mature HDL. The production amount of preβ1-HDL was calculated using preβ1-HDL content inhibited by a 37°C water bath for 12 h + DTNB minus the basic value (0 h). The amount of conversion is the value of the production amount plus the net content (12 h). The amount of serum preβ1-HDL produced in the water bath over the 12 h was less than the amount of transformation, and its net content decreased

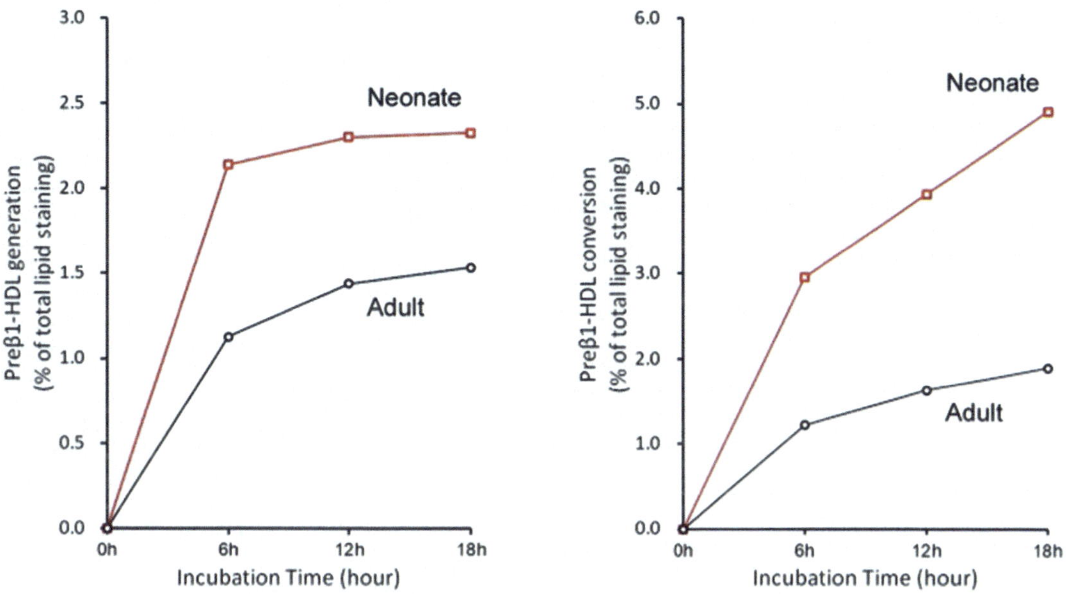

Fig. 12.3 *Reconstruction curve of nHDL particles.* The generation (left) and transformation (right) curves show that serum nHDL particle reconstruction is in a state of dynamic equilibrium. After the water bath experiments, the production and conversion rates of preβ1-HDL were calculated from the base values

a key mechanism by which HDL resists atherogenesis and reverses plaque. If the extracellular apo A-I or cell membrane ABCA1 is mutated or modified, it will cause dysfunction of the cholesterol efflux from the cell and accumulation of the lipid-free apo A-I, making it difficult to produce preβ1-HDL. There is a dynamic balance between production and transformation of preβ1-HDL in plasma or serum (Fig. 12.1). LCAT promotes the esterification of free cholesterol, and HDL is transformed from small particles to large particles. Hepatic lipase catalyzes the hydrolysis of lipids, and large particles of HDL are converted to small particles. Both CETP and phospholipid transporter are involved in lipid transfer between lipoprotein particles and in particle remodeling.

Neonatal umbilical cord blood is rich in preβ1-HDL. In vitro water bath experiments result in the inhibition of LCAT, and the metabolic activity of preβ1-HDL particles remodeling could be measured by detecting the rate of change in the production and conversion of preβ1-HDL (Fig. 12.2). The preβ1-HDL content and metabolic activity in neonatal cord blood are about twice of those in adults (Fig. 12.3). We speculate that preβ1-HDL plays a key role in cholesterol reverse transport and plaque reversal and this protective effect may diminish with age. It has been proposed [32] that blood lipid levels in newborns may be an ideal target for lipid-lowering therapy in patients with coronary heart disease. Neonatal blood lipids are characterized by extremely low LDL-C levels (<1.0 mmol/L) which is lower than HDL-C [33], whereas preβ1-HDL levels as newborns are significantly higher than those in adults. At present, the combined application of statins and PCSK9 inhibitors can achieve extremely low LDL-C levels in most patients with coronary heart disease. Regulating preβ1-HDL levels in newborns and its function of promoting cholesterol efflux from cells may be a promising way to prevent atherosclerotic cardiovascular disease and reduce residual cardiovascular risk in the future.

References

1. Arnett DK, Blumenthal RS, Albert MA, Buroker AB, Goldberger ZD, Hahn EJ et al (2019) 2019 ACC/AHA guideline on the primary prevention of cardiovascular disease: executive summary: a report of the American College of Cardiology/American Heart Association task force on clinical practice guidelines. J Am Coll Cardiol 74(10):1376–1414
2. Sabatine MS, Giugliano RP, Keech AC, Honarpour N, Wiviott SD, Murphy SA et al (2017) Evolocumab and clinical outcomes in patients with cardiovascular disease. N Engl J Med 376(18):1713–1722
3. Fruchart JC, Sacks F, Hermans MP, Assmann G, Brown WV, Ceska R et al (2008) The residual risk reduction initiative: a call to action to reduce residual vascular risk in patients with dyslipidemia. Am J Cardiol 102(10Suppl):1K–34K
4. Rader DJ, Hovingh GK (2014) HDL and cardiovascular disease. Lancet 384(9943):618–625
5. Tall AR, Rader DJ (2018) Trials and tribulations of CETP inhibitors. Circ Res 122(1):106–112
6. Lincoff AM, Nicholls SJ, Riesmeyer JS, Barter PJ, Brewer HB, Fox KAA et al (2017) Evacetrapib and cardiovascular outcomes in high-risk vascular disease. N Engl J Med 376(20):1933–1942
7. Barter P, Genest J (2019) HDL cholesterol and ASCVD risk stratification: a debate. Atherosclerosis 283:7–12
8. Sacks FM, Jensen MK (2018) From high-density lipoprotein cholesterol to measurements of function: prospects for the development of tests for high-density lipoprotein functionality in cardiovascular disease. Arterioscler Thromb Vasc Biol 38(3):487–499
9. Gillard BK, Rosales C, Xu B, Gotto AM Jr, Pownall HJ (2018) Rethinking reverse cholesterol transport and dysfunctional high-density lipoproteins. J Clin Lipidol 12(4):849–856
10. Toth PP, Barter PJ, Rosenson RS, Boden WE, Chapman MJ, Cuchel M et al (2013) High-density lipoproteins: a consensus statement from the National Lipid Association. J Clin Lipidol 7(5):484–525
11. Pourmousa M, Song HD, He Y, Heinecke JW, Segrest JP, Pastor RW (2018) Tertiary structure of apolipoprotein A-I in nascent high-density lipoproteins. Proc Natl Acad Sci U S A 115(20):5163–5168
12. Chen Y, Dong J, Chen X, Jiang H, Bakillah A, Zhang X et al (2016) Human serum preβ1-high density lipoprotein levels are independently and negatively associated with coronary artery diseases. Nutr Metab (Lond) 13:36
13. Chau P, Nakamura Y, Fielding CJ, Fielding PE (2006) Mechanism of prebeta-HDL formation and activation. Biochemistry 45(12):3981–3987
14. Gursky O (2015) Structural stability and functional remodeling of high-density lipoproteins. FEBS Lett 589(19 Pt A):2627–2639
15. Kunitake ST, La Sala KJ, Kane JP (1985) Apolipoprotein A-I-containing lipoproteins with pre-beta electrophoretic mobility. J Lipid Res 26(5):549–555
16. Castro GR, Fielding CJ (1988) Early incorporation of cell-derived cholesterol into pre-beta-migrating high-density lipoprotein. Biochemistry 27(1):25–29
17. Asztalos BF, Sloop CH, Wong L, Roheim PS (1993) Two-dimensional electrophoresis of plasma lipoproteins: recognition of new apo A-I-containing subpopulations. Biochim Biophys Acta 1169 (3):291–300
18. O'Connor PM, Naya-Vigne JM, Duchateau PN, Ishida BY, Mazur M, Schoenhaus SA et al (1997) Measurement of prebeta-1 HDL in human plasma by an ultrafiltration-isotope dilution technique. Anal Biochem 251(2):234–240
19. Nanjee MN, Brinton EA (2000) Very small apolipoprotein A-I-containing particles from human plasma: isolation and quantification by high-performance size-exclusion chromatography. Clin Chem 46(2):207–223
20. Miyazaki O, Kobayashi J, Fukamachi I, Miida T, Bujo H, Saito Y (2000) A new sandwich enzyme immunoassay for measurement of plasma pre-beta1-HDL levels. J Lipid Res 41(12):2083–2088
21. Miyazaki O, Ogihara J, Fukamachi I, Kasumi T (2014) Evidence for the presence of lipid-free monomolecular apolipoprotein A-1 in plasma. J Lipid Res 55 (2):214–225
22. Kane JP, Malloy MJ (2012) Prebeta-1 HDL and coronary heart disease. Curr Opin Lipidol 23(4):367–371
23. Guey LT, Pullinger CR, Ishida BY, O'Connor PM, Zellner C, Francone OL et al (2011) Relation of increased prebeta-1 high-density lipoprotein levels to risk of coronary heart disease. Am J Cardiol 108 (3):360–366
24. Sethi AA, Sampson M, Warnick R, Muniz N, Vaisman B, Nordestgaard BG et al (2010) High pre-beta1 HDL concentrations and low lecithin: cholesterol acyltransferase activities are strong positive risk markers for ischemic heart disease and independent of HDL-cholesterol. Clin Chem 56(7):1128–1137
25. Nofer JR, Remaley AT (2005) Tangier disease: still more questions than answers. Cell Mol Life Sci 62 (19–20):2150–2160
26. Hattori H, Kujiraoka T, Egashira T, Saito E, Fujioka T, Takahashi S et al (2004) Association of coronary heart disease with pre-beta-HDL concentrations in Japanese men. Clin Chem 50(3):589–595
27. Shiu SW, Wong Y, Tan KC (2017) Pre-β1 HDL in type 2 diabetes mellitus. Atherosclerosis 263:24–28
28. Nicholls SJ, Ray KK, Ballantyne CM, Beacham LA, Miller DL, Ruotolo G, ACCENTUATE Investigators et al (2017) Comparative effects of cholesteryl ester transfer protein inhibition, statin or ezetimibe on lipid factors: the ACCENTUATE trial. Atherosclerosis 261:12–18
29. Chen Y, Dong J, Zhang X, Chen X, Wang L, Chen H et al (2019) Evacetrapib reduces preβ-1 HDL in

patients with atherosclerotic cardiovascular disease or diabetes. Atherosclerosis 285:147–152
30. Nicholls SJ, Ruotolo G, Brewer HB, Kane JP, Wang MD, Krueger KA et al (2015) Cholesterol efflux capacity and pre-beta-1 HDL concentrations are increased in dyslipidemic patients treated with evacetrapib. J Am Coll Cardiol 66(20):2201–2210
31. Tall AR, Yvan-Charvet L, Terasaka N, Pagler T, Wang N (2008) HDL, ABC transporters, and cholesterol efflux: implications for the treatment of atherosclerosis. Cell Metab 7(5):365–375
32. Ge J, Wang Z (2016) Lower the low-density lipoprotein cholesterol to the level when you born. Cardiol Plus 1:1–6
33. Chen H, Chen Y, Jin X, Zhang X, Zhou J, Chen B et al (2011) A survey on blood lipid levels among newborns and healthy inhabitants in urban Shanghai (2008–2009). J Clin Lipidol 5(5):380–386

CGI-58: Versatile Regulator of Intracellular Lipid Droplet Homeostasis

13

Liqing Yu, Yi Li, Alison Grisé, and Huan Wang

Abstract

Comparative gene identification-58 (CGI-58), also known as α/β-hydrolase domain-containing 5 (ABHD5), is a member of a large family of proteins containing an α/β-hydrolase-fold. CGI-58 is well-known as the co-activator of adipose triglyceride lipase (ATGL), which is a key enzyme initiating cytosolic lipid droplet lipolysis. Mutations in either the human CGI-58 or ATGL gene cause an autosomal recessive neutral lipid storage disease, characterized by the excessive accumulation of triglyceride (TAG)-rich lipid droplets in the cytoplasm of almost all cell types. CGI-58, however, has ATGL-independent functions. Distinct phenotypes associated with CGI-58 deficiency commonly include ichthyosis (scaly dry skin), nonalcoholic steatohepatitis, and hepatic fibrosis. Through regulated interactions with multiple protein families, CGI-58 controls many metabolic and signaling pathways, such as lipid and glucose metabolism, energy balance, insulin signaling, inflammatory responses, and thermogenesis. Recent studies have shown that CGI-58 regulates the pathogenesis of common metabolic diseases in a tissue-specific manner. Future studies are needed to molecularly define ATGL-independent functions of CGI-58, including the newly identified serine protease activity of CGI-58. Elucidation of these versatile functions of CGI-58 may uncover fundamental cellular processes governing lipid and energy homeostasis, which may help develop novel approaches that counter against obesity and its associated metabolic sequelae.

Keywords

CGI-58 · ATGL · Lipid droplet · Lipolysis

13.1 Introduction

The human *comparative gene identification-58 (CGI-58)* gene was identified through comparative gene identification studies using the *Caenorhabditis elegans* proteome and human expressed sequence tag (EST) nucleotide database [109]. Human *CGI-58* gene is located at the chromosome 3p21.33 locus, spanning about 32kb and producing several splice variants. The full-length human *CGI-58* cDNA is transcribed from seven exons and encodes a 349 amino acid

L. Yu (✉) · Y. Li · H. Wang
Division of Endocrinology, Diabetes and Nutrition, Department of Medicine, University of Maryland School of Medicine, Baltimore, MD, USA
e-mail: Lyu@som.umaryland.edu

A. Grisé
College of Computer, Math, and Natural Sciences, College of Behavioral and Social Sciences, University of Maryland, College Park, MD, USA

X.-C. Jiang (ed.), *Lipid Transfer in Lipoprotein Metabolism and Cardiovascular Disease*, Advances in Experimental Medicine and Biology 1276, https://doi.org/10.1007/978-981-15-6082-8_13

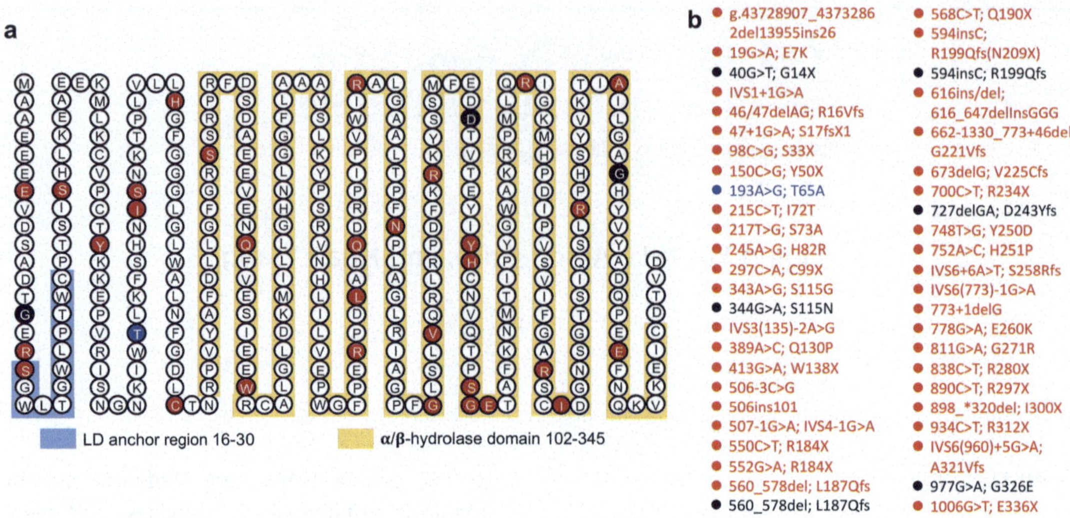

Fig. 13.1 (**a**) The amino acid sequence of human CGI-58 protein. The amino acids in red circles highlight those mutated in patients with CDS. Some altered splice donor or acceptor sites are not highlighted. According to the two studies using the mouse CGI-58 protein [16, 74], the amino acids 16–30 in the human CGI-58 protein are likely required for LD anchoring. (**b**) CGI-58 mutations reported in humans before March 2020. Biallelic mutations in red color are associated with the full phenotypes of CDS, biallelic mutations in blue color are associated with the partial phenotypes (no ichthyosis) of CDS, and those in black color denote monoallelic mutations associated with nonalcoholic fatty liver disease

protein of ~39 kDa (Fig. 13.1a). CGI-58 is also known as α/β hydrolase domain-containing 5 (ABHD5). The ABHD subfamily belongs to a large protein family defined by an α/β hydrolase fold [146, 258]. The α/β hydrolase fold has a highly conserved catalytic triad containing a nucleophile (serine, cysteine, or aspartic acid), an acidic residue, and histidine that are close in 3D structure, though apart from each other in sequence [116, 258]. The ABHD subfamily has a total of 19 members in humans and 15 members in mice [128, 202], yet the functions of most remain unknown. CGI-58 differs from other members in this subfamily in that the critical serine in the catalytic triad is substituted by asparagine [116].

Mutations in the human CGI-58 gene were identified as the cause of Chanarin-Dorfman syndrome (CDS, OMIM 275630) (Fig. 13.1), an autosomal recessive neutral lipid storage disease (NLSD) with ichthyosis (thickened dry skin) [58, 116]. CDS is characterized by the accumulation of triglyceride (TAG)-rich cytoplasmic lipid droplets (LDs) in most cell types, including leukocytes (Jordans' anomaly) [96], hepatocytes, myocytes, and cells in the epidermis, dermis, and intestinal mucosa [33, 46, 183, 205]. Patients with CDS often manifest hepatomegaly (hepatic steatosis and steatohepatitis), myopathy, microcephaly, cataracts, hearing loss, ataxia, mild mental retardation, and short stature [33, 46, 90, 183]. Since the initial description of the disease by Dorfman and Chanarin [33, 46], about 130 cases with more than 40 different mutations spanning the entire protein sequence have been reported worldwide [7, 49]. Types of mutations include deletion, insertion, missense, nonsense, and frameshift mutations (Fig. 13.1b) [1, 3, 5, 7, 9, 12, 22–24, 49, 52, 54, 89, 92, 116, 130, 150, 151, 169, 174, 181, 187, 192, 208, 219, 224, 243, 255]. While loss-of-function mutations cause CDS (Fig. 13.1), it is currently unknown whether gain-of-function exists for CGI-58 gene.

CGI-58 is ubiquitously expressed in mammals [18, 112, 211]. It is predicted to be cytosolic [116]. Interest in the scientific community regarding the functions of CGI-58 started in the early 2000s when three laboratories simultaneously reported that CGI-58 localizes at cytosolic LDs [121, 211, 244]. This was the time when biomedical scientists started to appreciate the cytosolic LD as an organelle that dynamically regulates energy storage and mobilization, rather than as an inert liposome-like structure that passively stores excess energy. The conceptual innovation placed cytosolic LDs at the center of cellular energy metabolism whose dysregulation is a hallmark of metabolic diseases, such as obesity, insulin resistance, type II diabetes, fatty liver, and cardiovascular disease. It was believed that excessive deposition of cytosolic lipid droplets would cause lipotoxicity, a biochemical mechanism that was widely used to explain impairments of cellular metabolism, cell signaling transduction, and redox imbalance associated with overnutrition-driven metabolic diseases [221]. Mutations in the human *CGI-58* gene were known to cause LD deposition in almost all cell types examined, which provided the biomedical research community an excellent opportunity to test how LD accumulation promotes lipotoxicity. Over the past 15 years, we have learned a great deal about the pros and cons of cytosolic LDs by studying the biochemistry, cell biology, and tissue-specific pathophysiology of CGI-58. This chapter summarizes the current knowledge about the role of CGI-58 in LD lipolysis (*i.e.*, hydrolysis of TAGs stored in cytosolic LDs) and discusses how CGI-58-dependent metabolic and signaling pathways regulate the pathogenesis of common metabolic diseases.

13.2 CGI-58 Interacts with Lipolysis-Regulatory Proteins

13.2.1 The PAT (Perilipin, Adipophilin, TIP47) Protein Family

Biochemical and cell biology studies have demonstrated that CGI-58 binds to cytosolic LDs and interacts with perilipin 1 (PLIN1), adipose differentiation-related protein (ADRP, also known as adipophilin or PLIN2), TIP47 (PLIN3), and muscle LD protein (MldP or PLIN5) [18, 63, 121, 161, 211, 244, 245]. These are members of the PAT (perilipin, adipophilin, TIP47) family that also includes S3-12 (or PLIN4) [105, 126, 133, 236]. The PAT family proteins share a highly conserved N-terminal structure. They localize at the surface of intracellular LDs of different lipid compositions and sizes, regulating energy storage and mobilization in response to nutritional fluctuations and various stimuli [126]. Using the two frame shift mutants (Leu-404fs and Val-398fs) that cause partial lipodystrophy in humans, Savage and associates have shown that the C-terminal region of human PLIN1 is essential for binding to CGI-58, and this interaction stabilizes CGI-58 localization on the LDs [63].

13.2.2 The PNPLA (Patatin-Like Phospholipase Domain Containing) Protein Family

The process that mobilizes the energy (mainly as TAGs) stored in intracellular LDs for utilization is called intracellular lipolysis (Fig. 13.2) [253]. During LD lipolysis, the three fatty acyl chains in a TAG molecule are sequentially cleaved into diacylglycerol (DAG), monoacylglycerol (MAG), and glycerol, releasing a fatty acid molecule at each step. The first enzyme that was discovered to catalyze hydrolysis of cytosolic LD-embedded TAGs is hormone-sensitive lipase (HSL) [86, 111, 177, 223]. The substrate spectrum of HSL appears to be quite broad, including DAGs, TAGs, MAGs, cholesteryl esters, and retinyl esters [40, 113, 234]. Monoacylglycerol lipase (MAGL) was reported, shortly after HSL, as a lipase that specifically hydrolyzes MAGs [98, 223]. Both HSL and MGL belong to the α/β-hydrolase fold family. For years, HSL was thought to be responsible for hydrolyzing TAGs in adipocyte LDs. However, HSL-null mice showed the accumulation of DAGs rather than TAGs in multiple tissues [77, 157, 179, 227],

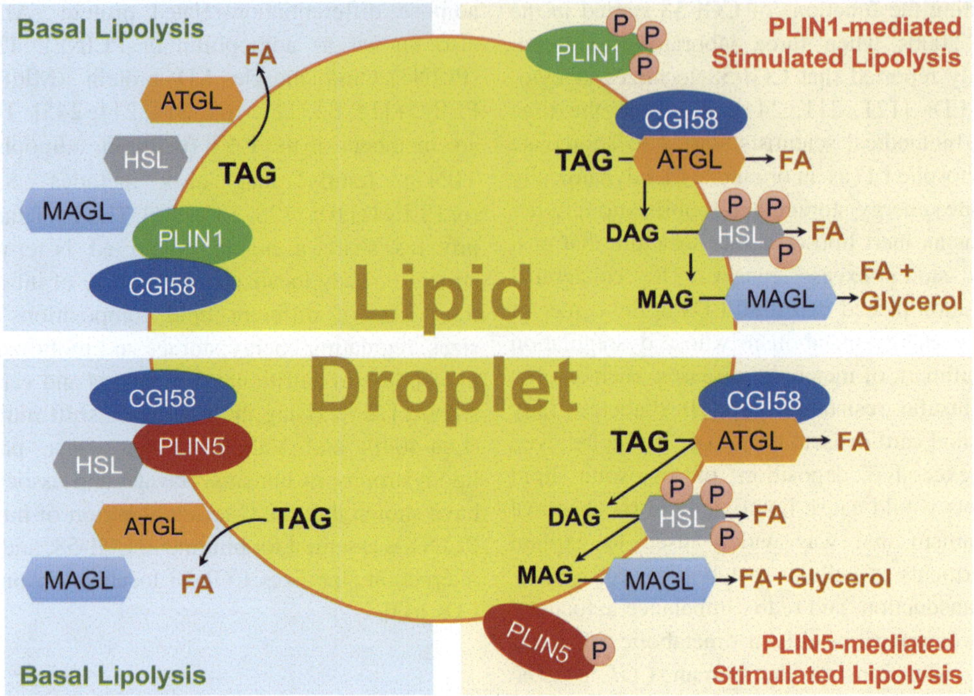

Fig. 13.2 Proposed model for CGI-58 regulation of cytosolic lipid droplet lipolysis. Lipolysis regulation differs between basal and stimulated conditions. Under the basal conditions, CGI-58 binds to PLIN1 in adipocytes or PLIN5 in oxidative nonadipocytes, preventing its interaction with ATGL. Thus, the lipolytic activity of ATGL is limited. After stimulation, perilipins are phosphorylated, resulting in the dissociation of CGI-58 from perilipins. CGI-58 then interacts with ATGL and substantially activates ATGL's TAG hydrolase activity to stimulate lipolysis, producing DAGs and fatty acids (FAs). The DAGs are then hydrolyzed to produce MAGs and FAs by HSL that was phosphorylated and recruited to the LDs during the lipolytic stimulation. Finally, the MAGs are hydrolyzed by MAGL to release the last fatty acyl chain from the glycerol backbone of a TAG molecule

indicating that other enzyme(s) are involved in the TAG hydrolysis. In 2004, three groups independently reported a new lipase possessing abundant TAG hydrolase activity, and the enzyme was named calcium-independent phospholipase $A_2\zeta$ ($iPLA_2\zeta$), desnutrin, or adipose triglyceride lipase (ATGL), respectively [95, 226, 264]. This newly discovered lipase turned out to be the rate-limiting enzyme of cytosolic LD lipolysis (Fig. 13.2), and, thus, the name ATGL became more popular than the other names. ATGL is also known as patatin-like phospholipase domain containing 2 (PNPLA2). The PNPLA protein family consists of a total of nine members, including PNPLA1 through PNPLA9, all of which seem to be implicated in lipid metabolism through their phospholipase or lipase activities, or other functions [104, 144]. Comparative studies of ATGL and CGI-58 in the context of adipose lipolysis have resulted in a major breakthrough regarding the biochemical function of CGI-58. In 2016, Dr. Rudolf Zechner and associates reported that CGI-58 functions as a coactivator of ATGL to promote in vitro TAG hydrolysis [112]. Subsequent studies were consistent with this original finding [70, 71, 161, 228, 233, 250]. Furthermore, CGI-58 was shown to release from perilipin proteins following lipolytic stimulation, which allowed CGI-58 to interact with ATGL and activate TAG hydrolysis [70, 71, 211, 228]. In this scenario, the interaction of CGI-58 and with perilipins functions as a brake of lipolysis (Fig. 13.2), though its efficiency may be cell-type specific due to distinct perilipin compositions and the different abilities of individual perilipins in sequestering CGI-58 in various cell types [161].

Although CGI-58 activates ATGL's TAG hydrolase activity [112], mutations in CGI-58 and ATGL cause distinct phenotypes in humans and mice [58, 65, 75, 78, 116, 135, 172, 237]. For example, human CGI-58 mutations cause NLSD with ichthyosis [33, 46, 90, 116, 183], whereas human ATGL mutations cause NSLD without skin defects but with mild myopathy [58]. Global CGI-58 knockout mice die ~16h after birth due to a skin barrier defect [172], yet global ATGL knockout mice are viable [78]. These phenotypic differences associated with mutations of the two genes indicate that CGI-58 must have ATGL-independent functions. Recently, CGI-58 was shown to interact with PNPLA1, another member of the PNPLA protein family, to stimulate PNPLA1-mediated ω-O-acylceramide production in skin [102, 154], providing a potential mechanism for skin barrier defect seen in patients with CDS. CGI-58 was also shown to functionally interact with the wild-type PNPLA3, the fatty liver-promoting PNPLA3(I148M) variant [180], and a lipase dead PNPLA3 mutant [32], suggesting that CGI-58 may coordinate with PNPLA3 and other lipases to regulate LD turnover independently of PNPLA3's lipase activity. Consistent with this scenario, two laboratories reported that PNPLA3, the fatty liver-causing PNPLA3(I148M) variant, in particular, competes with ATGL (PNPLA2) to bind with CGI-58, reducing TAG hydrolysis in the liver and brown adipocytes [233, 250]. These observations provided a mechanism for how the PNPLA3 (I148M) variant promotes fat deposition. However, such observations cannot explain why PNPLA3, including PNPLA3(I148M) but not a lipase dead mutant, retains its ability to reduce LD sizes when co-expressed with CGI-58 in the absence of ATGL [32]. It remains possible that PNPLA3 displays an in vivo lipase or transacylase activity toward specific substrates under some pathophysiological or nutritional conditions. In addition to PNPLA1-3, other members of PNPLA protein family may also interact with CGI-58 to fulfill unique functions under specific pathophysiological and nutritional conditions.

13.2.3 The Fatty Acid-Binding Protein (FABP) Family

Another group of proteins that interact with CGI-58 is the fatty acid binding protein (FABP) family members [85]. It was hypothesized that FABP interacts with CGI-58 to promote ATGL-mediated intracellular lipolysis by serving as an acceptor of free fatty acids released from TAG hydrolysis [85]. This was an important finding because it provided a mechanism for the handling of lipolytic products. Intriguingly, the lipolytic product long-chain acyl CoA was shown to bind CGI-58 and promote CGI-58 interactions with perilipins to suppress lipolysis [188]. This phenomenon seems to be an end-product feedback mechanism that fine-tunes hydrolysis of TAGs stored in intracellular LDs. The interactions of CGI-58 with LD coat proteins, lipases, and lipids suggest that CGI-58 likely play a key role in organizing major components of LD lipolysis into a functional "lipolysome" [85].

13.3 CGI-58 as a Serine Protease

The latest member of the CGI-58 interactome is histone deacetylase 4 (HDAC4). Backs and associates reported that CGI-58 functions in vitro and in vivo as a serine protease that cleaves HDAC4 in the heart in response to catecholamine stimulation, generating an N-terminal polypeptide of HDAC4 (HDAC4-NT) to protect cardiac functions [94]. This study was conceptually paradigm shifting, because it was the first to demonstrate that CGI-58 can function as a serine protease. CGI-58 was previously shown to function as a coactivator of a lipase that promotes lipolysis, and it was never thought to be a protease that promotes proteolysis. Perhaps, CGI-58 is a protein of dual function that promotes both lipolysis and proteolysis. This novel function of CGI-58 raises many new and exciting questions regarding the core function of the protein. For example, does CGI-58 cleave other proteins interacting with it? If yes, is this proteolytic function required for CGI-58 to activate LD lipolysis? Does a lipase require proteolytic cleavage prior to digesting a lipid molecule? Answers to these questions are expected to provide fundamental insights into the molecular and biochemical bases of lipolysis and its potential crosstalk with proteolysis.

13.4 Molecular Basis for CGI-58 Activation of ATGL-Dependent Lipolysis

The cellular, structural, and biochemical bases for CGI-58 and ATGL interaction to promote TAG hydrolysis remain incompletely understood. The N-terminal amino acids 1–30 of mouse CGI-58 were shown to form a lipophilic tryptophan-rich stretch, which is essential for CGI-58 to localize at the LD and activate ATGL in cultured cells [74]. This tryptophan-rich stretch appears to anchor CGI-58 to the LD surface through its three tryptophan residues serving as the left and right anchor arms [16]. A comparative study of mouse ABHD5 (CGI-58) and ABHD4, an ABHD family member that is closely related to ABHD5 but does not activate ATGL, identified R299 and G328 as essential residues for activating ATGL's TAG hydrolase activity. However, these two amino acids of ABHD5 did not affect ATGL translocation to LDs or ABHD5 binding to PLIN1 [189]. These studies collectively suggest that the LD localization is a prerequisite for a functional CGI-58 to activate ATGL in vivo.

Studies with ATGL mutants associated with NLSD have showed that the mutations result in the expression of either enzymatically inactive proteins localizing to LDs or active TAG hydrolase lacking LD localization [196]. Whereas CGI-58 was identified as a coactivator of ATGL [112], G0/G1 switch gene 2 (G0S2) was subsequently discovered as an inhibitor of ATGL function [246, 247]. It was further demonstrated that G0S2 and CGI-58 do not appear to compete with each other for binding to ATGL in cultured cells transfected with tag-proteins [131]. The 254 N-terminal amino acids of mouse ATGL were reported to be the minimal domain that can be activated by CGI-58 and inhibited by G0S2 [41]. Interestingly, deleting ~220 amino acids from the C-terminus of human ATGL protein increases its interaction and activation by CGI-58 in vitro in the test tube, despite defective LD localization in vivo in cultured cells [196]. This finding indicates that the C-terminal region of ATGL is required for its targeting to LDs and plays a regulatory role in ATGL activation by CGI-58. Considering the newly identified protease function of CGI-58 [94], it would be interesting to test whether CGI-58 activates ATGL by a two-step process. In the first step, CGI-58 may cleave ATGL to release the suppressive role of ATGL's C-terminal region on its enzymatic activity, which would be consistent with the observation that ATGL protein levels are often increased in the absence of CGI-58 [75, 242, 263]. The second step may involve conformational changes of the two proteins, resulting in tight interactions and correct positioning of "lipolysome" components on the surface of LD for hydrolysis of TAG in vivo.

13.5 CGI-58 Regulation of Autophagy and Lipophagy

The role of CGI-58 as the coactivator of ATGL to promote intracellular lipolysis has been established and reproduced in a series of in vitro and in vivo studies. ATGL is a cytosolic neutral lipase that initiates cytosolic/neutral lipolysis by cleaving a fatty acyl chain from a TAG molecule stored in cytosolic LDs, thus playing a critical role in intracellular lipolysis [95, 226, 257, 264]. Recently, the lipid-specific macroautophagy (lipophagy) was shown to also digest cytosolic LDs by delivering LD-associated fat to lysosomes for degradation by lysosomal acidic lipase (lysosomal/acidic lipolysis) [203]. Autophagy refers to the "self-eating" of a cell in response to starvation or nutrient deprivation for generating energy essential for its survival [155]. It is also a catabolic pathway for recycling of excessive or damaged organelles, such as mitochondria (mitophagy) [217]. In humans, insulin resistance suppresses CGI-58 mRNA expression in liver [99]. The nutritional and hormonal regulations of "neutral" lipolysis and lipophagy ("acidic" lipolysis) are strikingly similar. Both are induced by nutrient deprivation [45], and both are activated by glucagon or inhibited by insulin [45, 57]. It is currently unknown if CGI-58 promotes fat lipolysis by mediating lipophagy in addition to activating ATGL. It was demonstrated that ATGL, a lipase target of CGI-58, promotes autophagy and lipophagy in a sirtuin 1 (SIRT1)-dependent manner and that lipophagy is required for ATGL to promote LD catabolism and associated fatty acid oxidation in hepatocytes [190]. The crosstalk between ATGL-dependent lipolysis and autophagy was also seen in macrophages, though this crosstalk may be indirect or compensatory [66]. Some studies appear to suggest a role of CGI-58 in regulating autophagy and lipophagy. For example, in C2C12 muscle cells, CGI-58 overexpression increases, whereas CGI-58 knockdown decreases, autophagy and mitophagy through regulation of AMPK and mTORC1 signaling pathways [259]. CGI-58 was shown to bind Beclin1, a major regulator of autophagy [159, 163]. Many autophagy components can localize to LDs under some conditions [48, 51, 100, 160, 200, 203], though it is not known whether they interact with CGI-58 or other LD proteins to specifically regulate lipophagy. PLIN2, a major LD coat protein interacting with CGI-58 [244, 245], also binds the heat shock cognate protein of 70 kDa (Hsc70) for degradation via chaperone-mediated autophagy (CMA) [100]. The inhibition of CMA reduces both neutral and acidic lipolysis [100]. Hepatic CMA deficiency, like CGI-58 deletion, induces severe hepatic steatosis with liver damage and inflammation [220]. More studies are needed before the direct role of CGI-58 in the mediation of autophagy and lipophagy can be established.

13.6 Tissue-Specific Roles of CGI-58 in Energy and Lipid Metabolism

13.6.1 Adipose CGI-58 in Thermoregulation and Metabolic Health

CGI-58 is ubiquitously expressed in mammals, with the highest expression in adipose tissue. Adipose tissue is classically divided into white adipose tissue (WAT) and brown adipose tissue (BAT) that have distinct locations and opposite functions in energy balance. In general, WAT stores excess energy as TAGs in the large unilocular LD of white adipocytes, whereas BAT dissipates metabolic energy as heat for adaptive nonshivering thermogenesis in multilocular LD-containing brown adipocytes.

During prolonged fasting or increased energy demand, such as exercise and inflammation, the stored energy in WAT is mobilized via adipose LD lipolysis for utilization by cell types and pathways critical in sustaining life, meeting energetic demand, clearing infectious agents, or resolving inflammation. This process is generally defined as the stimulated adipose lipolysis, because it involves activation of a cell membrane receptor and its downstream signal transduction by neural and humoral factors in response to various stimuli [47, 108, 213, 253]. The classical signal-stimulating adipose lipolysis is the activation of β-adrenergic receptors by catecholamines released from the sympathetic nerves innervating adipose tissue. Binding of a catecholamine to the β receptor activates adenylate cyclase, which is an enzyme that uses ATP as the substrate to produce cAMP [55]. Elevation in cellular cAMP activates protein kinase A (PKA), which then phosphorylates several lipolytic components, such as PLIN1 and HSL, to stimulate lipolysis (Fig. 13.2) [213]. Thus, any stimulus that activates PKA or increases cellular cAMP levels is thus expected to stimulate adipose lipolysis. Phosphorylation of a perilipin, perhaps together with phosphorylation of CGI-58 on S239 [185], causes CGI-58 disassociation from the perilipin for CGI-58 to interact with ATGL (Fig. 13.2) [70, 71, 211, 213, 244, 251]. It was shown that the in vitro TAG hydrolase activity of ATGL can be increased up to 20-fold with CGI-58 interaction [112]. The in vivo significance of CGI-58 as an essential mediator of the stimulated lipolysis was demonstrated in a study showing that adipose-specific inactivation of CGI-58 abolishes the isoproterenol-stimulated increase in plasma levels of free fatty acids in mice [201].

The nonshivering thermogenesis in BAT is mainly mediated by uncoupling protein 1 (UCP-1), which resides in the inner membrane of a mitochondrion, uncoupling chemical energy from ATP synthesis and dissipating the energy as heat [27]. Under some environmental and pathophysiological conditions, such as cold exposure and β-adrenergic receptor activation, a cell type with features of both white and brown adipocytes appears in the classically white fat depots. This type of adipocytes is named brite or beige adipocytes that often express UCP-1 and produce heat [165, 238]. The process that drives the appearance of brite/beige adipocytes in WAT is called WAT browning or beigeing [97]. The origin of beige adipocytes may include mature white adipocyte transdifferentiation and/or de novo adipogenesis, depending on the condition that induces WAT browning [39, 83, 114, 115, 182, 229, 230].

Cytosolic LD lipolysis was thought to be central in nonshivering thermogenesis [27]. Several animal and human studies suggested the essential role of brown fat lipolysis in thermogenesis, though the genetic or pharmacological manipulation of adipose lipolysis employed in the studies inhibited intracellular lipolysis in both BAT and WAT [4, 15, 44, 78, 107]. We created mice deficient in CGI-58 in UCP1-positive brown and beige adipocytes (BAT-KO mice) and mice lacking CGI-58 in all adipocytes (FAT-KO mice), which allowed us to directly test the role of brown adipocyte LD lipolysis in thermoregulation. To our surprise, BAT-KO mice were not cold sensitive even when food was unavailable [201]. The mice became cold sensitive only when the following two conditions were met simultaneously: (1) deletion of CGI-58 in both WAT and BAT and (2) removal of food. Similar phenotypes were observed in mice lacking ATGL in BAT or the total adipose tissue [195]. When CGI-58 or ATGL was deleted in the total adipose tissue in mice, the in vivo lipolysis (fatty acid release from the tissue to the blood circulation) stimulated by isoproterenol, a β-adrenergic receptor agonist, was completely abolished in mice [195, 201]. The results demonstrated the indispensable role of CGI-58 or ATGL in mediating the stimulated lipolysis in the whole animal. These two animal studies also demonstrated a key role of WAT in regulating adaptive nonshivering thermogenesis, likely by providing the heat-producing cells with the metabolic fuels and/or by exposing the temperature sensors in the body to the thermogenically important adipokines or signaling molecules. It is currently unclear how food rescues the cold sensitivity of mice lacking CGI-58 or ATGL in the total adipose tissue [195, 201]. A simple explanation is that food serves as another source of metabolic fuels that may energize the heat-generating cells with glucose, fatty acids, and/or amino acids. However, we observed that only gastric gavage, but not intraperitoneal injections, of glucose can efficiently slow down hypothermia in mice lacking CGI-58 in both WAT and BAT (Wang H et al. unpublished data). This finding strongly supports a critical role of the gastrointestinal track in regulating the diet-induced thermogenesis. The gastrointestinal track is abundantly innervated and has special endocrine cells that secrete various incretins, which are important in local environment sensing and whole-body energy metabolism. Interestingly, secretin, a gut hormone that is derived from the S cells in the duodenum and jejunum of small intestine, was shown to mediate postprandial thermogenesis by activating its receptor in brown adipocytes to stimulate lipolysis and energy expenditure and to subsequently suppress satiation through the brain [119]. However, mice lacking CGI-58 or ATGL in BAT are defective in brown adipocyte lipolysis, yet they are capable of producing heat after a meal, suggesting that either other gastrointestinal factors or non-lipolytic pathways also mediate the postprandial thermogenesis. Nonetheless, it would be interesting to test whether secretin mediates postprandial heat production in mice lacking CGI-58 or ATGL in BAT and, if not, what other gastrointestinal factors are involved.

CGI-58 deletion in UCP1-positive cells in mice increases sympathetic innervation in both BAT and WAT. The animals also exhibit enhanced WAT browning, especially after cold exposure or β3-adrenergic receptor activation [201]. This observation implies that some signals and/or BATokines (factors secreted by BAT) are generated as a result of BAT CGI-58 deficiency. These signals and batokines can somehow be sensed by the central nervous system in the brain, thereby increasing the sympathetic outflow to activate compensatory thermogenic mechanisms. It is currently unknown what these signals and batokines are and whether they work locally or remotely or must be secreted into the blood circulation, which represents an important area for future research in BAT biology. It is important to note that BAT lipolysis deficiency induced by ATGL deletion in UCP1-positive cells does not increase WAT browning as evidenced by unaltered expression levels of UCP-1 protein in the inguinal subcutaneous fat [195], suggesting that deficiency of CGI-58's ATGL-independent functions in BAT promotes browning in WAT.

Genetic deletion of BAT CGI-58 in mice improves several fat-induced metabolic disorders, such as glucose intolerance, insulin resistance, and hepatic steatosis [201]. This improvement is more profound when CGI-58 is deleted in both BAT and WAT (our unpublished data). ATGL deletion in whole body or adipose tissue also protects mice from fat-induced metabolic abnormalities [4, 87, 194, 239]. These observations indicate that inhibiting adipose lipolysis may improve whole-body glucose handling as a result of failed mobilization of fatty acids for utilization, which would be consistent with the glucose fatty acid cycle (or the Randle cycle) theory [173].

13.6.2 Epidermis CGI-58 and Skin Barrier Function

A major phenotypic distinction of human patients with CGI-58 mutations from those with ATGL mutations is ichthyosis (scaly dry skin) [58, 116]. In mice, whole body ablation of CGI-58, but not ATGL, causes skin barrier defects [78, 172]. Using whole body and cell type-specific transgenic and knockout mouse models, it was shown that CGI-58 promotes the biosynthesis of the skin barrier lipids, ω-*O*-acylceramides, locally in the keratinocytes of suprabasal epidermal layers, and such function is ATGL independent [73]. It was further shown that CGI-58 interacts directly with PNPLA1 and recruits PNPLA1 to LDs where it functions as the coactivator of PNPLA1 for the biosynthesis of ω-*O*-acylceramides [102, 154]. Like CGI-58, PNPLA1 mutations in humans also cause ichthyosis [69]. Using biochemical approaches, cell cultures, and tissue-specific PNPLA1 knockout mice, several groups have demonstrated that PNPLA1 has transacylase or acyltransferase activity, which utilizes TAGs as an acyl donor and catalyzes the esterification of ω-hydroxy ceramides with linoleic acid to synthesize ω-*O*-acylceramides [73, 84, 102, 153]. Collectively, these studies strongly suggest that the defective activation of PNPLA1 is the molecular mechanism underlying CGI-58 mutation-induced ichthyosis in humans.

13.6.3 Muscle CGI-58 in Cardiomyopathy and Insulin Sensitivity

Patients with CDS accumulate neutral lipids in their skeletal muscle [138]. Heart murmurs, muscle weakness, and mild myopathy were reported in some CDS patients [90, 138, 235]. Two laboratories have generated muscle-specific CGI-58 knockout mice using MCK-cre transgenic mice [242, 263]. MCK-cre transgenic mice express cre recombinase in both skeletal and cardiac muscles, thereby deleting a loxP-floxed gene in both tissues [21]. Muscle CGI-58 knockout mice display intramyocellular deposition of neutral lipids in both cardiac and oxidative skeletal muscles [242, 263], implying that muscle fat deposition in human patients with CDS likely results from local CGI-58 deficiency in muscle. Neutral lipid deposition was not detected in the glycolytic skeletal muscle fibers in these animals [242]. The restriction of LD accumulation to the cardiac and oxidative (soleus) muscles highlights an essential role of CGI-58 in fatty acid oxidation in oxidative muscle types, which is consistent with other studies [10, 72].

CGI-58 deficiency in all muscles induces cardiac fibrosis, cardiac remodeling, and heart failure. The similar phenotypes were observed in muscle ATGL knockout mice [79]. In cardiac and oxidative skeletal muscles, CGI-58 interacts with PLIN3 and PLIN5, and this interaction regulates its association with ATGL [132, 167, 228]. These observations collectively suggest that CGI-58 may function through ATGL, promoting intracellular TAG hydrolysis in the muscle fibers. It was shown that cardiac ATGL-dependent TAG hydrolysis sustains mitochondrial functions by activating the PPAR-α pathway through the generation of endogenous ligands for PPAR-α [79]. CGI-58 may facilitate this pathway by activating ATGL in the cardiac muscle. Interestingly, CGI-58 was recently shown to function as a serine protease to protect heart failure by generating an N-terminal polypeptide from histone deacetylase 4 (HDAC4) through proteolysis [94]. The cardiac protective role of the HDAC4's N-terminal polypeptide generated by CGI-58 was not associated with reduction in cardiac TAG content. Although it is currently unclear whether similar mechanisms operate in other cell types, this study nonetheless uncovered a completely novel function of CGI-58 and emphasized a future direction for CGI-58 research.

Intramyocellular fat deposition in skeletal muscle is often associated with systemic insulin resistance due to accumulation of insulin signaling-suppressing lipids, such as diacylglycerols and ceramides that cause lipotoxicity [186, 222]. Despite intramyocellular accumulation of neutral lipids, mice lacking CGI-58 or ATGL in muscle are not glucose intolerant or insulin resistant [103, 204, 242]. This dissociation of cellular lipid deposition from insulin resistance suggests that how versus how much lipids are accumulated may be more important in driving tissue insulin resistance, which may be due to the differences in the molecular species of lipids deposited. Alternatively, cytosolic LD deposition, if not extremely excessive, may sequester insulin signaling-suppressing metabolites, protecting cells against lipotoxicity. Such scenario would be consistent with an observation that unsaturated fatty acids promote TAG accumulation, yet protect cells against lipotoxicity [120]. In addition, lipid deposition in different skeletal muscle fiber types may lead to different metabolic consequences [118, 123]. Mice overexpressing diacylglycerol acyltransferase 2 (DGAT2) in glycolytic (type II) muscle accumulate TAG in muscle and are insulin resistant [118]. However, mice overexpressing diacylglycerol acyltransferase 1 (DGAT1), another TG synthesis enzyme, in muscle accumulate TAG in the soleus, and these animals are not insulin resistant [122]. Endurance-trained athletes display increased fat content in their skeletal muscle, and they have enhanced insulin sensitivity ("athlete paradox") [67]. It seems that fat deposition in the glycolytic muscle is more problematic than in the oxidative muscle.

13.6.4 Liver CGI-58 in Nonalcoholic Fatty Liver Disease

Non-alcoholic fatty liver disease (NAFLD) is the most common liver disease in the United States and worldwide [254]. Patients with CDS (CGI-58 mutations) almost always display characteristics of advanced NAFLD, including severe hepatic steatosis, NASH, fibrosis, and cirrhosis [3, 24, 38, 76, 90, 139, 181, 205, 208, 215]. The CDS-causative mutations span the entire human CGI-58 protein sequence (Fig. 13.1). Interestingly, monoallelic mutations in the human *CGI-58* gene are also associated with NAFLD (Fig. 13.1b). The prevalence of CGI-58 monoallelic mutations that are associated

NAFLD was estimated to be 1 in 1,131 individuals in a normal population [255]. This study highlights an important role of CGI-58 in the pathogenesis of NAFLD in the general population. More importantly it was recently demonstrated that CGI-58 interacts with PNPLA3 [233, 250], a variant (I148M) of which is a major risk factor for fatty liver disease in all populations examined [149, 180, 207, 216]. CGI-58's association with PNPLA3 interferes with its ATGL interaction, thus inhibiting LD lipolysis [11, 233, 250]. CGI-58 is required for wildtype PNPLA3 and the PNPLA3(148M) variant to localize to hepatic LDs and for the overexpressed PNPLA3(148M) to promote hepatic steatosis [233]. It was shown that PNPLA3 accumulation on LDs, not its catalytic activity, is responsible for PNPLA3(148M)-induced hepatic steatosis [11]. While these studies provided an important mechanism for how CGI-58 coordinates with PNPLA3 and PNPLA2 (ATGL) to control cytosolic LD turnover, more research on the PNPLA3/CGI-58 interaction is needed to address why PNPLA3, including the PNPLA3(I148M) variant but not a lipase dead mutant, can substantially reduce LD size when co-expressed with CGI-58 in the absence of ATGL [32].

Antisense oligonucleotide (ASO)-mediated knockdown of CGI-58 in adult mice induced severe hepatic steatosis, though this study cannot establish a causal relationship between hepatic CGI-58 and fatty liver disease due to knockdown of CGI-58 in multiple tissues, including liver, adipose tissue, and macrophages [19, 28, 127, 129]. Selective inactivation of CGI-58 or ATGL in the liver of mice causes hepatic steatosis [75, 237], implying that fatty liver disease seen in patients with NLSD induced by CGI-58 or ATGL mutations is likely a local effect of hepatic CGI-58 or ATGL deficiency. These studies unequivocally demonstrated an important role of LD lipolysis in controlling lipid homeostasis in the liver. Besides TAGs, other species of lipids, such as DAGs, are also accumulated in mouse livers lacking CGI-58, especially when a high fat diet is used [19, 28, 75]. Although hepatic steatosis is often associated with insulin resistance and DAG accumulation is well-known to suppress insulin signaling [186], liver CGI-58 deficiency-induced hepatic steatosis and DAG accumulation are not associated with insulin resistance in mice [19, 28, 75]. One study demonstrated that this dissociation results from the sequestration of DAGs to LDs and ER, rather than the cell membrane, which prevented PKCε translocation to the plasma membrane to inhibit insulin-receptor kinase activity [28]. The dissociation of hepatic steatosis and insulin resistance is not restricted to the CGI-58 deficiency-induced fatty liver. For instance, hepatic overexpression of DGAT2 or liver-specific deletion of histone deacetylase 3 (HDAC3) in mice induces severe hepatic accumulation of lipids including TAGs, DAGs, and ceramides without causing insulin resistance [141, 212]. In humans, a genetic variant (I148M) of PNPLA3 confers susceptibility to NAFLD in multiple populations without affecting the index of insulin resistance [149, 180, 207, 216]. African-American descendants have significantly less hepatic steatosis despite a relatively high prevalence of obesity and diabetes, while Hispanic-American descendants are the opposite [175, 193]. The variation in correlation between hepatic steatosis and insulin resistance among ethnicities suggests that other factors should also be considered. It should be emphasized that clinical studies of NAFLD only found the association between insulin resistance and hepatic steatosis whereas the relationship between insulin resistance and other liver pathologies, such as NASH and hepatic fibrosis, has yet to be established.

It is currently unknown how liver CGI-58 deficiency induces NASH and hepatic fibrosis in addition to hepatic steatosis. The albumin-cre transgenic mice (Stock #: 003574; The Jackson Laboratory) used for liver-specific inactivation of CGI-58 and ATGL can delete a gene floxed by loxP sites in hepatocytes, biliary epithelial cells (cholangiocytes), and hepatic stellate cells [50, 64, 148, 168, 171, 184, 206, 214]. Each of these cell types has distinct physiological and pathological functions. For instance, injuries of hepatocytes and other liver cells stimulate inflammatory responses, causing NASH [20, 60]. Liver damage and inflammation often trigger ductular reaction (increases in the number of small biliary ductules lined by cholangiocytes) that may contribute to hepatic fibrogenesis to some extent [59, 176, 191]. Hepatic stellate cells increase collagen production after activation by various liver injuries, and this cell type is well accepted to be the major source of hepatic fibrosis [81, 82, 142]. Given that liver ATGL deficiency induced by the same albumin-cre transgenic mouse line does not cause these advanced pathological changes in liver [237], the mechanism underlying liver CGI-58 deficiency-induced NASH and hepatic fibrosis cannot be the inhibition of ATGL-mediated LD lipolysis in hepatocytes, cholangiocytes, or hepatic stellate cells. Consistently, patients with ATGL mutations do not develop NASH and hepatic fibrosis [6, 25, 58, 166, 197]. CGI-58, therefore, must have ATGL-independent functions in the liver. One of such functions may be its interaction with PNPLA3 [233, 250]. Like CGI-58 mutations, the PNPLA3(I148M) variant is also associated with NASH [180]. Another distinct function of CGI-58 is its interactions with almost all perilipins. This interaction may be needed for cellular processes, such as autophagy and lipophagy, besides activation of ATGL. Perilipins are coat proteins of cytosolic LDs. They are required for the biogenesis and turnover of cytosolic LDs. It has been shown that perilipins play an important role in the pathogenesis of hepatic steatosis, NASH, and hepatic fibrosis [29, 30, 34, 35, 61, 88, 91, 145, 162, 231]. Patients with NAFLD accumulate perilipins in the liver [61, 162, 209]. While perilipins may passively accumulate in the steatotic liver due to increased LDs, they may also actively increase to protect cells against lipotoxicity of free lipids. Other CGI-58 functions, such as its newly identified serine protease activity in the heart [94], may also exist in the liver and other tissues. This protease activity of CGI-58, like its lipase coactivator function, may target multiple proteins to regulate a variety of cellular processes important in lipid and energy metabolism.

Liver CGI-58 knockout mice on a regular low-energy chow diet develop a full spectrum of pathologies observed in human patients with advanced NAFLD [75]. The progression of these pathologies can be substantially facilitated by challenging the animals with a typical Western-type diet alone or in combination with fructose in drinking water (our unpublished data). Future studies are needed to discern whether CGI-58 needs to be deleted simultaneously in hepatocytes, cholangiocytes, and stellate cells or in a specific cell type to trigger NASH and fibrosis in liver. Studies are also needed to identify CGI-58's ATGL-independent mechanisms responsible for fatty liver progression, including testing the known ATGL-independent functions of CGI-58. Detailed comparative studies of liver CGI-58 and ATGL knockout mice may reveal mechanisms important in the etiology of NASH and hepatic fibrosis in general and shed light on novel drug targets against NAFLD progression.

13.6.5 Myeloid CGI-58 in Insulin Resistance, Inflammation, and Atherosclerosis

CGI-58 protein is expressed in mouse and human macrophages [13, 134]. It has been shown that myeloid cell-specific deletion of CGI-58 in mice worsens fat-induced tissue/systemic inflammation, proinflammatory activation of adipose tissue macrophages, glucose intolerance, and insulin resistance [134]. CGI-58-deficient macrophages accumulate cytosolic LDs and show reduced PPAR-γ signaling [134, 248]. Although the underlying mechanism remains unknown, sequestration of free fatty acids in cytosolic LDs may prevent these endogenous PPAR ligands from activating PPAR signaling as seen in ATGL-null cardiomyocytes [79]. As a result of PPARγ signaling suppression, CGI-58-null macrophages show mitochondrial dysfunction and accumulate reactive oxygen species, which activates NLRP3 inflammasome to promote secretion of proinflammatory cytokines [134]. Consistently, overexpression of CGI-58 in macrophages reduces inflammation in vitro and in vivo [241, 248]. The proinflammatory (M1-like) phenotype of CGI-58-null macrophages was also observed in other studies [65, 135]. In contrast, ATGL-deficient macrophages were shown to display the anti-inflammatory M2-like phenotype [2, 65, 110]. These collective observations indicate that CGI-58 also has ATGL-independent functions in myeloid cells, including macrophages.

The anti-inflammatory role of macrophage CGI-58 is expected to protect against atherosclerosis. One study with CGI-58 overexpression in macrophages did show such an atheroprotective role through the promotion of the PPAR/LXR-dependent cholesterol efflux without altering blood cholesterol levels [241]. However, the deletion of CGI-58 in myeloid cells of apoE knockout mice, or simultaneous knockdown of CGI-58 in multiple cell types including hepatocytes, adipocytes, and macrophages in LDLR-KO mice, does not worsen atherosclerosis or alter plasma cholesterol levels [65]. It is difficult to assess atherosclerosis risk in patients with CGI-58 mutations due to the rarity of disease, existence of other abnormalities, and relatively young subjects reported. The role of macrophage CGI-58 in atherogenesis has yet to be clarified. Macrophage CGI-58 deficiency causes foam cell formation [134]. Lipopolysaccharide (LPS) and saturated fatty acids downregulate CGI-58 expression in macrophages [134]. LPS and fatty acids are atherosclerosis risk factors, and many studies have shown that they promote foam cell formation and atherosclerosis [8, 14, 53, 56, 62, 101, 117, 137, 143, 156, 170]. Oxidized (ox)-LDL, a common atherosclerosis risk factor, inhibits CGI-58 expression in THP1 human macrophages (our unpublished data). These findings suggest a potential role of CGI-58 in modulating atherosclerosis risk factor-induced atherogenesis.

13.6.6 Intestine CGI-58 in Fat Absorption and Turnover

A major function of the small intestine is the absorption of nutrients including fats. Fat absorption occurs mainly in duodenum and jejunum. After digestion by pancreatic lipases, in the intestinal lumen, fat (mainly TAGs) becomes free fatty acids and monoacylglycerols (MAGs), which then enter the absorptive enterocytes and travel to the endoplasmic reticulum (ER) for re-esterification into TAGs for packaging into chylomicrons. Intestinal fat absorption is a very efficient process. Chylomicrons are quickly secreted into the lymphatic system heading to the blood circulation. Some of absorbed fat may be temporarily stored in the cytosolic LDs, especially after ingestion of a high fat diet [31, 164, 178, 262]. The TAGs stored in the cytosolic LDs have to be hydrolyzed before they can be assembled into primordial chylomicron particles in the ER lumen. CGI-58 and ATGL are expressed in the enterocytes. Genetic deletion of CGI-58 in these cells in mice induced the accumulation of cytosolic LDs predominantly in the nutrient absorptive segment of small intestine, regardless of dietary compositions and nutritional conditions [106, 240]. These observations demonstrated an important role of intestinal CGI-58 in mobilizing intestinal LDs for local and/or systemic utilization. Consistently, hepatic steatosis is attenuated in the intestine CGI-58 single or CGI-58/ATGL double knockout mice [106, 240]. Using intestine-specific CGI-58 knockout mice fed a synthetic diet containing 40% energy from lard and 0.2% (w/w) cholesterol, our laboratory has shown that intestinal absorption of total fat and long-chain fatty acids is significantly reduced, which is associated with reduced postprandial TAG section into the blood circulation and increased plasma concentrations of free and esterified cholesterol [240]. For reasons currently unknown, another group did not find similar changes in their intestine CGI-58 and ATGL single or double knockout mice fed a diet containing 60% energy from fat [34% (w/w) crude fat] and 1% (w/w) cholesterol. They instead showed a role of intestinal CGI-58 and ATGL in the turnover of lipids derived from the basolateral side of the absorptive enterocytes [106, 152]. More studies are clearly needed to address these controversial findings.

13.7 CGI-58 and Cancer

Cancer cells often accumulate LDs in the cytoplasm [17, 210]. The underlying mechanisms remain elusive. Sequestration of lipids in cytosolic LDs may protect cancer cells from lipotoxic stress [93]. Mutations in CGI-58 cause LD deposition in cells, which led to the first study exploring the role of CGI-58 in colorectal cancer development [158]. It was shown that CGI-58 deficiency promotes the epithelial-mesenchymal transition (EMT) and invasiveness of colorectal cancer cells by increasing aerobic glycolysis (the Warburg Effect) [158]. The increase in aerobic glycolysis in CGI-58-deficient cells may result from limited availability of fatty acids due to defective LD lipolysis. In addition, CGI-58 was shown to promote colorectal tumorigenesis by impairing Beclin1-mediated autophagy [163]. A subsequent study with prostate cancer cells was consistent with the tumor suppressor role of CGI-58 [37]. However, another group using the same prostate cancer cell line found that CGI-58 sustains cancer cell growth by inhibiting cell apoptosis and death [140]. CGI-58 was recently shown to be oncogenic in endometrial cancer [261]. It was reported that CGI-58 in tumor-associated macrophages indirectly promotes colorectal cancer growth by suppressing spermidine synthesis [136]. The same group also reported that CGI-58 suppresses NFκB-dependent metalloproteinase production in macrophages to indirectly inhibit colorectal cancer cell metastasis [199]. Besides regulating tumorigenesis directly and indirectly, CGI-58 was reported to inhibit the sensitivity of colorectal cancer cells to the chemotherapy drug fluorouracil [159]. CGI-58 expression patterns and levels may serve as markers for differentiating benign and malignant tumors in some tissues [36, 158]. DNA methylation and deletion may influence CGI-58 expression in some cancer types, such as cervical cancer [198]. CGI-58 is not the only LD-associated protein that is implicated in cancer development and progression. It was shown that ATGL mediates cancer-associated cachexia [42], correlates with the risk of pancreatic ductal adenocarcinoma [68], and promotes malignancies of breast cancer and hepatocellular carcinoma [43, 124, 125, 232, 249]. It was reported that ATGL deletion is linked to the aggressiveness of A549 lung carcinoma cells [218]. Inhibition of ATGL by the lipolysis suppressor protein G0S2 or a small molecule Atglistatin was found to attenuate the growth of cancer cells [256]. G0S2 was also observed to suppress oncogenic transformation of immortalized mouse embryonic fibroblasts [252]. Interestingly, inhibition of ATGL by hypoxia-inducible gene 2 (HIG2), unlike G0S2, was demonstrated to promote survival of colorectal cancer and renal cell carcinoma cell lines in hypoxia [260]. The role of LD-associated proteins CGI-58, ATGL, G0S2, and HIG2 in tumorigenesis may be cell type-specific, depending on how each cell type handles energy metabolism and signal transduction under different pathophysiological conditions.

13.8 CGI-58 and HCV Infection

A large proportion of patients chronically infected with hepatitis C virus (HCV) manifest LD deposition in the liver in the absence of other steatotic factors [147]. It was shown that the HCV nucleocapsid core, which is the major structural component of HCV virions, localizes at the surface of LDs to inhibit LD turnover in cultured cells and mouse livers [80]. The same group further showed that the HCV core inhibits ATGL-dependent LD lipolysis, but it unexpectedly enhances ATGL interaction with CGI-58 and the recruitment of the ATGL/CGI-58 complex to LDs [26]. Interestingly, an siRNA-based screen identified CGI-58 as a host factor that assists HCV assembly and release without affecting virus entry and replication [225]. They showed that several CDS-causing mutants of CGI-58 fail to localize at the surface of LDs, and those mutants are unable to support HCV production. Moreover, they identified a tribasic motif (KRK233-235) that is required for CGI-58 to promote lipolysis and HCV production, though not essential for CGI-58 localization to LDs. While this study may suggest that it is its lipase coactivator function that mediates HCV assembly and release, it remains unknown whether the newly identified serine protease function of CGI-58 is implicated in HCV production [94].

13.9 Concluding Remarks

Patients with CDS accumulate TAG-rich LDs in all cell types examined. Since the discovery of *CGI-58* gene mutations as the cause of CDS in 2001, enormous interest on the function of CGI-58 has been generated in the scientific community of lipid and energy metabolism. It has been well established that CGI-58 is a LD-associated protein that promotes intracellular LD lipolysis by activating ATGL's TAG hydrolase activity. In addition to ATGL, CGI-58 interacts with many other proteins and regulates LD dynamics and functions in a cell type-specific manner. Such broad protein-protein interactions of CGI-58 have provided important insights into the biochemical basis for its ATGL-independent functions. Future studies are needed to dissect the molecular itineraries of these interactions in regulation of intracellular LD biogenesis and turnover. As a versatile regulator of intracellular LD homeostasis, CGI-58 plays a central role in governing cellular and whole-body energy balance. Genetic deletion of CGI-58 in mice has uncovered distinct effects of LD deposition in different cell types on the pathogenesis of metabolic disease. CGI-58 was recently identified to possess the serine protease activity in the heart. It is unknown if CGI-58 has this protease activity in other tissue. If yes, what are the substrates and functional significance? Is the serine protease activity of CGI-58 coordinate with its lipase coactivator function to activate intracellular lipolysis? Clearly, more studies are needed to answer these exciting new questions.

Acknowledgment This work was partially supported by the grants DK111052 (to L.Y.) and DK116496 (to L. Y.) from the National Institute of Diabetes and Digestive and Kidney Diseases, the grant 1-18-IBS-346 (to L.Y.) from the American Diabetes Association Research Foundation, and the grant 17GRNT33670590 (to L.Y.) from the American Heart Association.

References

1. Adant I, Declercq M, Bird M, Bauters M, Boeckx N, Devriendt K, Cassiman D, Witters P (2020) Two cases of non-alcoholic fatty liver disease caused by biallelic ABHD5 mutations. J Hepatol 72 (5):1030–1032
2. Aflaki E, Balenga NA, Luschnig-Schratl P, Wolinski H, Povoden S, Chandak PG, Bogner-Strauss JG, Eder S, Konya V, Kohlwein SD, Heinemann A, Kratky D (2011) Impaired Rho GTPase activation abrogates cell polarization and migration in macrophages with defective lipolysis. Cell Mol Life Sci 68:3933–3947
3. Aggarwal S, Maras JS, Alam S, Khanna R, Gupta SK, Ahuja A (2012) Novel nonsense mutation of ABHD5 in Dorfman-Chanarin syndrome with unusual findings: a challenge for genotype-phenotype correlation. Eur J Med Genet 55:173–177
4. Ahmadian M, Abbott MJ, Tang T, Hudak CS, Kim Y, Bruss M, Hellerstein MK, Lee HY, Samuel VT, Shulman GI, Wang Y, Duncan RE, Kang C, Sul HS (2011) Desnutrin/ATGL is regulated by AMPK and is required for a brown adipose phenotype. Cell Metab 13:739–748
5. Akiyama M, Sawamura D, Nomura Y, Sugawara M, Shimizu H (2003) Truncation of CGI-58 protein causes malformation of lamellar granules resulting in ichthyosis in Dorfman-Chanarin syndrome. J Invest Dermatol 121:1029–1034
6. Akiyama M, Sakai K, Ogawa M, McMillan JR, Sawamura D, Shimizu H (2007) Novel duplication mutation in the patatin domain of adipose triglyceride lipase (PNPLA2) in neutral lipid storage disease with severe myopathy. Muscle Nerve 36:856–859
7. Al-Hage J, Abbas O, Nemer G, Kurban M (2019) Chanarin-Dorfman syndrome: a novel homozygous mutation in the ABHD5 gene. Clin Exp Dermatol 45 (2):257–259
8. Ameziane N, Beillat T, Verpillat P, Chollet-Martin S, Aumont MC, Seknadji P, Lamotte M, Lebret D, Ollivier V, de Prost D (2003) Association of the Toll-like receptor 4 gene Asp299Gly polymorphism with acute coronary events. Arterioscler Thromb Vasc Biol 23:e61–e64
9. Badeloe S, van Geel M, Nagtzaam I, Rubio-Gozalbo ME, Oei RL, Steijlen PM, van Steensel MA (2008) Chanarin-Dorfman syndrome caused by a novel splice site mutation in ABHD5. Br J Dermatol 158:1378–1380
10. Badin PM, Loubiere C, Coonen M, Louche K, Tavernier G, Bourlier V, Mairal A, Rustan AC, Smith SR, Langin D, Moro C (2012) Regulation of skeletal muscle lipolysis and oxidative metabolism by the co-lipase CGI-58. J Lipid Res 53:839–848
11. BasuRay S, Wang Y, Smagris E, Cohen JC, Hobbs HH (2019) Accumulation of PNPLA3 on lipid droplets is the basis of associated hepatic steatosis. Proc Natl Acad Sci U S A 116:9521–9526

12. Ben Selma Z, Yilmaz S, Schischmanoff PO, Blom A, Ozogul C, Laroche L, Caux F (2007) A novel S115G mutation of CGI-58 in a Turkish patient with Dorfman-Chanarin syndrome. J Invest Dermatol 127:2273–2276
13. Bergman R, Aviram M, Bitterman-Deutsch O, Oiknine Y, Shemer A, Srebnik A, Brook JG, Friedman-Birnbaum R (1991) Neutral lipid storage disease with ichthyosis: serum apolipoprotein levels and cholesterol metabolism in monocyte-derived macrophages. J Inherit Metab Dis 14:241–246
14. Bjorkbacka H (2006) Multiple roles of Toll-like receptor signaling in atherosclerosis. Curr Opin Lipidol 17:527–533
15. Blondin DP, Frisch F, Phoenix S, Guerin B, Turcotte EE, Haman F, Richard D, Carpentier AC (2017) Inhibition of intracellular triglyceride lipolysis suppresses cold-induced brown adipose tissue metabolism and increases shivering in humans. Cell Metab 25:438–447
16. Boeszoermenyi A, Nagy HM, Arthanari H, Pillip CJ, Lindermuth H, Luna RE, Wagner G, Zechner R, Zangger K, Oberer M (2015) Structure of a CGI-58 motif provides the molecular basis of lipid droplet anchoring. J Biol Chem 290:26361–26372
17. Bozza PT, Viola JP (2010) Lipid droplets in inflammation and cancer. Prostaglandins Leukot Essent Fatty Acids 82:243–250
18. Brown JM, Chung S, Das A, Shelness GS, Rudel LL, Yu L (2007) CGI-58 facilitates the mobilization of cytoplasmic triglyceride for lipoprotein secretion in hepatoma cells. J Lipid Res 48:2295–2305
19. Brown JM, Betters JL, Lord C, Ma Y, Han X, Yang K, Alger HM, Melchior J, Sawyer J, Shah R, Wilson MD, Liu X, Graham MJ, Lee R, Crooke R, Shulman GI, Xue B, Shi H, Yu L (2010) CGI-58 knockdown in mice causes hepatic steatosis but prevents diet-induced obesity and glucose intolerance. J Lipid Res 51:3306–3315
20. Browning JD, Horton JD (2004) Molecular mediators of hepatic steatosis and liver injury. J Clin Invest 114:147–152
21. Bruning JC, Michael MD, Winnay JN, Hayashi T, Horsch D, Accili D, Goodyear LJ, Kahn CR (1998) A muscle-specific insulin receptor knockout exhibits features of the metabolic syndrome of NIDDM without altering glucose tolerance. Mol Cell 2:559–569
22. Bruno C, Bertini E, Di Rocco M, Cassandrini D, Ruffa G, De Toni T, Seri M, Spada M, Li Volti G, D'Amico A, Trucco F, Arca M, Casali C, Angelini C, Dimauro S, Minetti C (2008) Clinical and genetic characterization of Chanarin-Dorfman syndrome. Biochem Biophys Res Commun 369:1125–1128
23. Cakir M, Bruno C, Cansu A, Cobanoglu U, Erduran E (2010) Liver cirrhosis in an infant with Chanarin-Dorfman syndrome caused by a novel splice-site mutation in ABHD5. Acta Paediatr 99:1592–1594
24. Cakmak E, Alagozlu H, Yonem O, Ataseven H, Citli S, Ozer H (2012) Steatohepatitis and liver cirrhosis in Chanarin-Dorfman syndrome with a new ABDH5 mutation. Clin Res Hepatol Gastroenterol 36:e34–e37
25. Campagna F, Nanni L, Quagliarini F, Pennisi E, Michailidis C, Pierelli F, Bruno C, Casali C, DiMauro S, Arca M (2008) Novel mutations in the adipose triglyceride lipase gene causing neutral lipid storage disease with myopathy. Biochem Biophys Res Commun 377:843–846
26. Camus G, Schweiger M, Herker E, Harris C, Kondratowicz AS, Tsou CL, Farese RV Jr, Herath K, Previs SF, Roddy TP, Pinto S, Zechner R, Ott M (2014) The hepatitis C virus core protein inhibits adipose triglyceride lipase (ATGL)-mediated lipid mobilization and enhances the ATGL interaction with comparative gene identification 58 (CGI-58) and lipid droplets. J Biol Chem 289:35770–35780
27. Cannon B, Nedergaard J (2004) Brown adipose tissue: function and physiological significance. Physiol Rev 84:277–359
28. Cantley JL, Yoshimura T, Camporez JP, Zhang D, Jornayvaz FR, Kumashiro N, Guebre-Egziabher F, Jurczak MJ, Kahn M, Guigni BA, Serr J, Hankin J, Murphy RC, Cline GW, Bhanot S, Manchem VP, Brown JM, Samuel VT, Shulman GI (2013) CGI-58 knockdown sequesters diacylglycerols in lipid droplets/ER-preventing diacylglycerol-mediated hepatic insulin resistance. Proc Natl Acad Sci U S A 110:1869–1874
29. Carr RM, Ahima RS (2016) Pathophysiology of lipid droplet proteins in liver diseases. Exp Cell Res 340:187–192
30. Carr RM, Patel RT, Rao V, Dhir R, Graham MJ, Crooke RM, Ahima RS (2012) Reduction of TIP47 improves hepatic steatosis and glucose homeostasis in mice. Am J Physiol Regul Integr Comp Physiol 302:R996–R1003
31. Cartwright IJ, Plonne D, Higgins JA (2000) Intracellular events in the assembly of chylomicrons in rabbit enterocytes. J Lipid Res 41:1728–1739
32. Chamoun Z, Vacca F, Parton RG, Gruenberg J (2013) PNPLA3/adiponutrin functions in lipid droplet formation. Biol Cell 105:219–233
33. Chanarin I, Patel A, Slavin G, Wills EJ, Andrews TM, Stewart G (1975) Neutral-lipid storage disease: a new disorder of lipid metabolism. Br Med J 1:553–555
34. Chang BH, Li L, Paul A, Taniguchi S, Nannegari V, Heird WC, Chan L (2006) Protection against fatty liver but normal adipogenesis in mice lacking adipose differentiation-related protein. Mol Cell Biol 26:1063–1076
35. Chen W, Chang B, Wu X, Li L, Sleeman M, Chan L (2013a) Inactivation of Plin4 downregulates Plin5 and reduces cardiac lipid accumulation in mice. Am J Physiol Endocrinol Metab 304:E770–E779
36. Chen WS, Chen PL, Li J, Lind AC, Lu D (2013b) Lipid synthesis and processing proteins ABHD5,

PGRMC1 and squalene synthase can serve as novel immunohistochemical markers for sebaceous neoplasms and differentiate sebaceous carcinoma from sebaceoma and basal cell carcinoma with clear cell features. J Cutan Pathol 40:631–638

37. Chen G, Zhou G, Aras S, He Z, Lucas S, Podgorski I, Skar W, Granneman JG, Wang J (2017) Loss of ABHD5 promotes the aggressiveness of prostate cancer cells. Sci Rep 7:13021
38. Ciesek S, Hadem J, Fischer J, Manns MP, Strassburg CP (2006) A rare cause of nonalcoholic fatty liver disease. Ann Intern Med 145:154–155
39. Cinti S (2002) Adipocyte differentiation and transdifferentiation: plasticity of the adipose organ. J Endocrinol Invest 25:823–835
40. Cook KG, Yeaman SJ, Stralfors P, Fredrikson G, Belfrage P (1982) Direct evidence that cholesterol ester hydrolase from adrenal cortex is the same enzyme as hormone-sensitive lipase from adipose tissue. Eur J Biochem 125:245–249
41. Cornaciu I, Boeszoermenyi A, Lindermuth H, Nagy HM, Cerk IK, Ebner C, Salzburger B, Gruber A, Schweiger M, Zechner R, Lass A, Zimmermann R, Oberer M (2011) The minimal domain of adipose triglyceride lipase (ATGL) ranges until leucine 254 and can be activated and inhibited by CGI-58 and G0S2, respectively. PLoS One 6:e26349
42. Das SK, Eder S, Schauer S, Diwoky C, Temmel H, Guertl B, Gorkiewicz G, Tamilarasan KP, Kumari P, Trauner M, Zimmermann R, Vesely P, Haemmerle G, Zechner R, Hoefler G (2011) Adipose triglyceride lipase contributes to cancer-associated cachexia. Science 333:233–238
43. Di Leo L, Vegliante R, Ciccarone F, Salvatori I, Scimeca M, Bonanno E, Sagnotta A, Grazi GL, Aquilano K, Ciriolo MR (2019) Forcing ATGL expression in hepatocarcinoma cells imposes glycolytic rewiring through PPAR-alpha/p300-mediated acetylation of p53. Oncogene 38:1860–1875
44. Doi K, Ohno T, Kurahashi M, Kuroshima A (1979) Thermoregulatory nonshivering thermogenesis in men, with special reference to lipid metabolism. Jpn J Physiol 29:359–372
45. Dong H, Czaja MJ (2011) Regulation of lipid droplets by autophagy. Trends Endocrinol Metab 22:234–240
46. Dorfman ML, Hershko C, Eisenberg S, Sagher F (1974) Ichthyosiform dermatosis with systemic lipidosis. Arch Dermatol 110:261–266
47. Duncan RE, Ahmadian M, Jaworski K, Sarkadi-Nagy E, Sul HS (2007) Regulation of lipolysis in adipocytes. Annu Rev Nutr 27:79–101
48. Dupont N, Chauhan S, Arko-Mensah J, Castillo EF, Masedunskas A, Weigert R, Robenek H, Proikas-Cezanne T, Deretic V (2014) Neutral lipid stores and lipase PNPLA5 contribute to autophagosome biogenesis. Curr Biol 24:609–620
49. Durdu M, Missaglia S, Moro L, Tavian D (2018) Clinical and genetic characterization of a Chanarin Dorfman Syndrome patient born to diseased parents. BMC Med Genet 19:88
50. Dutton JR, Chillingworth NL, Eberhard D, Brannon CR, Hornsey MA, Tosh D, Slack JM (2007) Beta cells occur naturally in extrahepatic bile ducts of mice. J Cell Sci 120:239–245
51. Eid N, Ito Y, Maemura K, Otsuki Y (2013) Elevated autophagic sequestration of mitochondria and lipid droplets in steatotic hepatocytes of chronic ethanol-treated rats: an immunohistochemical and electron microscopic study. J Mol Histol 44:311–326
52. Emre S, Unver N, Evans SE, Yuzbasioglu A, Gurakan F, Gumruk F, Karaduman A (2010) Molecular analysis of Chanarin-Dorfman syndrome (CDS) patients: Identification of novel mutations in the ABHD5 gene. Eur J Med Genet 53:141–144
53. Epstein SE, Zhu J, Burnett MS, Zhou YF, Vercellotti G, Hajjar D (2000) Infection and atherosclerosis: potential roles of pathogen burden and molecular mimicry. Arterioscler Thromb Vasc Biol 20:1417–1420
54. Eskiocak AH, Missaglia S, Moro L, Durdu M, Tavian D (2019) A novel mutation of ABHD5 gene in a Chanarin Dorfman patient with unusual dermatological findings. Lipids Health Dis 18:232
55. Fain JN, Garcija-Sainz JA (1983) Adrenergic regulation of adipocyte metabolism. J Lipid Res 24:945–966
56. Feingold KR, Shigenaga JK, Kazemi MR, McDonald CM, Patzek SM, Cross AS, Moser A, Grunfeld C (2012) Mechanisms of triglyceride accumulation in activated macrophages. J Leukoc Biol 92:829–839
57. Finn PF, Dice JF (2006) Proteolytic and lipolytic responses to starvation. Nutrition 22:830–844
58. Fischer J, Lefevre C, Morava E, Mussini JM, Laforet P, Negre-Salvayre A, Lathrop M, Salvayre R (2007) The gene encoding adipose triglyceride lipase (PNPLA2) is mutated in neutral lipid storage disease with myopathy. Nat Genet 39:28–30
59. Friedman SL (2008) Mechanisms of hepatic fibrogenesis. Gastroenterology 134:1655–1669
60. Friedman SL, Neuschwander-Tetri BA, Rinella M, Sanyal AJ (2018) Mechanisms of NAFLD development and therapeutic strategies. Nat Med 24:908–922
61. Fujii H, Ikura Y, Arimoto J, Sugioka K, Iezzoni JC, Park SH, Naruko T, Itabe H, Kawada N, Caldwell SH, Ueda M (2009) Expression of perilipin and adipophilin in nonalcoholic fatty liver disease; relevance to oxidative injury and hepatocyte ballooning. J Atheroscler Thromb 16:893–901
62. Funk JL, Feingold KR, Moser AH, Grunfeld C (1993) Lipopolysaccharide stimulation of RAW 264.7 macrophages induces lipid accumulation and foam cell formation. Atherosclerosis 98:67–82
63. Gandotra S, Lim K, Girousse A, Saudek V, O'Rahilly S, Savage DB (2011) Human frame shift mutations affecting the carboxyl terminus of perilipin increase lipolysis by failing to sequester the adipose triglyceride lipase (ATGL) coactivator AB-

hydrolase-containing 5 (ABHD5). J Biol Chem 286:34998–35006
64. Geisler F, Nagl F, Mazur PK, Lee M, Zimber-Strobl-U, Strobl LJ, Radtke F, Schmid RM, Siveke JT (2008) Liver-specific inactivation of Notch2, but not Notch1, compromises intrahepatic bile duct development in mice. Hepatology 48:607–616
65. Goeritzer M, Schlager S, Radovic B, Madreiter CT, Rainer S, Thomas G, Lord CC, Sacks J, Brown AL, Vujic N, Obrowsky S, Sachdev V, Kolb D, Chandak PG, Graier WF, Sattler W, Brown JM, Kratky D (2014) Deletion of CGI-58 or adipose triglyceride lipase differently affects macrophage function and atherosclerosis. J Lipid Res 55:2562–2575
66. Goeritzer M, Vujic N, Schlager S, Chandak PG, Korbelius M, Gottschalk B, Leopold C, Obrowsky S, Rainer S, Doddapattar P, Aflaki E, Wegscheider M, Sachdev V, Graier WF, Kolb D, Radovic B, Kratky D (2015) Active autophagy but not lipophagy in macrophages with defective lipolysis. Biochim Biophys Acta 1851:1304–1316
67. Goodpaster BH, He J, Watkins S, Kelley DE (2001) Skeletal muscle lipid content and insulin resistance: evidence for a paradox in endurance-trained athletes. J Clin Endocrinol Metab 86:5755–5761
68. Grace SA, Meeks MW, Chen Y, Cornwell M, Ding X, Hou P, Rutgers JK, Crawford SE, Lai JP (2017) Adipose triglyceride lipase (ATGL) expression is associated with adiposity and tumor stromal proliferation in patients with pancreatic ductal adenocarcinoma. Anticancer Res 37:699–703
69. Grall A, Guaguere E, Planchais S, Grond S, Bourrat E, Hausser I, Hitte C, Le Gallo M, Derbois C, Kim GJ, Lagoutte L, Degorce-Rubiales F, Radner FP, Thomas A, Kury S, Bensignor E, Fontaine J, Pin D, Zimmermann R, Zechner R, Lathrop M, Galibert F, Andre C, Fischer J (2012) PNPLA1 mutations cause autosomal recessive congenital ichthyosis in golden retriever dogs and humans. Nat Genet 44:140–147
70. Granneman JG, Moore HP, Granneman RL, Greenberg AS, Obin MS, Zhu Z (2007) Analysis of lipolytic protein trafficking and interactions in adipocytes. J Biol Chem 282:5726–5735
71. Granneman JG, Moore HP, Krishnamoorthy R, Rathod M (2009a) Perilipin controls lipolysis by regulating the interactions of AB-hydrolase containing 5 (Abhd5) and adipose triglyceride lipase (Atgl). J Biol Chem 284:34538–34544
72. Granneman JG, Moore HP, Mottillo EP, Zhu Z (2009b) Functional interactions between Mldp (LSDP5) and Abhd5 in the control of intracellular lipid accumulation. J Biol Chem 284:3049–3057
73. Grond S, Radner FPW, Eichmann TO, Kolb D, Grabner GF, Wolinski H, Gruber R, Hofer P, Heier C, Schauer S, Rulicke T, Hoefler G, Schmuth M, Elias PM, Lass A, Zechner R, Haemmerle G (2017) Skin barrier development depends on CGI-58 protein expression during late-stage keratinocyte differentiation. J Invest Dermatol 137:403–413
74. Gruber A, Cornaciu I, Lass A, Schweiger M, Poeschl M, Eder C, Kumari M, Schoiswohl G, Wolinski H, Kohlwein SD, Zechner R, Zimmermann R, Oberer M (2010) The N-terminal region of comparative gene identification-58 (CGI-58) is important for lipid droplet binding and activation of adipose triglyceride lipase. J Biol Chem 285:12289–12298
75. Guo F, Ma Y, Kadegowda AK, Xie P, Liu G, Liu X, Miao H, Ou J, Su X, Zheng Z, Xue B, Shi H, Yu L (2013) Deficiency of liver Comparative Gene Identification-58 causes steatohepatitis and fibrosis in mice. J Lipid Res 54:2109–2120
76. Gurakan F, Kaymaz F, Kocak N, Ors U, Yuce A, Atakan N (1999) A cause of fatty liver: neutral lipid storage disease with ichthyosis--electron microscopic findings. Dig Dis Sci 44:2214–2217
77. Haemmerle G, Zimmermann R, Hayn M, Theussl C, Waeg G, Wagner E, Sattler W, Magin TM, Wagner EF, Zechner R (2002) Hormone-sensitive lipase deficiency in mice causes diglyceride accumulation in adipose tissue, muscle, and testis. J Biol Chem 277:4806–4815
78. Haemmerle G, Lass A, Zimmermann R, Gorkiewicz G, Meyer C, Rozman J, Heldmaier G, Maier R, Theussl C, Eder S, Kratky D, Wagner EF, Klingenspor M, Hoefler G, Zechner R (2006) Defective lipolysis and altered energy metabolism in mice lacking adipose triglyceride lipase. Science 312:734–737
79. Haemmerle G, Moustafa T, Woelkart G, Buttner S, Schmidt A, van de Weijer T, Hesselink M, Jaeger D, Kienesberger PC, Zierler K, Schreiber R, Eichmann T, Kolb D, Kotzbeck P, Schweiger M, Kumari M, Eder S, Schoiswohl G, Wongsiriroj N, Pollak NM, Radner FP, Preiss-Landl K, Kolbe T, Rulicke T, Pieske B, Trauner M, Lass A, Zimmermann R, Hoefler G, Cinti S, Kershaw EE, Schrauwen P, Madeo F, Mayer B, Zechner R (2011) ATGL-mediated fat catabolism regulates cardiac mitochondrial function via PPAR-alpha and PGC-1. Nat Med 17:1076–1085
80. Harris C, Herker E, Farese RV Jr, Ott M (2011) Hepatitis C virus core protein decreases lipid droplet turnover: a mechanism for core-induced steatosis. J Biol Chem 286:42615–42625
81. Hernandez-Gea V, Friedman SL (2011) Pathogenesis of liver fibrosis. Annu Rev Pathol 6:425–456
82. Higashi T, Friedman SL, Hoshida Y (2017) Hepatic stellate cells as key target in liver fibrosis. Adv Drug Deliv Rev 121:27–42
83. Himms-Hagen J, Melnyk A, Zingaretti MC, Ceresi E, Barbatelli G, Cinti S (2000) Multilocular fat cells in WAT of CL-316243-treated rats derive directly from white adipocytes. Am J Physiol Cell Physiol 279: C670–C681

84. Hirabayashi T, Anjo T, Kaneko A, Senoo Y, Shibata A, Takama H, Yokoyama K, Nishito Y, Ono T, Taya C, Muramatsu K, Fukami K, Munoz-Garcia A, Brash AR, Ikeda K, Arita M, Akiyama M, Murakami M (2017) PNPLA1 has a crucial role in skin barrier function by directing acylceramide biosynthesis. Nat Commun 8:14609
85. Hofer P, Boeszoermenyi A, Jaeger D, Feiler U, Arthanari H, Mayer N, Zehender F, Rechberger G, Oberer M, Zimmermann R, Lass A, Haemmerle G, Breinbauer R, Zechner R, Preiss-Landl K (2015) Fatty acid-binding proteins interact with comparative gene identification-58 linking lipolysis with lipid ligand shuttling. J Biol Chem 290:18438–18453
86. Holm C, Kirchgessner TG, Svenson KL, Fredrikson G, Nilsson S, Miller CG, Shively JE, Heinzmann C, Sparkes RS, Mohandas T et al (1988) Hormone-sensitive lipase: sequence, expression, and chromosomal localization to 19 cent-q13.3. Science 241:1503–1506
87. Hoy AJ, Bruce CR, Turpin SM, Morris AJ, Febbraio MA, Watt MJ (2011) Adipose triglyceride lipase-null mice are resistant to high-fat diet-induced insulin resistance despite reduced energy expenditure and ectopic lipid accumulation. Endocrinology 152:48–58
88. Hsieh K, Lee YK, Londos C, Raaka BM, Dalen KT, Kimmel AR (2012) Perilipin family members preferentially sequester to either triacylglycerol-specific or cholesteryl-ester-specific intracellular lipid storage droplets. J Cell Sci 125:4067–4076
89. Huigen MC, van der Graaf M, Morava E, Dassel AC, van Steensel MA, Seyger MM, Wevers RA, Willemsen MA (2015) Cerebral lipid accumulation in Chanarin-Dorfman Syndrome. Mol Genet Metab 114:51–54
90. Igal RA, Rhoads JM, Coleman RA (1997) Neutral lipid storage disease with fatty liver and cholestasis. J Pediatr Gastroenterol Nutr 25:541–547
91. Imai Y, Varela GM, Jackson MB, Graham MJ, Crooke RM, Ahima RS (2007) Reduction of hepatosteatosis and lipid levels by an adipose differentiation-related protein antisense oligonucleotide. Gastroenterology 132:1947–1954
92. Israeli S, Pessach Y, Sarig O, Goldberg I, Sprecher E (2012) Beneficial effect of acitretin in Chanarin-Dorfman syndrome. Clin Exp Dermatol 37:31–33
93. Jarc E, Eichmann TO, Zimmermann R, Petan T (2018) Lipidomic data on lipid droplet triglyceride remodelling associated with protection of breast cancer cells from lipotoxic stress. Data Brief 18:234–240
94. Jebessa ZH, Shanmukha Kumar D, Dewenter M, Lehmann LH, Xu C, Schreiter F, Siede D, Gong XM, Worst BC, Federico G, Sauer SW, Fischer T, Wechselberger L, Muller OJ, Sossalla S, Dieterich C, Most P, Grone HJ, Moro C, Oberer M, Haemmerle G, Katus HA, Tyedmers J, Backs J (2019) The lipid droplet-associated protein ABHD5 protects the heart through proteolysis of HDAC4. Nat Metab 1:1157–1167
95. Jenkins CM, Mancuso DJ, Yan W, Sims HF, Gibson B, Gross RW (2004) Identification, cloning, expression, and purification of three novel human calcium-independent phospholipase A2 family members possessing triacylglycerol lipase and acylglycerol transacylase activities. J Biol Chem 279:48968–48975
96. Jordans GH (1953) The familial occurrence of fat containing vacuoles in the leukocytes diagnosed in two brothers suffering from dystrophia musculorum progressiva (ERB.). Acta Med Scand 145:419–423
97. Kajimura S, Spiegelman BM, Seale P (2015) Brown and beige fat: physiological roles beyond heat generation. Cell Metab 22:546–559
98. Karlsson M, Contreras JA, Hellman U, Tornqvist H, Holm C (1997) cDNA cloning, tissue distribution, and identification of the catalytic triad of monoglyceride lipase. Evolutionary relationship to esterases, lysophospholipases, and haloperoxidases. J Biol Chem 272:27218–27223
99. Kato M, Higuchi N, Enjoji M (2008) Reduced hepatic expression of adipose tissue triglyceride lipase and CGI-58 may contribute to the development of non-alcoholic fatty liver disease in patients with insulin resistance. Scand J Gastroenterol 43:1–2
100. Kaushik S, Cuervo AM (2015) Degradation of lipid droplet-associated proteins by chaperone-mediated autophagy facilitates lipolysis. Nat Cell Biol 17:759–770
101. Kiechl S, Lorenz E, Reindl M, Wiedermann CJ, Oberhollenzer F, Bonora E, Willeit J, Schwartz DA (2002) Toll-like receptor 4 polymorphisms and atherogenesis. N Engl J Med 347:185–192
102. Kien B, Grond S, Haemmerle G, Lass A, Eichmann TO, Radner FPW (2018) ABHD5 stimulates PNPLA1-mediated omega-O-acylceramide biosynthesis essential for a functional skin permeability barrier. J Lipid Res 59:2360–2367
103. Kienesberger PC, Lee D, Pulinilkunnil T, Brenner DS, Cai L, Magnes C, Koefeler HC, Streith IE, Rechberger GN, Haemmerle G, Flier JS, Zechner R, Kim YB, Kershaw EE (2009a) Adipose triglyceride lipase deficiency causes tissue-specific changes in insulin signaling. J Biol Chem 284:30218–30229
104. Kienesberger PC, Oberer M, Lass A, Zechner R (2009b) Mammalian patatin domain containing proteins: a family with diverse lipolytic activities involved in multiple biological functions. J Lipid Res 50(Suppl):S63–S68
105. Kimmel AR, Brasaemle DL, McAndrews-Hill M, Sztalryd C, Londos C (2010) Adoption of PERILIPIN as a unifying nomenclature for the mammalian PAT-family of intracellular lipid storage droplet proteins. J Lipid Res 51:468–471
106. Korbelius M, Vujic N, Sachdev V, Obrowsky S, Rainer S, Gottschalk B, Graier WF, Kratky D (2019) ATGL/CGI-58-dependent hydrolysis of a

lipid storage pool in murine enterocytes. Cell Rep 28:1923-34 e4

107. Labbe SM, Caron A, Bakan I, Laplante M, Carpentier AC, Lecomte R, Richard D (2015) In vivo measurement of energy substrate contribution to cold-induced brown adipose tissue thermogenesis. FASEB J 29:2046–2058
108. Lafontan M, Langin D (2009) Lipolysis and lipid mobilization in human adipose tissue. Prog Lipid Res 48:275–297
109. Lai CH, Chou CY, Ch'ang LY, Liu CS, Lin W (2000) Identification of novel human genes evolutionarily conserved in Caenorhabditis elegans by comparative proteomics. Genome Res 10:703–713
110. Lammers B, Chandak PG, Aflaki E, Van Puijvelde GH, Radovic B, Hildebrand RB, Meurs I, Out R, Kuiper J, Van Berkel TJ, Kolb D, Haemmerle G, Zechner R, Levak-Frank S, Van Eck M, Kratky D (2011) Macrophage adipose triglyceride lipase deficiency attenuates atherosclerotic lesion development in low-density lipoprotein receptor knockout mice. Arterioscler Thromb Vasc Biol 31:67–73
111. Langin D, Laurell H, Holst LS, Belfrage P, Holm C (1993) Gene organization and primary structure of human hormone-sensitive lipase: possible significance of a sequence homology with a lipase of Moraxella TA144, an antarctic bacterium. Proc Natl Acad Sci U S A 90:4897–4901
112. Lass A, Zimmermann R, Haemmerle G, Riederer M, Schoiswohl G, Schweiger M, Kienesberger P, Strauss JG, Gorkiewicz G, Zechner R (2006) Adipose triglyceride lipase-mediated lipolysis of cellular fat stores is activated by CGI-58 and defective in Chanarin-Dorfman Syndrome. Cell Metab 3:309–319
113. Lee FT, Adams JB, Garton AJ, Yeaman SJ (1988) Hormone-sensitive lipase is involved in the hydrolysis of lipoidal derivatives of estrogens and other steroid hormones. Biochim Biophys Acta 963:258–264
114. Lee YH, Petkova AP, Mottillo EP, Granneman JG (2012) In vivo identification of bipotential adipocyte progenitors recruited by beta3-adrenoceptor activation and high-fat feeding. Cell Metab 15:480–491
115. Lee YH, Petkova AP, Konkar AA, Granneman JG (2015) Cellular origins of cold-induced brown adipocytes in adult mice. FASEB J 29:286–299
116. Lefevre C, Jobard F, Caux F, Bouadjar B, Karaduman A, Heilig R, Lakhdar H, Wollenberg A, Verret JL, Weissenbach J, Ozguc M, Lathrop M, Prud'homme JF, Fischer J (2001) Mutations in CGI-58, the gene encoding a new protein of the esterase/lipase/thioesterase subfamily, in Chanarin-Dorfman syndrome. Am J Hum Genet 69:1002–1012
117. Lehr HA, Sagban TA, Ihling C, Zahringer U, Hungerer KD, Blumrich M, Reifenberg K, Bhakdi S (2001) Immunopathogenesis of atherosclerosis: endotoxin accelerates atherosclerosis in rabbits on hypercholesterolemic diet. Circulation 104:914–920
118. Levin MC, Monetti M, Watt MJ, Sajan MP, Stevens RD, Bain JR, Newgard CB, Farese RV Sr, Farese RV Jr (2007) Increased lipid accumulation and insulin resistance in transgenic mice expressing DGAT2 in glycolytic (type II) muscle. Am J Physiol Endocrinol Metab 293:E1772–E1781
119. Li Y, Schnabl K, Gabler SM, Willershauser M, Reber J, Karlas A, Laurila S, Lahesmaa M, Din U, Bast-Habersbrunner MA, Virtanen KA, Fromme T, Bolze F, O'Farrell LS, Alsina-Fernandez J, Coskun T, Ntziachristos V, Nuutila P, Klingenspor M (2018) Secretin-activated brown fat mediates prandial thermogenesis to induce satiation. Cell 175:1561-74 e12
120. Listenberger LL, Han X, Lewis SE, Cases S, Farese RV Jr, Ory DS, Schaffer JE (2003) Triglyceride accumulation protects against fatty acid-induced lipotoxicity. Proc Natl Acad Sci U S A 100:3077–3082
121. Liu P, Ying Y, Zhao Y, Mundy DI, Zhu M, Anderson RG (2004) Chinese hamster ovary K2 cell lipid droplets appear to be metabolic organelles involved in membrane traffic. J Biol Chem 279:3787–3792
122. Liu L, Shi X, Bharadwaj KG, Ikeda S, Yamashita H, Yagyu H, Schaffer JE, Yu YH, Goldberg IJ (2009a) DGAT1 expression increases heart triglyceride content but ameliorates lipotoxicity. J Biol Chem 284:36312–36323
123. Liu L, Shi X, Choi CS, Shulman GI, Klaus K, Nair KS, Schwartz GJ, Zhang Y, Goldberg IJ, Yu YH (2009b) Paradoxical coupling of triglyceride synthesis and fatty acid oxidation in skeletal muscle overexpressing DGAT1. Diabetes 58:2516–2524
124. Liu X, Liang Y, Song R, Yang G, Han J, Lan Y, Pan S, Zhu M, Liu Y, Wang Y, Meng F, Cui Y, Wang J, Zhang B, Song X, Lu Z, Zheng T, Liu L (2018) Long non-coding RNA NEAT1-modulated abnormal lipolysis via ATGL drives hepatocellular carcinoma proliferation. Mol Cancer 17:90
125. Liu M, Yu X, Lin L, Deng J, Wang K, Xia Y, Tang X, Hong H (2019) ATGL promotes the proliferation of hepatocellular carcinoma cells via the p-AKT signaling pathway. J Biochem Mol Toxicol 33:e22391
126. Londos C, Sztalryd C, Tansey JT, Kimmel AR (2005) Role of PAT proteins in lipid metabolism. Biochimie 87:45–49
127. Lord CC, Betters JL, Ivanova PT, Milne SB, Myers DS, Madenspacher J, Thomas G, Chung S, Liu M, Davis MA, Lee RG, Crooke RM, Graham MJ, Parks JS, Brasaemle DL, Fessler MB, Brown HA, Brown JM (2012) CGI-58/ABHD5-derived signaling lipids regulate systemic inflammation and insulin action. Diabetes 61:355–363
128. Lord CC, Thomas G, Brown JM (2013) Mammalian alpha beta hydrolase domain (ABHD) proteins: lipid metabolizing enzymes at the interface of cell signaling and energy metabolism. Biochim Biophys Acta 1831:792–802

129. Lord CC, Ferguson D, Thomas G, Brown AL, Schugar RC, Burrows A, Gromovsky AD, Betters J, Neumann C, Sacks J, Marshall S, Watts R, Schweiger M, Lee RG, Crooke RM, Graham MJ, Lathia JD, Sakaguchi TF, Lehner R, Haemmerle G, Zechner R, Brown JM (2016) Regulation of hepatic triacylglycerol metabolism by CGI-58 does not require ATGL Co-activation. Cell Rep 16:939–949
130. Louhichi N, Bahloul E, Marrakchi S, Othman HB, Triki C, Aloulou K, Trabelsi L, Mahfouth N, Ayadi-Mnif Z, Keskes L, Fakhfakh F, Turki H (2019) Thyroid involvement in Chanarin-Dorfman syndrome in adults in the largest series of patients carrying the same founder mutation in ABHD5 gene. Orphanet J Rare Dis 14:112
131. Lu X, Yang X, Liu J (2010) Differential control of ATGL-mediated lipid droplet degradation by CGI-58 and G0S2. Cell Cycle 9:2719–2725
132. MacPherson RE, Ramos SV, Vandenboom R, Roy BD, Peters SJ (2013) Skeletal muscle PLIN proteins, ATGL and CGI-58, interactions at rest and following stimulated contraction. Am J Physiol Regul Integr Comp Physiol 304:R644–R650
133. Martin S, Parton RG (2006) Lipid droplets: a unified view of a dynamic organelle. Nat Rev Mol Cell Biol 7:373–378
134. Miao H, Ou J, Ma Y, Guo F, Yang Z, Wiggins M, Liu C, Song W, Han X, Wang M, Cao Q, Chung BH, Yang D, Liang H, Xue B, Shi H, Gan L, Yu L (2014) Macrophage CGI-58 deficiency activates ROS-inflammasome pathway to promote insulin resistance in mice. Cell Rep 7:223–235
135. Miao H, Ou J, Zhang X, Chen Y, Xue B, Shi H, Gan L, Yu L, Liang H (2015) Macrophage CGI-58 deficiency promotes IL-1beta transcription by activating the SOCS3-FOXO1 pathway. Clin Sci (Lond) 128:493–506
136. Miao H, Ou J, Peng Y, Zhang X, Chen Y, Hao L, Xie G, Wang Z, Pang X, Ruan Z, Li J, Yu L, Xue B, Shi H, Shi C, Liang H (2016) Macrophage ABHD5 promotes colorectal cancer growth by suppressing spermidine production by SRM. Nat Commun 7:11716
137. Michelsen KS, Wong MH, Shah PK, Zhang W, Yano J, Doherty TM, Akira S, Rajavashisth TB, Arditi M (2004) Lack of Toll-like receptor 4 or myeloid differentiation factor 88 reduces atherosclerosis and alters plaque phenotype in mice deficient in apolipoprotein E. Proc Natl Acad Sci U S A 101:10679–10684
138. Miranda A, DiMauro S, Eastwood A, Hays A, Johnson WG, Olarte M, Whitlock R, Mayeux R, Rowland LP (1979) Lipid storage myopathy, ichthyosis, and steatorrhea. Muscle Nerve 2:1–13
139. Missaglia S, Valadares ER, Moro L, Faguntes ED, Quintao Roque R, Giardina B, Tavian D (2014) Early onset of Chanarin-Dorfman syndrome with severe liver involvement in a patient with a complex rearrangement of ABHD5 promoter. BMC Med Genet 15:32
140. Mitra R, Le TT, Gorjala P, Goodman OB Jr (2017) Positive regulation of prostate cancer cell growth by lipid droplet forming and processing enzymes DGAT1 and ABHD5. BMC Cancer 17:631
141. Monetti M, Levin MC, Watt MJ, Sajan MP, Marmor S, Hubbard BK, Stevens RD, Bain JR, Newgard CB, Farese RV Sr, Hevener AL, Farese RV Jr (2007) Dissociation of hepatic steatosis and insulin resistance in mice overexpressing DGAT in the liver. Cell Metab 6:69–78
142. Moreira RK (2007) Hepatic stellate cells and liver fibrosis. Arch Pathol Lab Med 131:1728–1734
143. Mullick AE, Tobias PS, Curtiss LK (2005) Modulation of atherosclerosis in mice by Toll-like receptor 2. J Clin Invest 115:3149–3156
144. Murakami M, Taketomi Y, Miki Y, Sato H, Hirabayashi T, Yamamoto K (2011) Recent progress in phospholipase A(2) research: from cells to animals to humans. Prog Lipid Res 50:152–192
145. Najt CP, Senthivinayagam S, Aljazi MB, Fader KA, Olenic SD, Brock JR, Lydic TA, Jones AD, Atshaves BP (2016) Liver-specific loss of Perilipin 2 alleviates diet-induced hepatic steatosis, inflammation, and fibrosis. Am J Physiol Gastrointest Liver Physiol 310:G726–G738
146. Nardini M, Dijkstra BW (1999) Alpha/beta hydrolase fold enzymes: the family keeps growing. Curr Opin Struct Biol 9:732–737
147. Negro F (2006) Mechanisms and significance of liver steatosis in hepatitis C virus infection. World J Gastroenterol 12:6756–6765
148. Newberry EP, Xie Y, Lodeiro C, Solis R, Moritz W, Kennedy S, Barron L, Onufer E, Alpini G, Zhou T, Blaner WS, Chen A, Davidson NO (2019) Hepatocyte and stellate cell deletion of liver fatty acid binding protein reveals distinct roles in fibrogenic injury. FASEB J 33:4610–4625
149. Nischalke HD, Berger C, Luda C, Berg T, Muller T, Grunhage F, Lammert F, Coenen M, Kramer B, Korner C, Vidovic N, Oldenburg J, Nattermann J, Sauerbruch T, Spengler U (2011) The PNPLA3 rs738409 148M/M genotype is a risk factor for liver cancer in alcoholic cirrhosis but shows no or weak association in hepatitis C cirrhosis. PLoS One 6: e27087
150. Nur BG, Gencpinar P, Yuzbasioglu A, Emre SD, Mihci E (2015) Chanarin-Dorfman syndrome: genotype-phenotype correlation. Eur J Med Genet 58:238–242
151. Oberer M, Boeszoermenyi A, Nagy HM, Zechner R (2011) Recent insights into the structure and function of comparative gene identification-58. Curr Opin Lipidol 22:149–158
152. Obrowsky S, Chandak PG, Patankar JV, Povoden S, Schlager S, Kershaw EE, Bogner-Strauss JG, Hoefler G, Levak-Frank S, Kratky D (2013) Adipose triglyceride lipase is a TG hydrolase of the small

intestine and regulates intestinal PPARalpha signaling. J Lipid Res 54:425–435
153. Ohno Y, Kamiyama N, Nakamichi S, Kihara A (2017) PNPLA1 is a transacylase essential for the generation of the skin barrier lipid omega-O-acylceramide. Nat Commun 8:14610
154. Ohno Y, Nara A, Nakamichi S, Kihara A (2018) Molecular mechanism of the ichthyosis pathology of Chanarin-Dorfman syndrome: stimulation of PNPLA1-catalyzed omega-O-acylceramide production by ABHD5. J Dermatol Sci 92:245–253
155. Ohsumi Y (2014) Historical landmarks of autophagy research. Cell Res 24:9–23
156. Ostos MA, Recalde D, Zakin MM, Scott-Algara D (2002) Implication of natural killer T cells in atherosclerosis development during a LPS-induced chronic inflammation. FEBS Lett 519:23–29
157. Osuga J, Ishibashi S, Oka T, Yagyu H, Tozawa R, Fujimoto A, Shionoiri F, Yahagi N, Kraemer FB, Tsutsumi O, Yamada N (2000) Targeted disruption of hormone-sensitive lipase results in male sterility and adipocyte hypertrophy, but not in obesity. Proc Natl Acad Sci U S A 97:787–792
158. Ou J, Miao H, Ma Y, Guo F, Deng J, Wei X, Zhou J, Xie G, Shi H, Xue B, Liang H, Yu L (2014) Loss of abhd5 promotes colorectal tumor development and progression by inducing aerobic glycolysis and epithelial-mesenchymal transition. Cell Rep 9:1798–1811
159. Ou J, Peng Y, Yang W, Zhang Y, Hao J, Li F, Chen Y, Zhao Y, Xie X, Wu S, Zha L, Luo X, Xie G, Wang L, Sun W, Zhou Q, Li J, Liang H (2019) ABHD5 blunts the sensitivity of colorectal cancer to fluorouracil via promoting autophagic uracil yield. Nat Commun 10:1078
160. Ouimet M, Franklin V, Mak E, Liao X, Tabas I, Marcel YL (2011) Autophagy regulates cholesterol efflux from macrophage foam cells via lysosomal acid lipase. Cell Metab 13:655–667
161. Patel S, Yang W, Kozusko K, Saudek V, Savage DB (2014) Perilipins 2 and 3 lack a carboxy-terminal domain present in perilipin 1 involved in sequestering ABHD5 and suppressing basal lipolysis. Proc Natl Acad Sci U S A 111:9163–9168
162. Pawella LM, Hashani M, Eiteneuer E, Renner M, Bartenschlager R, Schirmacher P, Straub BK (2014) Perilipin discerns chronic from acute hepatocellular steatosis. J Hepatol 60:633–642
163. Peng Y, Miao H, Wu S, Yang W, Zhang Y, Xie G, Xie X, Li J, Shi C, Ye L, Sun W, Wang L, Liang H, Ou J (2016) ABHD5 interacts with BECN1 to regulate autophagy and tumorigenesis of colon cancer independent of PNPLA2. Autophagy 12:2167–2182
164. Petit V, Arnould L, Martin P, Monnot MC, Pineau T, Besnard P, Niot I (2007) Chronic high-fat diet affects intestinal fat absorption and postprandial triglyceride levels in the mouse. J Lipid Res 48:278–287
165. Petrovic N, Walden TB, Shabalina IG, Timmons JA, Cannon B, Nedergaard J (2010) Chronic peroxisome proliferator-activated receptor gamma (PPARgamma) activation of epididymally derived white adipocyte cultures reveals a population of thermogenically competent, UCP1-containing adipocytes molecularly distinct from classic brown adipocytes. J Biol Chem 285:7153–7164
166. Piva E, Pajola R, Binotto G, Plebani M (2009) Jordans' anomaly in a new neutral lipid storage disease. Am J Hematol 84:254–255
167. Pollak NM, Jaeger D, Kolleritsch S, Zimmermann R, Zechner R, Lass A, Haemmerle G (2015) The interplay of protein kinase A and perilipin 5 regulates cardiac lipolysis. J Biol Chem 290:1295–1306
168. Postic C, Shiota M, Niswender KD, Jetton TL, Chen Y, Moates JM, Shelton KD, Lindner J, Cherrington AD, Magnuson MA (1999) Dual roles for glucokinase in glucose homeostasis as determined by liver and pancreatic beta cell-specific gene knockouts using Cre recombinase. J Biol Chem 274:305–315
169. Pujol RM, Gilaberte M, Toll A, Florensa L, Lloreta J, Gonzalez-Ensenat MA, Fischer J, Azon A (2005) Erythrokeratoderma variabilis-like ichthyosis in Chanarin-Dorfman syndrome. Br J Dermatol 153:838–841
170. Pussinen PJ, Mattila K (2004) Periodontal infections and atherosclerosis: mere associations? Curr Opin Lipidol 15:583–588
171. Pyzik M, Rath T, Kuo TT, Win S, Baker K, Hubbard JJ, Grenha R, Gandhi A, Kramer TD, Mezo AR, Taylor ZS, McDonnell K, Nienaber V, Andersen JT, Mizoguchi A, Blumberg L, Purohit S, Jones SD, Christianson G, Lencer WI, Sandlie I, Kaplowitz N, Roopenian DC, Blumberg RS (2017) Hepatic FcRn regulates albumin homeostasis and susceptibility to liver injury. Proc Natl Acad Sci U S A 114:E2862–E2E71
172. Radner FP, Streith IE, Schoiswohl G, Schweiger M, Kumari M, Eichmann TO, Rechberger G, Koefeler HC, Eder S, Schauer S, Theussl HC, Preiss-Landl K, Lass A, Zimmermann R, Hoefler G, Zechner R, Haemmerle G (2010) Growth retardation, impaired triacylglycerol catabolism, hepatic steatosis, and lethal skin barrier defect in mice lacking comparative gene identification-58 (CGI-58). J Biol Chem 285:7300–7311
173. Randle PJ, Garland PB, Hales CN, Newsholme EA (1963) The glucose fatty-acid cycle. Its role in insulin sensitivity and the metabolic disturbances of diabetes mellitus. Lancet 1:785–789
174. Redaelli C, Coleman RA, Moro L, Dacou-Voutetakis C, Elsayed SM, Prati D, Colli A, Mela D, Colombo R, Tavian D (2010) Clinical and genetic characterization of Chanarin-Dorfman syndrome patients: first report of large deletions in the ABHD5 gene. Orphanet J Rare Dis 5:33
175. Rich NE, Oji S, Mufti AR, Browning JD, Parikh ND, Odewole M, Mayo H, Singal AG (2018) Racial and ethnic disparities in nonalcoholic fatty liver disease

prevalence, severity, and outcomes in the United States: a systematic review and meta-analysis. Clin Gastroenterol Hepatol 16:198–210 e2
176. Richardson MM, Jonsson JR, Powell EE, Brunt EM, Neuschwander-Tetri BA, Bhathal PS, Dixon JB, Weltman MD, Tilg H, Moschen AR, Purdie DM, Demetris AJ, Clouston AD (2007) Progressive fibrosis in nonalcoholic steatohepatitis: association with altered regeneration and a ductular reaction. Gastroenterology 133:80–90
177. Rizack MA (1964) Activation of an epinephrine-sensitive lipolytic activity from adipose tissue by adenosine 3′,5′-Phosphate. J Biol Chem 239:392–395
178. Robertson MD, Parkes M, Warren BF, Ferguson DJ, Jackson KG, Jewell DP, Frayn KN (2003) Mobilisation of enterocyte fat stores by oral glucose in humans. Gut 52:834–839
179. Roduit R, Masiello P, Wang SP, Li H, Mitchell GA, Prentki M (2001) A role for hormone-sensitive lipase in glucose-stimulated insulin secretion: a study in hormone-sensitive lipase-deficient mice. Diabetes 50:1970–1975
180. Romeo S, Kozlitina J, Xing C, Pertsemlidis A, Cox D, Pennacchio LA, Boerwinkle E, Cohen JC, Hobbs HH (2008) Genetic variation in PNPLA3 confers susceptibility to nonalcoholic fatty liver disease. Nat Genet 40:1461–1465
181. Ronchetti A, Prati D, Pezzotta MG, Tavian D, Colombo R, Callea F, Colli A (2008) Severe steatohepatitis in a patient with a rare neutral lipid storage disorder due to ABHD5 mutation. J Hepatol 49:474–477
182. Rosenwald M, Perdikari A, Rulicke T, Wolfrum C (2013) Bi-directional interconversion of brite and white adipocytes. Nat Cell Biol 15:659–667
183. Rozenszajn L, Klajman A, Yaffe D, Efrati P (1966) Jordans' anomaly in white blood cells. report of case. Blood 28:258–265
184. Russell JO, Lu WY, Okabe H, Abrams M, Oertel M, Poddar M, Singh S, Forbes SJ, Monga SP (2019) Hepatocyte-specific beta-catenin deletion during severe liver injury provokes cholangiocytes to differentiate into hepatocytes. Hepatology 69:742–759
185. Sahu-Osen A, Montero-Moran G, Schittmayer M, Fritz K, Dinh A, Chang YF, McMahon D, Boeszoermenyi A, Cornaciu I, Russell D, Oberer M, Carman GM, Birner-Gruenberger R, Brasaemle DL (2015) CGI-58/ABHD5 is phosphorylated on Ser239 by protein kinase A: control of subcellular localization. J Lipid Res 56:109–121
186. Samuel VT, Shulman GI (2012) Mechanisms for insulin resistance: common threads and missing links. Cell 148:852–871
187. Samuelov L, Fuchs-Telem D, Sarig O, Sprecher E (2011) An exceptional mutational event leading to Chanarin-Dorfman syndrome in a large consanguineous family. Br J Dermatol 164:1390–1392
188. Sanders MA, Madoux F, Mladenovic L, Zhang H, Ye X, Angrish M, Mottillo EP, Caruso JA, Halvorsen G, Roush WR, Chase P, Hodder P, Granneman JG (2015) Endogenous and synthetic ABHD5 ligands regulate ABHD5-perilipin interactions and lipolysis in fat and muscle. Cell Metab 22:851–860
189. Sanders MA, Zhang H, Mladenovic L, Tseng YY, Granneman JG (2017) Molecular basis of ABHD5 lipolysis activation. Sci Rep 7:42589
190. Sathyanarayan A, Mashek MT, Mashek DG (2017) ATGL promotes autophagy/lipophagy via SIRT1 to control hepatic lipid droplet catabolism. Cell Rep 19:1–9
191. Sato K, Marzioni M, Meng F, Francis H, Glaser S, Alpini G (2019) Ductular reaction in liver diseases: pathological mechanisms and translational significances. Hepatology 69:420–430
192. Schleinitz N, Fischer J, Sanchez A, Veit V, Harle JR, Pelissier JF (2005) Two new mutations of the ABHD5 gene in a new adult case of Chanarin Dorfman syndrome: an uncommon lipid storage disease. Arch Dermatol 141:798–800
193. Schneider AL, Lazo M, Selvin E, Clark JM (2014) Racial differences in nonalcoholic fatty liver disease in the U.S. population. Obesity (Silver Spring) 22:292–299
194. Schoiswohl G, Stefanovic-Racic M, Menke MN, Wills RC, Surlow BA, Basantani MK, Sitnick MT, Cai L, Yazbeck CF, Stolz DB, Pulinilkunnil T, O'Doherty RM, Kershaw EE (2015) Impact of reduced ATGL-mediated adipocyte lipolysis on obesity-associated insulin resistance and inflammation in male mice. Endocrinology 156:3610–3624
195. Schreiber R, Diwoky C, Schoiswohl G, Feiler U, Wongsiriroj N, Abdellatif M, Kolb D, Hoeks J, Kershaw EE, Sedej S, Schrauwen P, Haemmerle G, Zechner R (2017) Cold-Induced thermogenesis depends on ATGL-mediated lipolysis in cardiac muscle, but not brown adipose tissue. Cell Metab 26:753-63 e7
196. Schweiger M, Schoiswohl G, Lass A, Radner FP, Haemmerle G, Malli R, Graier W, Cornaciu I, Oberer M, Salvayre R, Fischer J, Zechner R, Zimmermann R (2008) The C-terminal region of human adipose triglyceride lipase affects enzyme activity and lipid droplet binding. J Biol Chem 283:17211–17220
197. Schweiger M, Lass A, Zimmermann R, Eichmann TO, Zechner R (2009) Neutral lipid storage disease: genetic disorders caused by mutations in adipose triglyceride lipase/PNPLA2 or CGI-58/ABHD5. Am J Physiol Endocrinol Metab 297:E289–E296
198. Senchenko VN, Kisseljova NP, Ivanova TA, Dmitriev AA, Krasnov GS, Kudryavtseva AV, Panasenko GV, Tsitrin EB, Lerman MI, Kisseljov FL, Kashuba VI, Zabarovsky ER (2013) Novel tumor suppressor candidates on chromosome

3 revealed by NotI-microarrays in cervical cancer. Epigenetics 8:409–420
199. Shang S, Ji X, Zhang L, Chen J, Li C, Shi R, Xiang W, Kang X, Zhang D, Yang F, Dai R, Chen P, Chen S, Chen Y, Li Y, Miao H (2019) Macrophage ABHD5 suppresses NFkappaB-dependent matrix metalloproteinase expression and cancer metastasis. Cancer Res 79:5513–5526
200. Shibata M, Yoshimura K, Furuya N, Koike M, Ueno T, Komatsu M, Arai H, Tanaka K, Kominami E, Uchiyama Y (2009) The MAP1-LC3 conjugation system is involved in lipid droplet formation. Biochem Biophys Res Commun 382:419–423
201. Shin H, Ma Y, Chanturiya T, Cao Q, Wang Y, Kadegowda AKG, Jackson R, Rumore D, Xue B, Shi H, Gavrilova O, Yu L (2017) Lipolysis in brown adipocytes is not essential for cold-induced thermogenesis in mice. Cell Metab 26:764-77 e5
202. Simon GM, Cravatt BF (2006) Endocannabinoid biosynthesis proceeding through glycerophospho-N-acyl ethanolamine and a role for alpha/beta-hydrolase 4 in this pathway. J Biol Chem 281:26465–26472
203. Singh R, Kaushik S, Wang Y, Xiang Y, Novak I, Komatsu M, Tanaka K, Cuervo AM, Czaja MJ (2009) Autophagy regulates lipid metabolism. Nature 458:1131–1135
204. Sitnick MT, Basantani MK, Cai L, Schoiswohl G, Yazbeck CF, Distefano G, Ritov V, DeLany JP, Schreiber R, Stolz DB, Gardner NP, Kienesberger PC, Pulinilkunnil T, Zechner R, Goodpaster BH, Coen P, Kershaw EE (2013) Skeletal muscle triacylglycerol hydrolysis does not influence metabolic complications of obesity. Diabetes 62:3350–3361
205. Slavin G, Wills EJ, Richmond JE, Chanarin I, Andrews T, Stewart G (1975) Morphological features in a neutral lipid storage disease. J Clin Pathol 28:701–710
206. Sparks EE, Huppert KA, Brown MA, Washington MK, Huppert SS (2010) Notch signaling regulates formation of the three-dimensional architecture of intrahepatic bile ducts in mice. Hepatology 51:1391–1400
207. Speliotes EK, Butler JL, Palmer CD, Voight BF, Giant Consortium, M. IGen Consortium, Nash CRN, Hirschhorn JN (2010) PNPLA3 variants specifically confer increased risk for histologic nonalcoholic fatty liver disease but not metabolic disease. Hepatology 52:904–912
208. Srinivasan R, Hadzic N, Fischer J, Knisely AS (2004) Steatohepatitis and unsuspected micronodular cirrhosis in Dorfman-Chanarin syndrome with documented ABHD5 mutation. J Pediatr 144:662–665
209. Straub BK, Stoeffel P, Heid H, Zimbelmann R, Schirmacher P (2008) Differential pattern of lipid droplet-associated proteins and de novo perilipin expression in hepatocyte steatogenesis. Hepatology 47:1936–1946
210. Straub BK, Herpel E, Singer S, Zimbelmann R, Breuhahn K, Macher-Goeppinger S, Warth A, Lehmann-Koch J, Longerich T, Heid H, Schirmacher P (2010) Lipid droplet-associated PAT-proteins show frequent and differential expression in neoplastic steatogenesis. Mod Pathol 23:480–492
211. Subramanian V, Rothenberg A, Gomez C, Cohen AW, Garcia A, Bhattacharyya S, Shapiro L, Dolios G, Wang R, Lisanti MP, Brasaemle DL (2004) Perilipin A mediates the reversible binding of CGI-58 to lipid droplets in 3T3-L1 adipocytes. J Biol Chem 279:42062–42071
212. Sun Z, Miller RA, Patel RT, Chen J, Dhir R, Wang H, Zhang D, Graham MJ, Unterman TG, Shulman GI, Sztalryd C, Bennett MJ, Ahima RS, Birnbaum MJ, Lazar MA (2012) Hepatic Hdac3 promotes gluconeogenesis by repressing lipid synthesis and sequestration. Nat Med 18:934–942
213. Sztalryd C, Brasaemle DL (2017) The perilipin family of lipid droplet proteins: Gatekeepers of intracellular lipolysis. Biochim Biophys Acta Mol Cell Biol Lipids 1862:1221–1232
214. Taguchi K, Hirano I, Itoh T, Tanaka M, Miyajima A, Suzuki A, Motohashi H, Yamamoto M (2014) Nrf2 enhances cholangiocyte expansion in Pten-deficient livers. Mol Cell Biol 34:900–913
215. Takeda K, Tanaka K, Kumamoto T, Morioka D, Endo I, Togo S, Shimada H (2010) Living donor liver transplantation for Dorfman-Chanarin syndrome with 1 year follow-up: case report. Transplant Proc 42:3858–3861
216. Tian C, Stokowski RP, Kershenobich D, Ballinger DG, Hinds DA (2010) Variant in PNPLA3 is associated with alcoholic liver disease. Nat Genet 42:21–23
217. Tolkovsky AM (2009) Mitophagy. Biochim Biophys Acta 1793:1508–1515
218. Tomin T, Fritz K, Gindlhuber J, Waldherr L, Pucher B, Thallinger GG, Nomura DK, Schittmayer M, Birner-Gruenberger R (2018) Deletion of adipose triglyceride lipase links triacylglycerol accumulation to a more-aggressive phenotype in A549 lung carcinoma cells. J Proteome Res 17:1415–1425
219. Ujihara M, Nakajima K, Yamamoto M, Teraishi M, Uchida Y, Akiyama M, Shimizu H, Sano S (2010) Epidermal triglyceride levels are correlated with severity of ichthyosis in Dorfman-Chanarin syndrome. J Dermatol Sci 57:102–107
220. Umemura A, Park EJ, Taniguchi K, Lee JH, Shalapour S, Valasek MA, Aghajan M, Nakagawa H, Seki E, Hall MN, Karin M (2014) Liver damage, inflammation, and enhanced tumorigenesis after persistent mTORC1 inhibition. Cell Metab 20:133–144
221. Unger RH (1995) Lipotoxicity in the pathogenesis of obesity-dependent NIDDM. Genetic and clinical implications. Diabetes 44:863–870

222. Unger RH (2002) Lipotoxic diseases. Annu Rev Med 53:319–336
223. Vaughan M, Berger JE, Steinberg D (1964) Hormone-sensitive lipase and monoglyceride lipase activities in adipose tissue. J Biol Chem 239:401–409
224. Verma SB, Mittal A, Wollina U, Eckstein GH, Gohel K, Giehl K (2017) Chanarin-Dorfman syndrome with rare renal involvement. Br J Dermatol 176:545–548
225. Vieyres G, Welsch K, Gerold G, Gentzsch J, Kahl S, Vondran FW, Kaderali L, Pietschmann T (2016) ABHD5/CGI-58, the Chanarin-Dorfman syndrome protein, mobilises lipid stores for Hepatitis C virus production. PLoS Pathog 12:e1005568
226. Villena JA, Roy S, Sarkadi-Nagy E, Kim KH, Sul HS (2004) Desnutrin, an adipocyte gene encoding a novel patatin domain-containing protein, is induced by fasting and glucocorticoids: ectopic expression of desnutrin increases triglyceride hydrolysis. J Biol Chem 279:47066–47075
227. Wang SP, Laurin N, Himms-Hagen J, Rudnicki MA, Levy E, Robert MF, Pan L, Oligny L, Mitchell GA (2001) The adipose tissue phenotype of hormone-sensitive lipase deficiency in mice. Obes Res 9:119–128
228. Wang H, Bell M, Sreenevasan U, Hu H, Liu J, Dalen K, Londos C, Yamaguchi T, Rizzo MA, Coleman R, Gong D, Brasaemle D, Sztalryd C (2011) Unique regulation of adipose triglyceride lipase (ATGL) by perilipin 5, a lipid droplet-associated protein. J Biol Chem 286:15707–15715
229. Wang QA, Tao C, Gupta RK, Scherer PE (2013) Tracking adipogenesis during white adipose tissue development, expansion and regeneration. Nat Med 19:1338–1344
230. Wang W, Kissig M, Rajakumari S, Huang L, Lim HW, Won KJ, Seale P (2014) Ebf2 is a selective marker of brown and beige adipogenic precursor cells. Proc Natl Acad Sci U S A 111:14466–14471
231. Wang C, Zhao Y, Gao X, Li L, Yuan Y, Liu F, Zhang L, Wu J, Hu P, Zhang X, Gu Y, Xu Y, Wang Z, Li Z, Zhang H, Ye J (2015) Perilipin 5 improves hepatic lipotoxicity by inhibiting lipolysis. Hepatology 61:870–882
232. Wang YY, Attane C, Milhas D, Dirat B, Dauvillier S, Guerard A, Gilhodes J, Lazar I, Alet N, Laurent V, Le Gonidec S, Biard D, Herve C, Bost F, Ren GS, Bono F, Escourrou G, Prentki M, Nieto L, Valet P, Muller C (2017) Mammary adipocytes stimulate breast cancer invasion through metabolic remodeling of tumor cells. JCI Insight 2:e87489
233. Wang Y, Kory N, BasuRay S, Cohen JC, Hobbs HH (2019) PNPLA3, CGI-58, and inhibition of hepatic triglyceride hydrolysis in mice. Hepatology 69:2427–2441
234. Wei S, Lai K, Patel S, Piantedosi R, Shen H, Colantuoni V, Kraemer FB, Blaner WS (1997) Retinyl ester hydrolysis and retinol efflux from BFC-1beta adipocytes. J Biol Chem 272:14159–14165
235. Williams ML, Koch TK, O'Donnell JJ, Frost PH, Epstein LB, Grizzard WS, Epstein CJ (1985) Ichthyosis and neutral lipid storage disease. Am J Med Genet 20:711–726
236. Wolins NE, Brasaemle DL, Bickel PE (2006) A proposed model of fat packaging by exchangeable lipid droplet proteins. FEBS Lett 580:5484–5491
237. Wu JW, Wang SP, Alvarez F, Casavant S, Gauthier N, Abed L, Soni KG, Yang G, Mitchell GA (2011) Deficiency of liver adipose triglyceride lipase in mice causes progressive hepatic steatosis. Hepatology 54:122–132
238. Wu J, Bostrom P, Sparks LM, Ye L, Choi JH, Giang AH, Khandekar M, Virtanen KA, Nuutila P, Schaart G, Huang K, Tu H, van Marken Lichtenbelt WD, Hoeks J, Enerback S, Schrauwen P, Spiegelman BM (2012a) Beige adipocytes are a distinct type of thermogenic fat cell in mouse and human. Cell 150:366–376
239. Wu JW, Wang SP, Casavant S, Moreau A, Yang GS, Mitchell GA (2012b) Fasting energy homeostasis in mice with adipose deficiency of desnutrin/adipose triglyceride lipase. Endocrinology 153:2198–2207
240. Xie P, Guo F, Ma Y, Zhu H, Wang F, Xue B, Shi H, Yang J, Yu L (2014a) Intestinal Cgi-58 deficiency reduces postprandial lipid absorption. PLoS One 9: e91652
241. Xie P, Zeng X, Xiao J, Sun B, Yang D (2014b) Transgenic CGI-58 expression in macrophages alleviates the atherosclerotic lesion development in ApoE knockout mice. Biochim Biophys Acta 1841:1683–1690
242. Xie P, Kadegowda AK, Ma Y, Guo F, Han X, Wang M, Groban L, Xue B, Shi H, Li H, Yu L (2015) Muscle-specific deletion of comparative gene identification-58 (CGI-58) causes muscle steatosis but improves insulin sensitivity in male mice. Endocrinology 156:1648–1658
243. Yamaguchi T, Osumi T (2009) Chanarin-Dorfman syndrome: deficiency in CGI-58, a lipid droplet-bound coactivator of lipase. Biochim Biophys Acta 1791:519–523
244. Yamaguchi T, Omatsu N, Matsushita S, Osumi T (2004) CGI-58 interacts with perilipin and is localized to lipid droplets. Possible involvement of CGI-58 mislocalization in Chanarin-Dorfman syndrome. J Biol Chem 279:30490–30497
245. Yamaguchi T, Omatsu N, Morimoto E, Nakashima H, Ueno K, Tanaka T, Satouchi K, Hirose F, Osumi T (2007) CGI-58 facilitates lipolysis on lipid droplets but is not involved in the vesiculation of lipid droplets caused by hormonal stimulation. J Lipid Res 48:1078–1089
246. Yang X, Lu X, Lombes M, Rha GB, Chi YI, Guerin TM, Smart EJ, Liu J (2010) The G(0)/G(1) switch gene 2 regulates adipose lipolysis through association

with adipose triglyceride lipase. Cell Metab 11:194–205
247. Yang X, Zhang X, Heckmann BL, Lu X, Liu J (2011) Relative contribution of adipose triglyceride lipase and hormone-sensitive lipase to tumor necrosis factor-alpha (TNF-alpha)-induced lipolysis in adipocytes. J Biol Chem 286:40477–40485
248. Yang D, Chen H, Zeng X, Xie P, Wang X, Liu C (2016) Macrophage CGI-58 attenuates inflammatory responsiveness via promotion of PPARgamma signaling. Cell Physiol Biochem 38:696–713
249. Yang D, Li Y, Xing L, Tan Y, Sun J, Zeng B, Xiang T, Tan J, Ren G, Wang Y (2018) Utilization of adipocyte-derived lipids and enhanced intracellular trafficking of fatty acids contribute to breast cancer progression. Cell Commun Signal 16:32
250. Yang A, Mottillo EP, Mladenovic-Lucas L, Zhou L, Granneman JG (2019) Dynamic interactions of ABHD5 with PNPLA3 regulate triacylglycerol metabolism in brown adipocytes. Nat Metab 1:560–569
251. Yen CL, Farese RV Jr (2006) Fat breakdown: a function for CGI-58 (ABHD5) provides a new piece of the puzzle. Cell Metab 3:305–307
252. Yim CY, Sekula DJ, Hever-Jardine MP, Liu X, Warzecha JM, Tam J, Freemantle SJ, Dmitrovsky E, Spinella MJ (2016) G0S2 suppresses oncogenic transformation by repressing a MYC-regulated transcriptional program. Cancer Res 76:1204–1213
253. Young SG, Zechner R (2013) Biochemistry and pathophysiology of intravascular and intracellular lipolysis. Genes Dev 27:459–484
254. Younossi ZM, Koenig AB, Abdelatif D, Fazel Y, Henry L, Wymer M (2016) Global epidemiology of nonalcoholic fatty liver disease-Meta-analytic assessment of prevalence, incidence, and outcomes. Hepatology 64:73–84
255. Youssefian L, Vahidnezhad H, Saeidian AH, Pajouhanfar S, Sotoudeh S, Mansouri P, Amirkashani D, Zeinali S, Levine MA, Peris K, Colombo R, Uitto J (2019) Inherited non-alcoholic fatty liver disease and dyslipidemia due to monoallelic ABHD5 mutations. J Hepatol 71:366–370
256. Zagani R, El-Assaad W, Gamache I, Teodoro JG (2015) Inhibition of adipose triglyceride lipase (ATGL) by the putative tumor suppressor G0S2 or a small molecule inhibitor attenuates the growth of cancer cells. Oncotarget 6:28282–28295
257. Zechner R, Strauss JG, Haemmerle G, Lass A, Zimmermann R (2005) Lipolysis: pathway under construction. Curr Opin Lipidol 16:333–340
258. Zhang L, Godzik A, Skolnick J, Fetrow JS (1998) Functional analysis of the Escherichia coli genome for members of the alpha/beta hydrolase family. Fold Des 3:535–548
259. Zhang J, Xu D, Nie J, Han R, Zhai Y, Shi Y (2014) Comparative gene identification-58 (CGI-58) promotes autophagy as a putative lysophosphatidylglycerol acyltransferase. J Biol Chem 289 (47):33044–33053
260. Zhang X, Saarinen AM, Hitosugi T, Wang Z, Wang L, Ho TH, Liu J (2017) Inhibition of intracellular lipolysis promotes human cancer cell adaptation to hypoxia. Elife 6:e31132
261. Zhou Q, Wang F, Zhou K, Huang K, Zhu Q, Luo X, Yu J, Shi Z (2019) Oncogenic role of ABHD5 in endometrial cancer. Cancer Manag Res 11:2139–2150
262. Zhu J, Lee B, Buhman KK, Cheng JX (2009) A dynamic, cytoplasmic triacylglycerol pool in enterocytes revealed by ex vivo and in vivo coherent anti-Stokes Raman scattering imaging. J Lipid Res 50:1080–1089
263. Zierler KA, Jaeger D, Pollak NM, Eder S, Rechberger GN, Radner FP, Woelkart G, Kolb D, Schmidt A, Kumari M, Preiss-Landl K, Pieske B, Mayer B, Zimmermann R, Lass A, Zechner R, Haemmerle G (2013) Functional cardiac lipolysis in mice critically depends on comparative gene identification-58. J Biol Chem 288:9892–9904
264. Zimmermann R, Strauss JG, Haemmerle G, Schoiswohl G, Birner-Gruenberger R, Riederer M, Lass A, Neuberger G, Eisenhaber F, Hermetter A, Zechner R (2004) Fat mobilization in adipose tissue is promoted by adipose triglyceride lipase. Science 306:1383–1386

Mfsd2a: A Physiologically Important Lysolipid Transporter in the Brain and Eye

14

Bernice H. Wong and David L. Silver

Abstract

Lipids and essential fatty acids are required for normal brain development and continued photoreceptor membrane biogenesis for the maintenance of vision. The blood-brain barrier and blood-eye barriers prohibit the free diffusion of solutes into the brain and eye so that transporter-mediated uptake predominates at these barriers. The major facilitator superfamily of transporters constitutes one of the largest families of facilitative transporters across all domains of life. A unique family member, major facilitator superfamily domain containing 2a (Mfsd2a) is a lysophosphatidylcholine (LPC) transporter expressed at the blood-brain and blood-retinal barriers and demonstrated to be the major pathway for brain and eye accretion of docosahexaenoic acid (DHA) as an LPC. In addition to LPC-DHA, Mfsd2a can transport other LPCs containing mono- and polyunsaturated fatty acids. Mfsd2a deficiency in mouse and humans results in severe microcephaly, underscoring the importance of LPC transport in brain development. Beyond its role in brain development, LPC-DHA uptake in the brain and eye negatively regulates de novo lipogenesis. This review focuses on the current understanding of the physiological roles of Mfsd2a in the brain and eye and the proposed transport mechanism of Mfsd2a.

Keywords

Major facilitator superfamily (MFS) · Major facilitator superfamily domain containing 2a (Mfsd2a) · Lipid transfer activity · Lysolipids · Brain · Eye

14.1 Major Facilitator Superfamily

Lipids are organic compounds that are essential in living cells. Mammalian cell membranes are largely made up of glycerolipids, phospholipids, and cholesterol, organized into a lipid bilayer. The transport of molecules across this hydrophobic membrane is vital for cell growth, metabolism, and signal transduction, and are facilitated by transport proteins such as channels and transporters. Primary active transporters like ATP-binding cassette transporters are fueled by energy released from ATP hydrolysis. Conversely, secondary facilitative transporters do not utilize ATP hydrolysis for transport and can be generally categorized as facilitative or active facilitate. The former transports solutes down their concentration gradients across membranes, while active facilitative transporters transport solutes against their concentration gradients

B. H. Wong · D. L. Silver (✉)
Signature Research Program in Cardiovascular and Metabolic Disorders, Duke-NUS Medical School, Singapore, Singapore
e-mail: David.Silver@Duke-NUS.edu.sg

X.-C. Jiang (ed.), *Lipid Transfer in Lipoprotein Metabolism and Cardiovascular Disease*, Advances in Experimental Medicine and Biology 1276, https://doi.org/10.1007/978-981-15-6082-8_14

either as symporters or antiporters. Facilitative active transporters derive their energy for transport through coupling solute transport with the transport of ions such as sodium or protons down their concentration gradients across membranes [45, 60, 75].

The major facilitator superfamily (MFS) is one of the largest families of secondary transporters. The vast majority of characterized MFS transporters transport minimally polar hydrophilic substrates such as mono- and disaccharides, amino acids, and nucleosides [45, 53]. However, there are three MFS transporters that are exceptions to this general feature of the MFS family, namely, Spns2, Mfsd2a, and Mfsd2b [39, 52, 73]. Spns2 and Mfsd2b are sphingosine-1-phosphate transporters [39, 73], while Mfsd2a, the subject of this review article, is a lysophosphatidylcholine transporter.

MFS proteins have a highly conserved fold that is composed of 12 transmembrane alpha helices separated into two 6-transmembrane units that exhibit a pseudo twofold symmetry about an axis perpendicular to the membrane plane [45, 53]. Within each of these 6-transmembrane domains, the two 3-transmembrane units are organized as inverted repeats [45, 60, 75].

The majority of crystal structures of MFS proteins are bacterial proteins, with some exceptions being human glucose transporters [26, 27, 75], where the first MFS structures to be elucidated are that of *Escherichia coli* Lactose:H+ symporter (LacY) [1] and *Escherichia coli* Glycerol-3-phosphate:Pi antiporter (GlpT) [36]. A common transport mechanism that has been proposed as a result of these structural and biochemical studies is that substrates are transported in a rocker-switch, alternating access mechanism [75]. The N- and C- terminal domains rotate about the central substrate binding site and open exclusively to either the cytoplasm or extracellular space at any one time, rocking between an inward open or outward open state [1, 36, 45]. The four important transmembrane helices that surround the central pocket and essential for transport activity are domains 1, 4, 7, and 10. Transmembrane domains consisting of 2, 5, 8, and 11 or 3, 6, 9, and 12 are positioned just outside the core helices and mediate the interference between the N- and C- domains and support the structural integrity of the transporter, respectively [75].

14.2 Major Facilitator Superfamily Containing 2a

Major facilitator superfamily domain containing 2a (MFSD2A) was first identified by Angers et al. as an orphan transporter to be significantly induced in brown adipose tissue (BAT) of mice lacking both nuclear receptors retinoid-related orphan receptor alpha and gamma (RORα and RORγ) [5]. The *Mfsd2a* gene is approximately 14.3kb long, with 14 exons and 13 introns. Analysis of the amino acid sequence indicates it is most closely related to the bacterial-sodium melibiose symporter MelB at a 43–37% similarity [29]. Importantly, amino acid sequence of both mouse and human MFSD2A proteins is approximately 85% identical and is highly conserved from fish to human [11].

Mfsd2a is expressed in the brain, spinal cord, BAT, liver, kidney, lung, placenta, testes [11], and eye [74]. mRNA expression of Mfsd2a is greatly induced in murine liver and BAT during fasting and follows an oscillatory expression profile consistent with a circadian rhythm, with peak expression at circadian time 12 [5]. Additionally, Mfsd2a mRNA was also significantly upregulated exclusively in BATs by cold exposure and β-adrenergic receptor signaling pathway [5]. Berger et al. identified Mfsd2a to be induced by fasting and regulated by both peroxisome proliferator-activated receptor alpha (PPARα) and glucagon signaling in the liver, which turns over rapidly in liver upon refeeding [11]. While mRNA can be detected in BAT, Mfsd2a protein level is extremely low [11]. The function of Mfsd2a in BAT has not been determined.

14.3 Lipids and Essential Fatty Acids Are Important for Brain Growth

The brain is made up of glycerophospholipids, cholesterol, and sphingolipids, making it one of the most lipid-rich organs in the body [41]. Prenatal brain development is a complex developmental process that begins with the development of the neural tube, which ultimately differentiates into the brain and spinal cord. This is also the time where hundreds of specialized cell types come together, organizing a network of synaptic connectivity and a functioning blood-brain barrier (BBB) (Fig. 14.1) [6, 24, 61]. The BBB separates the brain from blood and serves to maintain a tightly controlled environment where toxins and pathogens are prevented from freely entering or leaving the brain by diffusion. The BBB is governed by tight junctions of endothelial cells of blood vessels, supported by astrocytes and pericytes [6, 24]. This is followed by postnatal brain growth, which is accompanied by the proliferation of astrocytes and oligodendrocytes [14, 28, 42] and myelination of axons and synaptogenesis [8, 28, 51]. Massive amounts of membrane phospholipids are therefore required for brain growth, where it has been postulated that lipids are derived exclusively from de novo biosynthesis within cells of the brain.

De novo lipogenic gene expression is controlled by sterol regulatory element-binding proteins (Srebp-1 and Srebp-2). In support of the vital role of de novo lipogenesis in brain development, the genetic deficiency of Scap, an essential chaperone protein for Srebp, in neurons in the developing central nervous system resulted in microcephaly and early postnatal lethality [65]. In addition, deficiency of Scap in mature astrocytes and oligodendrocytes have profound effects on myelination [71].

14.4 LPC-DHA Transport into the Brain

Docosahexaenoic acid (DHA) is an omega-3 fatty acid composed of 22 carbons and 6 double bonds. DHA can be synthesized by the liver through chain extension and desaturation of the essential fatty acid linolenic acid. DHA is highly enriched in brain phospholipids, particularly in the phosphatidylethanolamine (PE), phosphatidylserine (PS), and to a lesser extent, phosphatidylcholine (PC) pools within membranes, and comprises up to 15% or more of the total fatty acid composition of the prefrontal cortex [13, 50]. In humans, DHA is rapidly taken up as early as the end of the second trimester, coinciding with the development of the BBB where considerable amounts of membrane phospholipids are required for the growing brain [17, 38, 61]. DHA is continuously acquired from early postnatal days until approximately 2 years of age [23, 47, 66]. While DHA supplementation studies in term infants or pregnant and lactating women have been inconclusive for enhancing cognitive development [25, 31, 55], DHA supplementation in preterm infants has shown some benefit to cognitive development, presumably because preterm infants might have lower brain DHA levels [7, 18]. Likewise, decreased levels of DHA in the developing brain have been associated with negative effects on cognitive function [33, 49] and neurodevelopmental disorders [19, 35, 48]. Importantly, DHA itself cannot be de novo synthesized and must be transported across the BBB into brain.

The form by which DHA gets taken up into brain, either as unesterified DHA or DHA esterified as lysophosphatidylcholine-DHA (LPC-DHA), has been a point of debate. LPCs circulate in blood bound to albumin [20, 56, 67] where it was first shown by Illingworth and Portman to be taken up and reacylated readily in brains of squirrel monkeys [37]. As early as 1965, it was hypothesized by Switzer and Eder that plasma LPCs serve as precursors for the renewal of cellular membranes [67]. Importantly, Thiès et al. reported a preference for unsaturated fatty acids esterified as 2-acyl-LPC in young rat brains where LPC-DHA was transported 12 fold more than unesterified DHA, suggesting that LPCs might be an efficient delivery of polyunsaturated fatty acids (PUFAs) into the developing brain [69, 70]. Moreover, Lagarde et al. was the first

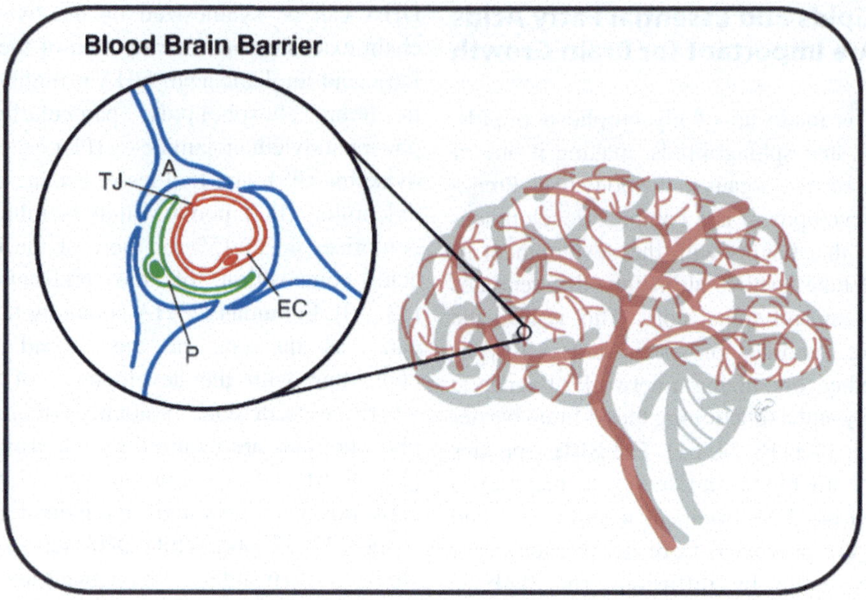

Fig. 14.1 *Blood-brain barrier*. The BBB is governed by tight junctions (TJ) of endothelial cells (EC) of blood vessels, supported by astrocytes (A) and pericytes (P)

to propose that LPC is the preferred carrier of PUFAs like DHA or arachidonic acid (AA) to the brain [43]. As will be further discussed below, Mfsd2a is the LPC transporter that explains the LPC transport activity first described by Lagarde and co-workers. More recently, it was demonstrated that supplementing adult mice with dietary LPC-DHA, but not unesterified DHA, were able to increase brain DHA levels twofold [64]. Collectively, these findings support the conclusion that LPC-DHA, and not unesterified DHA, is the primary carrier of DHA delivery to the brain. However, Mfsd2a KO mice have residual phospholipid containing DHA in the brain and eye, indicating the possibility of either compensatory de novo biosynthesis, other transport mechanisms, or acquisition of DHA during embryogenesis in the brain and eye prior to blood-barrier formation. It is important to note that single cell sequencing projects and bulk RNA-seq of the blood-brain barrier in mice [68, 72, 79] have shown that mRNA expression for proteins proposed to be involved in the uptake of unesterified DHA by the BBB endothelium, such as LPL, and its essential chaperone GPIHBP1 [78], CD36, and FATP1-6 (Slc27a1-6), and ACSL6 are not expressed by the endothelium of the BBB.

14.5 Mfsd2a Deficiency in the Brain

Importantly, Nguyen et al. and Ben-Zvi et al. discovered Mfsd2a to be highly expressed at the endothelium of the BBB [10, 52]. Through targeted lipidomic analysis, Mfsd2a was found to be the major pathway for brain DHA accretion, where a significant 60–70% reduction in steady-state levels of total percentage DHA-containing phospholipids was observed in brains of 2aKO mice relative to wild-type controls [15, 52]. Conversely, brains of 2aKO mice had a modest 35% increase in steady-state levels of total percentage AA-containing phospholipids [52], a phenomenon commonly observed in rodent models of DHA deficiency [62].

More recently, using endothelial-specific and inducible endothelial-specific Mfsd2a deletion mouse models, Chan et al. showed that Mfsd2a deficiency results in a unique form of postnatal microcephaly, with DHA deficiency preceding the onset of microcephaly [15]. Only adult

2aKO mice exhibit a minor loss of Purkinje cells in the cerebellum and a decrease in neuronal cell density in the CA1 and CA3 regions of the hippocampus [52]. Because the brains of 2aKO embryos are deficient in DHA but are not microcephalic until postnatal life, these cell loss phenotypes are secondary events. These findings also indicate that DHA deficiency is an unlikely cause underlying microcephaly, but rather the absence of bulk LPC transport, where LPCs are phospholipid membrane building blocks.

Recently, transcriptomic and lipidomic analysis in Mfsd2a deficiency mouse models was used as a tool to understand how the brain adapts to DHA deficiency, thus revealing functions of DHA in the brain [15]. It was discovered that Mfsd2a deficiency resulted in a de-repression of the Srebp1 and Srebp2 pathways leading to an increase in de novo synthesis of unsaturated fatty acids in phospholipids. It was shown that Mfsd2a is expressed in neural stem cells (NSCs) isolated from early postnatal mice and that NSCs treated with LPC-DHA and other LPC-PUFAs can acutely downregulate Srebp1 and Srebp2 target gene expression in an Mfsd2a-dependent fashion and that the mechanism is in part through inhibition of Srebp-1 receptor processing [15]. Moreover, Mfsd2a itself is regulated by Srebp, forming a negative feedback loop on Srebp processing that can balance de novo lipogenesis with exogenous uptake of LPC-DHA. The regulation of brain Srebp function by LPC-DHA transported by Mfsd2a might serve the purpose of fine-tuning membrane phospholipid saturation and hence biophysical properties during brain development [15].

Another reported feature of Mfsd2a deficiency in the brain and eye is that Mfsd2a knockout mice have increased transcytosis resulting in increased BBB permeability [4, 10]. It has been suggested that the microcephaly and DHA deficiency in 2aKO mice could be due to a leaky BBB, but it is unclear how a leaky BBB would result in less DHA uptake and not more relative to wild-type (WT) mice. Nonetheless, this issue has been resolved in that BBB permeability, but not microcephaly and DHA deficiency, can be completely rescued in Mfsd2a-deficient mice by genetic deficiency of Cav1 [4]. Andreone et al. generated a transporter-dead Mfsd2a knockin mouse model bearing a D96A aspartate to alanine point mutation, a conserved residue with D97 in the human Mfsd2a constituting the sodium binding site, and showed that consistent with the lack of transport activity, Mfsd2a$^{D96A/D96A}$ mice exhibited microcephaly and DHA deficiency in the brain [4]. These findings indicate that microcephaly and DHA deficiency are primary phenotypes of Mfsd2a deficiency, and not a result of a leaky BBB [4], and that LPC transport via Mfsd2a is essential for DHA accretion and postnatal brain growth. Of note, the increased transcytosis phenotype in the BBB or blood-retina barrier of 2aKO mice reported by the Gu lab [4, 10, 16] has not been observed in other studies [46, 74].

14.6 Mfsd2a Deficiency in the Eye

The retina is a highly organized structure, with photoreceptors (PR), extensive retinal glial network, and retinal pigment epithelium (RPE) organized into distinct layers. Rods and cones are the two types of PR found in mammalian eyes, which make up 70% majority of cells in the retina. DHA, localized with rhodopsin [30], is found primarily in phospholipids of membrane discs that make up rod PR outer segments (OS), making the retina a tissue with the highest concentration of DHA per unit area in the body [58]. With daily daylight exposure, OS discs which are photosensitive, accumulate photo-damaged proteins and lipids [9] and must be synthesized continuously throughout one's lifetime for the maintenance of healthy vision [63, 77]. The villi-containing apical membrane of the RPE is particularly important for this renewal process, where through its interaction with the distal ends of the OS (Fig. 14.2) facilitates the daily phagocytosis of OS discs that make up one-tenth of the OS length. This

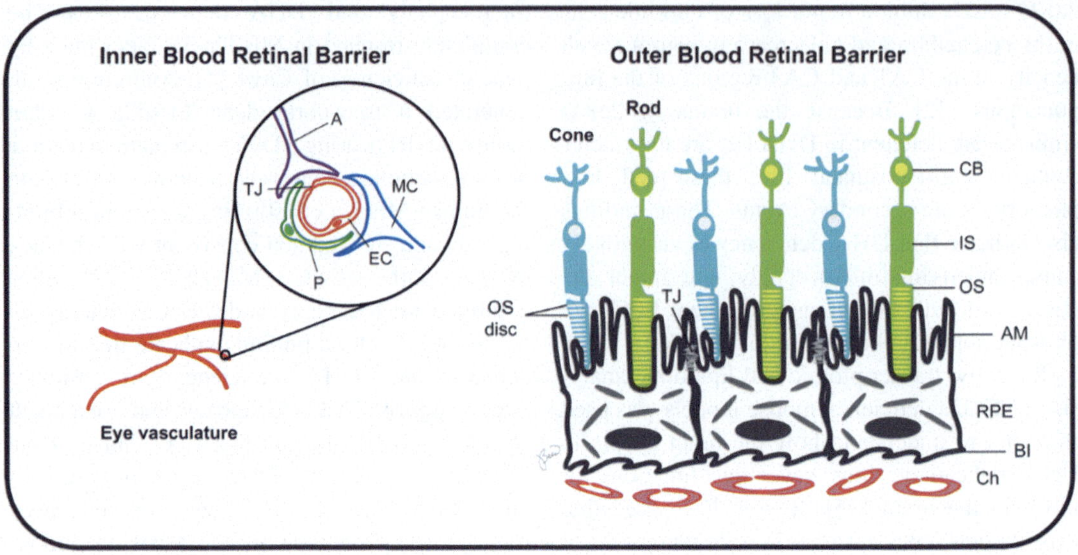

Fig. 14.2 *Blood-retinal barrier*. The BRB is made up of the inner BRB, formed by tight junctions (TJ) of the endothelium of retinal capillaries (EC), supported by pericytes (P), astrocytes (A) and Müller cells (MC). The outer BRB is governed by TJ of the retinal pigment epithelium (RPE). DHA is found primarily in phospholipids of the outer segment (OS) discs of rod and cones and interact closely with the apical membrane villi of the RPE. As photo-damaged discs need to undergo a constant renewal process for the maintenance of vision, inner segments that contain metabolic machinery synthesize new membrane discs that move along the length of the OS where they are eventually phagocytosed by the RPE. *CB* cell body, *IS* inner segments, *OS* outer segments, *AM* apical membrane, *RPE* retinal pigment epithelium, *BI* basal infoldings, *Ch* choroid

process of phagocytosis is balanced with an equal rate of disc regeneration, so that the OS length is maintained [76], thus highlighting the importance of lipids and essential fatty acids for membrane biogenesis and turnover.

Similar to the BBB of the brain, the eye contains cellular barriers that prevent the diffusion of blood-borne material or lipids from entering the retina freely. The eye contains two blood-eye barriers (Fig. 14.2), the inner blood-retina barrier (inner BRB) that is established by tight junctions between retinal endothelial cells, supported by the pericytes, astrocytes, and Müller cells [22, 59] and the outer BRB that is governed by tight junctions of the RPE [21].

Mfsd2a is expressed at the endothelium of the BRB and RPE. The RPE is the major site of Mfsd2a expression and is quantitatively important for DHA accretion into the retina via LPC-DHA transport [74] (Fig. 14.3). Whole-body Mfsd2a-deficient (2aKO) mice displayed a unique form of a slow, progressive retina degeneration [46, 74]. However, a 40% deficiency in phospholipids containing DHA in eyes of 2aKO mice did not result in the expected severe and rapid retinal degeneration nor significant visual dysfunction [46, 74]. Like the brain, upregulation of de novo lipogenesis pathways was observed in eyes of 2aKO mice which might serve, as a compensatory mechanism to synthesize new OS discs in the absence of Mfsd2a [46, 74]. In addition, the BRB was found to be intact in 2aKO mice [46, 74], which is inconsistent with a report that Mfsd2a is required to suppresses transcytosis for the development and maintenance of a functional BRB [16]. This discrepancy is not due to strain-specific differences as the strain used in the Lobanova study was the same as reported by Chow et al. [16]. The most remarkable finding from studying 2aKO retinas is that phototransduction tested by electroretinography [74] or light evoked potential recordings of single rods [46] indicated that phototransduction in 2aKO and WT was indistinguishable. These findings might suggest that the compensatory changes in lipid composition in 2aKO retinas of

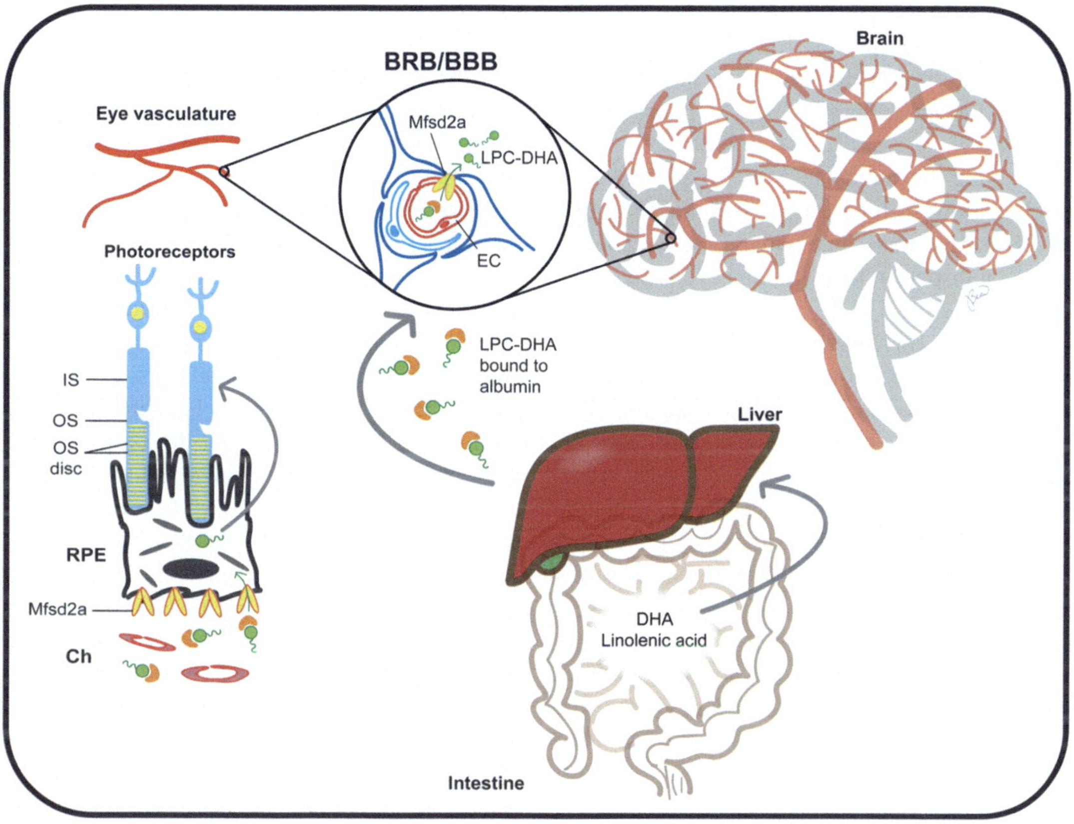

Fig. 14.3 *MFSD2A transports LPC-DHA across the BBB and BRB.* DHA can come preformed from the diet or its precursor linolenic acid, conjugated to LPC in the liver, and transported in blood plasma bound to albumin. At the BBB, Mfsd2a translocates LPC-DHA across the endothelial plasma membrane into the brain. Mfsd2a is expressed at both the inner and outer BRB, but Mfsd2a at the RPE is the major route by which LPC-DHA gets into the eye. *IS* inner segments, *OS* outer segments, *RPE* retinal pigment epithelium, *Ch* choroid

increased monounsaturated fatty acids and arachidonic acid in phospholipid pools might compensate for the severe reduction in DHA.

14.7 Inactivating Mutations of MFSD2A in Humans

To date, four unrelated consanguineous families with homozygous non-synonymous inactivating mutations in MFSD2A have been identified that presented with severe microcephaly and intellectual impairments [2, 32, 34]. The first two families, one from Libya and the other from Egypt, harbored a p.Thr159Met or p.Ser166Leu protein change [32]. Mutant Mfsd2a proteins were stably expressed and localized to the plasma membrane when expressed in HEK 293 cells, comparable to WT Mfsd2a, but had complete inactivation of transport activity [32]. The third family from Pakistan was a large pedigree, with ten affected family members harboring a p. Ser339Leu protein change that presented with severe non-lethal microcephaly [2]. Again, mutant Mfsd2a proteins were stably expressed and had proper membrane localization when expressed in HEK 293 cells but exhibited a partial inactivation of transport activity relative to WT protein [2]. A fourth family was identified in Israel that harbored a p.Pro402His protein change, with complete inactivation of transport activity, and presented with severe non-lethal microcephaly [34]. Consistent with reduced or complete inactivation of transport activity that would be expected to reduce brain and eye LPC uptake, increased plasma LPC levels have been observed in all affected family members [2, 32, 34]. In further support of this explanation for increased plasma LPC in patients with inactivating mutations in Mfsd2a, plasma LPC levels were also found to be increased by 40% in 2aKO mice, consistent with 85–90% reduction in LPC transport in the brain and eye using tracer studies [32, 52, 74].

Both p.Thr159Met and p.Ser166Leu mutations were found on transmembrane domain 4 of Mfsd2a, p.Ser339Leu was found on transmembrane domain 8, while p.Pro402His was found on the extracellular loop between transmembrane 10 and 11. A molecular explanation for the loss-of-function caused by Ser166Leu and Pro402His is not known. However, Thr159met is homologous to Thr121 in MelB, which is essential for establishing a hydrogen bond with conserved aspartate residues at the sodium binding pocket. Therefore, it can be predicted that Thr159Met inactivity is a consequence of absence of sodium binding [32].

14.8 Proposed Transport Mechanism of Mfsd2a

Mfsd2a does not transport unesterified PUFAs, but PUFAs esterified as a LPC [52]. It was determined through structure-activity relationship studies that lysophospholipid with a minimal acyl chain length of 14 carbons and a zwitterionic headgroup (e.g., PC, PE, and PS) is essential for transport by Mfsd2a [52]. More recently, Quek et al. showed that the acyl-carnitines can also be transported by Mfsd2a, again underscoring the importance of a zwitterionic headgroup and not strictly a phosphorylcholine headgroup as a necessary feature for lysolipid transport [57]. Notably, Mfsd2a has a higher transport capacity for LPCs having unsaturated fatty acids like DHA relative to LPCs with saturated fatty acids like palmitate [52]. This latter finding is important, because it indicates that LPC transport capacity is inversely correlated with the physiological levels of LPCs in human plasma, where LPC-palmitate is the most abundant [2, 32, 56]. Presumably, this preference for LPC-PUFA by Mfsd2a would allow the brain to obtain the lower abundant essential fatty acids diluted in a larger milieu of LPCs containing non-essential fatty acids.

Using homology modeling based upon crystal structures of MelB and LacY, and further refinement by site-directed mutagenesis and biochemical transport analysis, Quek et al. identified the following four important structural features of human Mfsd2a: a sodium binding site, a hydrophobic cleft, a lipid phosphate headgroup binding residue (Lys436), and ionic locks [57]. The hydrophobic cleft is likely involved in LPC acyl chain binding, while the Lys436 is involved in coordinating the LPC phosphate headgroup interaction. The ionic locks are presumably involved in stabilizing the outward open conformation during the transport cycle as previously proposed for similar ionic locks identified on MelB [29]. This proposed model of transport co-opts the standard rocker-switch model, with the exception that LPCs bound to albumin would first bind to the outer leaflet of the plasma membrane and diffuse laterally into Mfsd2a facing the outward open conformation until hydrophobic forces position the acyl chain of the LPC into the hydrophobic cleft and headgroup binding to Lys436 (Fig. 14.4). Sodium binding to its binding site comprising residues Asp93, Asp97, and Thr159 would drive a conformational change to an inward-open conformation that would push the LPC-DHA down along the hydrophobic cleft and flip over to the inner lipid leaflet, where it exits the transporter by diffusing laterally along the inner membrane [57]. This "flipping" activity would in theory allow LPCs to bypass the tight junctions of the BBB endothelium [12]. Once LPCs reach other cells at the BBB such as astrocytes, it could be converted to PC-DHA through activity of the LPCATs [44].

14.9 Concluding Remarks

Mfsd2a is a sodium-dependent lysophosphatidylcholine co-transporter highly expressed at the

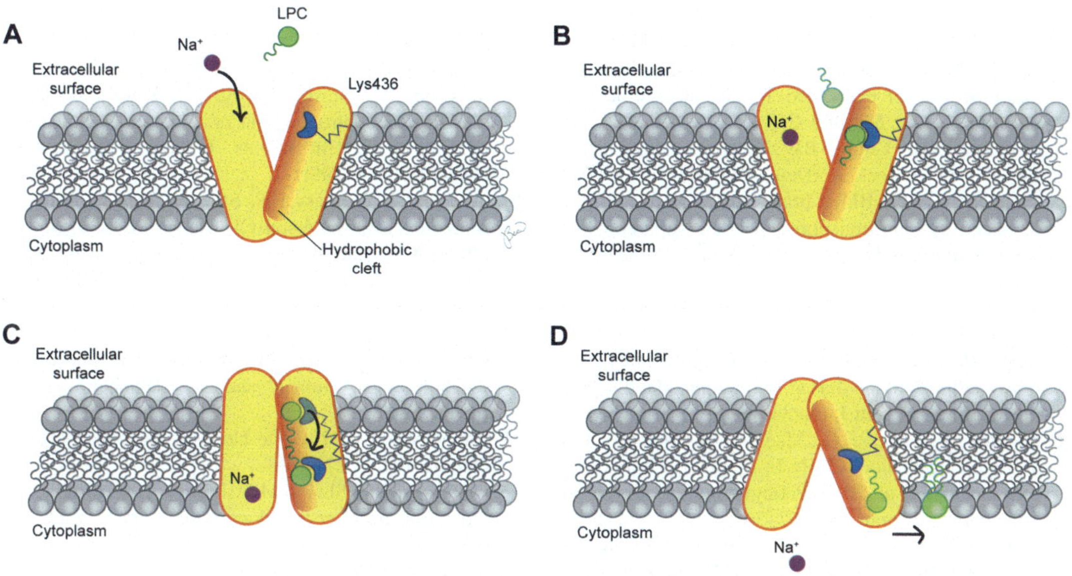

Fig. 14.4 *Proposed mechanism of LPC transport.* (**a**) Mfsd2a in the outward-open conformation facing the extracellular surface. (**b**) Sodium binds to sodium binding site, while LPC diffuses laterally into the central cavity and binds to Lys436. (**c**) In the presence of sodium, Mfsd2a undergoes a conformational change, and LPC "flips" along hydrophobic cleft. (**d**) LPC diffuses laterally into the inner leaflet and converts to PC-DHA

blood-brain barrier and blood-eye barriers that is essential for normal human brain development. Mfsd2a shows high specificity for the transport of LPCs with long chain and unsaturated fatty acyl chains. LPC-DHA in particular negatively regulates Srebp activity during brain development, and this function is likely important to maintain proper membrane phospholipid saturation. An important question that remains to be answered is a determination of the transport mechanism of LPCs by Mfsd2a. This determination awaits the development of new biochemical assays to reconstitute transport on purified Mfsd2a and the determination of atomic resolution structures. Interestingly, Mfsd2a is expressed by other cell types and tissues such as liver and a determination of the function of Mfsd2a outside of the brain and eye will likely reveal new biology into the function of LPCs. For example, Piccirillo and others have shown that Mfsd2a is required for the maintenance of memory T cells [54], perhaps in part through TOX, which might regulate Mfsd2a [3, 40]. Lastly, a word of caution, many recent papers have been published using non-validated Mfsd2a antibodies that are likely leading to erroneous conclusions on the regulation of, site of, expression of, and involvement of Mfsd2a in particular biological and pathophysiological processes. It is critical that Mfsd2a antibodies be validated using both cell-based overexpression and Mfsd2a deficiency cell or mouse models.

Acknowledgments This work was supported by grants from the National Research Foundation, Singapore (NRF-NRFI2017-05 to D.L.S.), and the Ministry of Health (MOH-000217-00).

References

1. Abramson J, Smirnova I, Kasho V, Verner G, Kaback HR, Iwata S (2003) Structure and mechanism of the lactose permease of Escherichia coli. Science 301:610–615
2. Alakbarzade V, Hameed A, Quek DQ, Chioza BA, Baple EL, Cazenave-Gassiot A, Nguyen LN, Wenk MR, Ahmad AQ, Sreekantan-Nair A et al (2015) A partially inactivating mutation in the sodium-dependent lysophosphatidylcholine transporter MFSD2A causes a non-lethal microcephaly syndrome. Nat Genet 47:814–817

3. Alfei F, Kanev K, Hofmann M, Wu M, Ghoneim HE, Roelli P, Utzschneider DT, von Hoesslin M, Cullen JG, Fan Y et al (2019) TOX reinforces the phenotype and longevity of exhausted T cells in chronic viral infection. Nature 571:265–269
4. Andreone BJ, Chow BW, Tata A, Lacoste B, Ben-Zvi A, Bullock K, Deik AA, Ginty DD, Clish CB, Gu C (2017) Blood-brain barrier permeability is regulated by lipid transport-dependent suppression of caveolae-mediated transcytosis. Neuron 94:581–594. e585
5. Angers M, Uldry M, Kong D, Gimble JM, Jetten AM (2008) Mfsd2a encodes a novel major facilitator superfamily domain-containing protein highly induced in brown adipose tissue during fasting and adaptive thermogenesis. Biochem J 416:347–355
6. Armulik A, Genove G, Mae M, Nisancioglu MH, Wallgard E, Niaudet C, He L, Norlin J, Lindblom P, Strittmatter K et al (2010) Pericytes regulate the blood-brain barrier. Nature 468:557–561
7. Baack ML, Puumala SE, Messier SE, Pritchett DK, Harris WS (2016) Daily enteral DHA supplementation alleviates deficiency in premature infants. Lipids 51:423–433
8. Baumann N, Pham-Dinh D (2001) Biology of oligodendrocyte and myelin in the mammalian central nervous system. Physiol Rev 81:871–927
9. Bazan NG (2006) Cell survival matters: docosahexaenoic acid signaling, neuroprotection and photoreceptors. Trends Neurosci 29:263–271
10. Ben-Zvi A, Lacoste B, Kur E, Andreone BJ, Mayshar Y, Yan H, Gu C (2014) Mfsd2a is critical for the formation and function of the blood-brain barrier. Nature 509:507–511
11. Berger JH, Charron MJ, Silver DL (2012) Major facilitator superfamily domain-containing protein 2a (MFSD2A) has roles in body growth, motor function, and lipid metabolism. PLoS One 7:e50629
12. Betsholtz C (2015) Lipid transport and human brain development. Nat Genet 47:699–701
13. Carver JD, Benford VJ, Han B, Cantor AB (2001) The relationship between age and the fatty acid composition of cerebral cortex and erythrocytes in human subjects. Brain Res Bull 56:79–85
14. Catalani A, Sabbatini M, Consoli C, Cinque C, Tomassoni D, Azmitia E, Angelucci L, Amenta F (2002) Glial fibrillary acidic protein immunoreactive astrocytes in developing rat hippocampus. Mech Ageing Dev 123:481–490
15. Chan JP, Wong BH, Chin CF, Galam DLA, Foo JC, Wong LC, Ghosh S, Wenk MR, Cazenave-Gassiot A, Silver DL (2018) The lysolipid transporter Mfsd2a regulates lipogenesis in the developing brain. PLoS Biol 16:e2006443
16. Chow BW, Gu C (2017) Gradual suppression of transcytosis governs functional blood-retinal barrier formation. Neuron 93:1325–1333. e1323
17. Clandinin MT, Chappell JE, Heim T, Swyer PR, Chance GW (1981) Fatty acid utilization in perinatal de novo synthesis of tissues. Early Hum Dev 5:355–366
18. Collins CT, Sullivan TR, McPhee AJ, Stark MJ, Makrides M, Gibson RA (2015) A dose response randomised controlled trial of docosahexaenoic acid (DHA) in preterm infants. Prostaglandins Leukot Essent Fat Acids 99:1–6
19. Colombo J, Kannass KN, Shaddy DJ, Kundurthi S, Maikranz JM, Anderson CJ, Blaga OM, Carlson SE (2004) Maternal DHA and the development of attention in infancy and toddlerhood. Child Dev 75:1254–1267
20. Croset M, Brossard N, Polette A, Lagarde M (2000) Characterization of plasma unsaturated lysophosphatidylcholines in human and rat. Biochem J 345 (Pt 1):61–67
21. Cunha-Vaz J (1979) The blood-ocular barriers. Surv Ophthalmol 23:279–296
22. Cunha-Vaz JG, Shakib M, Ashton N (1966) Studies on the permeability of the blood-retinal barrier. I. On the existence, development, and site of a blood-retinal barrier. Br J Ophthalmol 50:441–453
23. Cunnane SC, Francescutti V, Brenna JT, Crawford MA (2000) Breast-fed infants achieve a higher rate of brain and whole body docosahexaenoate accumulation than formula-fed infants not consuming dietary docosahexaenoate. Lipids 35:105–111
24. Daneman R, Zhou L, Kebede AA, Barres BA (2010) Pericytes are required for blood-brain barrier integrity during embryogenesis. Nature 468:562–566
25. Delgado-Noguera MF, Calvache JA, Bonfill Cosp X (2010) Supplementation with long chain polyunsaturated fatty acids (LCPUFA) to breastfeeding mothers for improving child growth and development Cochrane Database Syst Rev:CD007901
26. Deng D, Sun P, Yan C, Ke M, Jiang X, Xiong L, Ren W, Hirata K, Yamamoto M, Fan S et al (2015) Molecular basis of ligand recognition and transport by glucose transporters. Nature 526:391–396
27. Deng D, Xu C, Sun P, Wu J, Yan C, Hu M, Yan N (2014) Crystal structure of the human glucose transporter GLUT1. Nature 510:121–125
28. Dobbing J, Sands J (1979) Comparative aspects of the brain growth spurt. Early Hum Dev 3:79–83
29. Ethayathulla AS, Yousef MS, Amin A, Leblanc G, Kaback HR, Guan L (2014) Structure-based mechanism for Na(+)/melibiose symport by MelB. Nat Commun 5:3009
30. Fliesler SJ, Anderson RE (1983) Chemistry and metabolism of lipids in the vertebrate retina. Prog Lipid Res 22:79–131
31. Gould JF, Smithers LG, Makrides M (2013) The effect of maternal omega-3 (n-3) LCPUFA supplementation during pregnancy on early childhood cognitive and visual development: a systematic review and meta-analysis of randomized controlled trials. Am J Clin Nutr 97:531–544
32. Guemez-Gamboa A, Nguyen LN, Yang H, Zaki MS, Kara M, Ben-Omran T, Akizu N, Rosti RO, Rosti B,

Scott E et al (2015) Inactivating mutations in MFSD2A, required for omega-3 fatty acid transport in brain, cause a lethal microcephaly syndrome. Nat Genet 47:809–813
33. Guesnet P, Alessandri JM (2011) Docosahexaenoic acid (DHA) and the developing central nervous system (CNS) – implications for dietary recommendations. Biochimie 93:7–12
34. Harel T, Quek DQY, Wong BH, Cazenave-Gassiot A, Wenk MR, Fan H, Berger I, Shmueli D, Shaag A, Silver DL et al (2018) Homozygous mutation in MFSD2A, encoding a lysolipid transporter for docosahexanoic acid, is associated with microcephaly and hypomyelination. Neurogenetics 19:227–235
35. Heird WC, Lapillonne A (2005) The role of essential fatty acids in development. Annu Rev Nutr 25:549–571
36. Huang Y, Lemieux MJ, Song J, Auer M, Wang DN (2003) Structure and mechanism of the glycerol-3-phosphate transporter from Escherichia coli. Science 301:616–620
37. Illingworth DR, Portman OW (1972) The uptake and metabolism of plasma lysophosphatidylcholine in vivo by the brain of squirrel monkeys. Biochem J 130:557–567
38. Innis SM (2005) Essential fatty acid transfer and fetal development. Placenta 26(Suppl A):S70–S75
39. Kawahara A, Nishi T, Hisano Y, Fukui H, Yamaguchi A, Mochizuki N (2009) The sphingolipid transporter spns2 functions in migration of zebrafish myocardial precursors. Science 323:524–527
40. Khan O, Giles JR, McDonald S, Manne S, Ngiow SF, Patel KP, Werner MT, Huang AC, Alexander KA, Wu JE et al (2019) TOX transcriptionally and epigenetically programs CD8(+) T cell exhaustion. Nature 571:211–218
41. Korade Z, Kenworthy AK (2008) Lipid rafts, cholesterol, and the brain. Neuropharmacology 55:1265–1273
42. Kriegstein A, Alvarez-Buylla A (2009) The glial nature of embryonic and adult neural stem cells. Annu Rev Neurosci 32:149–184
43. Lagarde M, Bernoud N, Brossard N, Lemaitre-Delaunay D, Thies F, Croset M, Lecerf J (2001) Lysophosphatidylcholine as a preferred carrier form of docosahexaenoic acid to the brain. J Mol Neurosci 16:201–204; discussion 215-221
44. Lands WE (1960) Metabolism of glycerolipids. 2. The enzymatic acylation of lysolecithin. J Biol Chem 235:2233–2237
45. Law CJ, Maloney PC, Wang DN (2008) Ins and outs of major facilitator superfamily antiporters. Annu Rev Microbiol 62:289–305
46. Lobanova ES, Schuhmann K, Finkelstein S, Lewis TR, Cady MA, Hao Y, Keuthan C, Ash JD, Burns ME, Shevchenko A et al (2019) Disrupted blood-retina lysophosphatidylcholine transport impairs photoreceptor health but not visual signal transduction. J Neurosci 39:9689–9701
47. Martinez M (1992) Tissue levels of polyunsaturated fatty acids during early human development. J Pediatr 120:S129–S138
48. Martinez M (1996) Docosahexaenoic acid therapy in docosahexaenoic acid-deficient patients with disorders of peroxisomal biogenesis. Lipids 31(Suppl):S145–S152
49. McNamara RK (2010) DHA deficiency and prefrontal cortex neuropathology in recurrent affective disorders. J Nutr 140:864–868
50. McNamara RK, Liu Y, Jandacek R, Rider T, Tso P (2008) The aging human orbitofrontal cortex: decreasing polyunsaturated fatty acid composition and associated increases in lipogenic gene expression and stearoyl-CoA desaturase activity. Prostaglandins Leukot Essent Fat Acids 78:293–304
51. Micheva KD, Beaulieu C (1996) Quantitative aspects of synaptogenesis in the rat barrel field cortex with special reference to GABA circuitry. J Comp Neurol 373:340–354
52. Nguyen LN, Ma D, Shui G, Wong P, Cazenave-Gassiot A, Zhang X, Wenk MR, Goh EL, Silver DL (2014) Mfsd2a is a transporter for the essential omega-3 fatty acid docosahexaenoic acid. Nature 509:503–506
53. Pao SS, Paulsen IT, Saier MH Jr (1998) Major facilitator superfamily. Microbiol Mol Biol Rev 62:1–34
54. Piccirillo AR, Hyzny EJ, Beppu LY, Menk AV, Wallace CT, Hawse WF, Buechel HM, Wong BH, Foo JC, Cazenave-Gassiot A et al (2019) The lysophosphatidylcholine transporter MFSD2A is essential for CD8(+) memory T cell maintenance and secondary response to infection. J Immunol 203:117–126
55. Qawasmi A, Landeros-Weisenberger A, Leckman JF, Bloch MH (2012) Meta-analysis of long-chain polyunsaturated fatty acid supplementation of formula and infant cognition. Pediatrics 129:1141–1149
56. Quehenberger O, Armando AM, Brown AH, Milne SB, Myers DS, Merrill AH, Bandyopadhyay S, Jones KN, Kelly S, Shaner RL et al (2010) Lipidomics reveals a remarkable diversity of lipids in human plasma. J Lipid Res 51:3299–3305
57. Quek DQ, Nguyen LN, Fan H, Silver DL (2016) Structural insights into the transport mechanism of the human sodium-dependent lysophosphatidylcholine transporter Mfsd2a. J Biol Chem 291:9383–9394
58. SanGiovanni JP, Chew EY (2005) The role of omega-3 long-chain polyunsaturated fatty acids in health and disease of the retina. Prog Retin Eye Res 24:87–138
59. Shakib M, Cunha-Vaz JG (1966) Studies on the permeability of the blood-retinal barrier. IV. Junctional complexes of the retinal vessels and their role in the permeability of the blood-retinal barrier. Exp Eye Res 5:229–234
60. Shi Y (2013) Common folds and transport mechanisms of secondary active transporters. Annu Rev Biophys 42:51–72

61. Silbereis JC, Pochareddy S, Zhu Y, Li M, Sestan N (2016) The cellular and molecular landscapes of the developing human central nervous system. Neuron 89:248–268
62. Simopoulos AP (2008) The importance of the omega-6/omega-3 fatty acid ratio in cardiovascular disease and other chronic diseases. Exp Biol Med (Maywood) 233:674–688
63. Steinberg RH (1985) Interactions between the retinal pigment epithelium and the neural retina. Doc Ophthalmol 60:327–346
64. Sugasini D, Thomas R, Yalagala PCR, Tai LM, Subbaiah PV (2017) Dietary docosahexaenoic acid (DHA) as lysophosphatidylcholine, but not as free acid, enriches brain DHA and improves memory in adult mice. Sci Rep 7:11263
65. Suzuki R, Ferris HA, Chee MJ, Maratos-Flier E, Kahn CR (2013) Reduction of the cholesterol sensor SCAP in the brains of mice causes impaired synaptic transmission and altered cognitive function. PLoS Biol 11: e1001532
66. Svennerholm L (1968) Distribution and fatty acid composition of phosphoglycerides in normal human brain. J Lipid Res 9:570–579
67. Switzer S, Eder HA (1965) Transport of lysolecithin by albumin in human and rat plasma. J Lipid Res 6:506–511
68. Tabula Muris C, Overall c, Logistical c, Organ c, processing, Library, p., sequencing, Computational data a, Cell type a, Writing g et al (2018) Single-cell transcriptomics of 20 mouse organs creates a Tabula Muris. Nature 562:367–372
69. Thies F, Delachambre MC, Bentejac M, Lagarde M, Lecerf J (1992) Unsaturated fatty acids esterified in 2-acyl-l-lysophosphatidylcholine bound to albumin are more efficiently taken up by the young rat brain than the unesterified form. J Neurochem 59:1110–1116
70. Thies F, Pillon C, Moliere P, Lagarde M, Lecerf J (1994) Preferential incorporation of sn-2 lysoPC DHA over unesterified DHA in the young rat brain. Am J Phys 267:R1273–R1279
71. van Deijk AF, Camargo N, Timmerman J, Heistek T, Brouwers JF, Mogavero F, Mansvelder HD, Smit AB, Verheijen MH (2017) Astrocyte lipid metabolism is critical for synapse development and function in vivo. Glia 65:670–682
72. Vanlandewijck M, He L, Mae MA, Andrae J, Ando K, Del Gaudio F, Nahar K, Lebouvier T, Lavina B, Gouveia L et al (2018) A molecular atlas of cell types and zonation in the brain vasculature. Nature 560:475–480
73. Vu TM, Ishizu AN, Foo JC, Toh XR, Zhang F, Whee DM, Torta F, Cazenave-Gassiot A, Matsumura T, Kim S et al (2017) Mfsd2b is essential for the sphingosine-1-phosphate export in erythrocytes and platelets. Nature 550:524–528
74. Wong BH, Chan JP, Cazenave-Gassiot A, Poh RW, Foo JC, Galam DL, Ghosh S, Nguyen LN, Barathi VA, Yeo SW et al (2016) Mfsd2a is a transporter for the essential omega-3 fatty acid docosahexaenoic acid (DHA) in eye and is important for photoreceptor cell development. J Biol Chem 291:10501–10514
75. Yan N (2015) Structural biology of the major facilitator superfamily transporters. Annu Rev Biophys 44:257–283
76. Young RW (1967) The renewal of photoreceptor cell outer segments. J Cell Biol 33:61–72
77. Young RW (1976) Visual cells and the concept of renewal. Invest Ophthalmol Vis Sci 15:700–725
78. Young SG, Davies BS, Voss CV, Gin P, Weinstein MM, Tontonoz P, Reue K, Bensadoun A, Fong LG, Beigneux AP (2011) GPIHBP1, an endothelial cell transporter for lipoprotein lipase. J Lipid Res 52:1869–1884
79. Zhang Y, Chen K, Sloan SA, Bennett ML, Scholze AR, O'Keeffe S, Phatnani HP, Guarnieri P, Caneda C, Ruderisch N et al (2014) An RNA-sequencing transcriptome and splicing database of glia, neurons, and vascular cells of the cerebral cortex. J Neurosci 34:11929–11947

The manufacturer's authorised representative in the EU is Springer Nature Customer Service Centre GmbH, Europaplatz 3, 69115 Heidelberg, Germany. If you have any concerns regarding our products, please contact ProductSafety@springernature.com

Printed and bound by CPI Group (UK) Ltd, Croydon, CR0 4YY
30/09/2026
02225557-0020